AF413087

Topics in Lipid Research

From structural elucidation to biological function

Topics in Lipid Research

From structural elucidation to biological function

Edited by
Roger Klein and **Brigitte Schmitz**
Molteno Institute, University of Cambridge

The Proceedings of the 3rd International Conference on Lipid Chemistry organized by the Lipid Chemistry Group of the Perkin Division of The Royal Society of Chemistry, held 8th – 11th April 1986 in Cambridge, England.

British Library Cataloguing in Publication Data

Topics in lipid research : from structural
  elucidation to biological function
  1. Lipids
  I. Klein, Roger   II. Schmitz, Brigitte
  III. Royal Society of Chemistry
  547.7'7   QP751

ISBN 0-85186-353-1

Published by The Royal Society of Chemistry,
Burlington House, London, W1V 0BN

Printed in Great Britain by
Whitstable Litho Ltd., Whitstable, Kent

PREFACE

Over the last decade many lipid chemists and biochemists have made a conscious effort to move from purely structural investigations to those which throw more light on the biological function of lipid molecules. Lipid molecules of rather simple structure have potent biological activities in their own right and are not just the convenient adjuncts of the more exciting membrane proteins - a view not uncommon in the late sixties and earlier seventies!

For many years lipids were considered to be molecules containing just C, H and O, related to fatty acids and hydrocarbons, with high oil-water partition coefficients; more recently, however, it has been realized that the more complex structures such as glycolipids and lipoproteins, and indeed glycolipoproteins, are fascinating both as structural challenges and as functional elements of the cell.

The contributions to this book grew out of a meeting, held in Cambridge in April 1986 by the Royal Society of Chemistry, concerned with the inter-relationships between structure and function for lipid molecules, and their macromolecular ordered systems.

Our choice of individual topics was undoubtedly personal; we felt, however, that the areas chosen - platelet activating factor, eicosanoids, glycolipids, membrane structure and function and environmental adaptation - are some of those areas in which there has been a very successful coming together of physical and chemical work at the molecular level with investigations at supra-molecular level involving biophysicists, biochemists, physiologists and pharmacologists.

In each of the sections of the book we have contributions some of which are review papers and some of which are more detailed research papers. We have tried to present a leavened mixture of papers ranging from those with a chemical and structural bias to those with a more functional and biological basis.

Our overall intention was to present in this book a synopsis of current thinking in those areas of lipid research which we felt to be most interesting. This is especially true of the round table discussion on aspects of membrane probes and anaesthetics representing as it does the edited, and at that only slightly edited, verbatim recording of an unscripted verbal duel between the various participants.

We would like to express our thanks to all those who took part in the Round Table for being both coherent and incisive in what they had to say. Our thanks go also to all the other contributors to the book whose manuscripts were of a high standard thus making our editorial task that much easier.

We should like to dedicate this book to Alec Bangham on the occasion of his sixty-fifth birthday on 10th November 1986, in recognition of his contribution towards understanding the relationships between structure and function for membrane lipids.

Cambridge, August 1986   Roger Klein and Brigitte Schmitz

Alec Bangham, MD, FRS

CONTENTS

Contents

<u>Section</u> 1

PLATELET ACTIVATING FACTOR

CHEMISTRY OF PLATELET ACTIVATING FACTOR (PAF, PAF ACETHER, AGEPC) — 1-O-ALKYL-2-ACETYL-sn-GLYCERO-3-PHOSPHOCHOLINES — AND SOME PAF ANTAGONISTS

Helmut K. Mangold

Bundesanstalt für Fettforschung, Institut für Biochemie und Technologie -H.P. Kaufmann-Institut- , Piusallee 68, D-4400 Münster (Federal Republic of Germany)

INTRODUCTION

The platelet activating factor (PAF) was first isolated from rabbit blood cells (Benveniste et al., 1972) and later identified as a mixture of 1-O-alkyl-2-acetyl-sn-glycero-3-phosphocholines having predominantly saturated alkyl moieties with 16 and 18 carbon atoms (Benveniste et al., 1979; Demopoulos et al., 1979); unsaturated alkyl moieties were found recently (Mueller et al., 1984; Weintraub et al., 1985). Numerous biological effects of this potent lipid mediator and the structural requirements for such actions are described in several review articles (e.g., Benveniste and Vargaftig, 1983; Braquet et al., 1986; Pinckard et al., 1982; Snyder, 1985; Weber, 1986), the proceedings of two symposia (Benveniste and Arnoux, 1983; Winslow and Lee, 1986), and a monograph (Snyder, 1986).

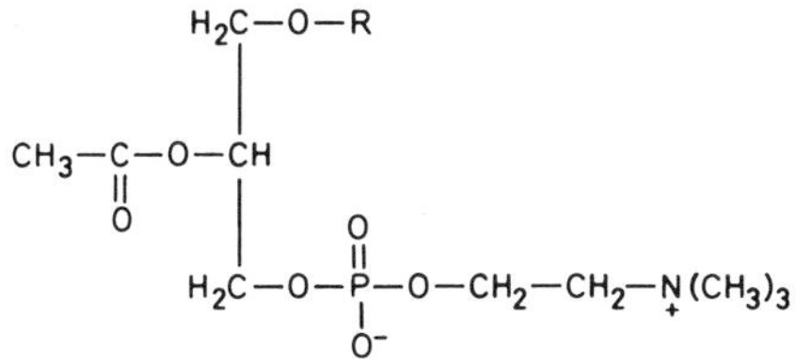

$$H_2C-O-R$$
$$CH_3-\underset{\underset{O}{\|}}{C}-O-CH$$
$$H_2C-O-\underset{\underset{O^-}{|}}{\overset{\overset{O}{\|}}{P}}-O-CH_2-CH_2-\overset{+}{N}(CH_3)_3$$

PAF

As an extension of an earlier survey (Mangold, 1983), the present article gives an account of work aimed at the synthesis and semi-synthesis of PAF including radioactively labelled preparations as well as structurally related compounds tagged with a photoreactive group. In addition, this review provides information on PAF antagonists that were synthesized or isolated from natural sources.

SYNTHESIS OF 1-O-ALKYL-2-ACETYL-sn-GLYCERO-3-PHOSPHO-CHOLINES

Efficient procedures for the synthesis of 1-O-alkyl-2-acylglycero-phospholipids with a satisfactory degree of positional and chiral purity are well worked out (Eibl, 1981, 1984; Eibl and Woolley, 1986; Paltauf, 1983). As a rule, the synthesis of 1-O-alkyl-2-acetyl-sn-glycero-3-phosphocholines

Synthesis of 1-*O*-alkyl-2-acetyl-*sn*-glycero-3-phosphocholines
from 1-*O*-alkyl-*sn*-glycerols

$$
\begin{array}{ccc}
\text{H}_2\text{C}-\text{O}-\text{R} & & \text{H}_2\text{C}-\text{O}-\text{R} \\
\text{HO}-\text{CH} & \xrightarrow[\text{Et}_3\text{N}]{\text{TrCl, Toluene}} & \text{HO}-\text{CH} \\
\text{H}_2\text{C}-\text{OH} & & \text{H}_2\text{C}-\text{O}-\text{Tr} \\
\text{I} & & \text{II}
\end{array}
$$

$$
\xrightarrow[\text{BzCl}]{\text{NaH, DMF}}
\quad
\begin{array}{c}
\text{H}_2\text{C}-\text{O}-\text{R} \\
\text{Bz}-\text{O}-\text{CH} \\
\text{H}_2\text{C}-\text{O}-\text{Tr} \\
\text{III}
\end{array}
\xrightarrow[\text{H}_3\text{BO}_3]{(\text{EtO})_3\text{B}}
\begin{array}{c}
\text{H}_2\text{C}-\text{O}-\text{R} \\
\text{Bz}-\text{O}-\text{CH} \\
\text{H}_2\text{C}-\text{OH} \\
\text{IV}
\end{array}
$$

V: $\text{H}_2\text{C}-\text{O}-\text{R}$; $\text{Bz}-\text{O}-\text{CH}$; $\text{H}_2\text{C}-\text{O}-\overset{\text{O}}{\underset{\text{OH}}{\text{P}}}-\text{O}-\text{CH}_2-\text{CH}_2-\text{Br}$ (from IV via $\text{Cl}_2\text{P(O)OCH}_2\text{CH}_2\text{Br}$, Hydrolysis)

VI: $\text{H}_2\text{C}-\text{O}-\text{R}$; $\text{Bz}-\text{O}-\text{CH}$; $\text{H}_2\text{C}-\text{O}-\overset{\text{O}}{\underset{\text{O}^-}{\text{P}}}-\text{O}-\text{CH}_2-\text{CH}_2-\overset{+}{\text{N}}(\text{CH}_3)_3$ (from V via Me_3N, Ag_2CO_3)

VI $\xrightarrow{\text{H}_2/\text{Pd}}$ **VII:** $\text{H}_2\text{C}-\text{O}-\text{R}$; $\text{HO}-\text{CH}$; $\text{H}_2\text{C}-\text{O}-\overset{\text{O}}{\underset{\text{O}^-}{\text{P}}}-\text{O}-\text{CH}_2-\text{CH}_2-\overset{+}{\text{N}}(\text{CH}_3)_3$

$$\text{VII} \xrightarrow[\text{Et}_3\text{N}]{\text{Ac}_2\text{O}} \text{PAF}$$

starts out from a 1-O-alkyl-sn-glycerol which is prepared by alkylation of 1,2-isopropylidene-sn-glycerol with an alkyl methanesulfonate followed by hydrolytic removal of the protecting group (Baumann and Mangold, 1964). In the latter reaction and in all following steps, experimental conditions must be employed that minimize racemization and avoid migration of substituents in positions 2 and 3 as far as possible. The consecutive introduction of the acetyl group in position 2 and of the phosphorylcholine moiety in position 3 of the glycerol backbone leads to mixtures of positional isomers due to acyl migration (Godfroid et al., 1980; Hirth et al., 1983). In order to avoid this isomerization reaction, it is advisable to introduce the phosphorylcholine moiety in position 3 before the most labile bond is formed by acetylation of the hydroxy group in position 2. This strategy is applied in the route of synthesis shown above (Heymans et al., 1981). The primary hydroxy group of a 1-O-alkyl-sn-glycerol, I, is protected by tritylation to yield II, and the secondary hydroxy group is protected by introduction of a benzyl group; detritylation of compound III affords IV.

1-O-Alkyl-2-O-benzyl-sn-glycerol, IV, the key intermediate in this synthesis, can be obtained by several other routes as well (Eibl, 1984; Fujita et al., 1982; Ohno et al., 1985; Tsuri and Kamata, 1985). The phosphorylation of IV with bromoethylphosphoric acid dichloride is well worked out. Nevertheless, it may be of advantage to use dimethylphosphoryl chloride for phosphorylation followed by removal of the protecting methyl groups and hydrolysis (Bittman et al., 1984). The yield of VI may be improved by preparing first the corresponding ethanolamine phospholipid - via a phospholane intermediate - followed by quaternization (H. Eibl, private communication, 1986). The catalytic hydrogenolysis of the protecting benzyl group in position 2 of VI does not lead to migration of the phosphorylcholine moiety and, therefore, acetylation of VII with acetic anhydride, preferably in the presence of dimethylaminopyridine, affords PAF of high chemical purity and high biological activity (Hirth and Barner, 1982; Muramatsu et al., 1981; Surles et al., 1985).

The removal of the protecting benzyl group in VI by catalytic hydrogenolysis excludes the preparation of PAF having unsaturated alkyl moieties by this route of synthesis. The synthesis of unsaturated PAF is possible by protecting the primary hydroxy function of an unsaturated 1-O-alkyl-sn-glycerol with a p-methoxytrityl group and the secondary hydroxy function with a benzoyl group. The p-methoxytrityl group is removed by acid hydrolysis under relatively mild conditions and thus a 1-O-alkyl-2-benzoyl-sn-glycerol becomes available as the key intermediate instead of IV. Following phosphorylation and amination, the protecting benzoyl group is removed by alkaline hydrolysis and unsaturated PAF is obtained by acetylation of the lyso compound thus formed (Hirth et al., 1983). Another route of synthesis makes use of the methoxyethoxymethyl group for protecting the secondary hydroxy function in an unsaturated rac-1-O-alkylglycerol. After introduction of the phosphorylcholine moiety, the protecting group is removed by reaction with zinc bromide in methylene chloride and the resulting lyso compound is acetylated to afford rac-1-O-alkyl-2-acetylglycerophosphocholine. The latter substance is treated with phospholipase A$_2$ (Wells and Hanahan, 1969) and the resulting 1-O-alkyl-sn-glycero-3-phosphocholine is separated from the unaffected 'unnatural' 3-O-alkyl-2-acetyl-sn-glycero-1-phosphocholine and reacetylated to yield the desired PAF (Surles et al., 1985).

Semi-synthesis of 1-*O*-alkyl-2-acetyl-*sn*-glycero-3-phosphocholines

Preparation of 1-*O*-alkyl-2-acetyl-*sn*-glycero-3-phosphocholines
from the choline plasmalogens of beef heart

1-*O*-(1'-Alkenyl)-2-acyl-*sn*-glycero-3-phosphocholines → Hydrogenation
(Choline plasmalogens)

1-*O*-Alkyl-2-acyl-*sn*-glycero-3-phosphocholines → Methanolysis

1-*O*-Alkyl-*sn*-glycero-3-phosphocholines → Acetylation

1-*O*-Alkyl-2-acetyl-*sn*-glycero-3-phosphocholines

Preparation of 1-*O*-alkyl-2-acetyl-*sn*-glycero-3-phosphocholines
from the neutral ether lipids of shark liver oils

1-*O*-Alkyl-2,3-diacyl-*sn*-glycerols → Enzymatic hydrolysis

1-*O*-Alkyl-2-acyl-*sn*-glycerols → 1. Hydrogenation
                                      2. Phosphorylation
                                      3. Amination

1-*O*-Alkyl-2-acyl-*sn*-glycero-3-phosphocholines → 1. Partial hydrolysis
                                                      2. Acetylation

1-*O*-Alkyl-2-acetyl-*sn*-glycero-3-phosphocholines

Preparation of 1-*O*-alkyl-2-acetyl-*sn*-glycero-3-phosphocholines
using cell suspension cultures of rape (*Brassica napus*)

1-*O*-Alkylglycerol → Rape cells

1-*O*-Alkyl-2-acyl-*sn*-glycero-3-phosphocholines → 1. Partial hydrolysis
                                                      2. Acetylation

1-*O*-Alkyl-2-acetyl-*sn*-glycero-3-phosphocholine

SEMI-SYNTHESIS OF 1-O-ALKYL-2-ACETYL-sn-GLYCERO-3-PHOS-PHOCHOLINES

Naturally occurring ether lipids such as the ethanolamine plasmalogens of bovine brain (Benveniste et al., 1979) or the choline plasmalogens of bovine heart (Demopoulos et al., 1979), the neutral ether lipids of shark liver oils (Muramatsu et al., 1981), and the 1-O-alkyl-2-acyl-sn-glycero-3-phosphoethanolamines of bovine erythrocytes (Kumar et al., 1984) offer themselves as starting materials for the preparation of 1-O-alkyl-2-acyl-sn-glycero-3-phosphocholines from which PAF can be prepared easily. The latter intermediates can also be obtained by utilizing the capability of cell suspension cultures of rape to acylate, phosphorylate, and aminate exogenous 1-O-alkyl-sn-glycerols (Weber et al., 1984; Weber and Mangold, 1985). Three procedures for the preparation of PAF are outlined on the previous page.

LABELLED COMPOUNDS

1-O-Alkyl-2-acetyl-sn-glycero-3-phosphocholines labelled with ^{3}H, ^{14}C or another radio-isotope are needed in numerous biochemical and biomedical studies. ^{3}H-Labelled preparations that are obtained by catalytic tritiation of the 1-alkenyl moieties of choline plasmalogens (Demopoulos et al., 1979) or the alkenyl moieties of 'unsaturated PAF' derived from shark liver oils (Muramatsu et al., 1981) are available from NEN, Boston, Mass., U.S.A. and Amersham International, Amersham, Bucks., U.K., respectively. ^{3}H-Labelled compounds prepared by tritiation of synthetic PAF containing (Z)-9-alkenyl moieties (Surles et al., 1985; Wyrick et al., 1985) are commercially available as well. Unsaturated PAF could, of course, also be labelled with ^{125}I or ^{131}I. [^{14}C]PAF labelled in the choline moiety can be obtained by reacting 1-O-alkyl-2-acetyl-sn-glycero-3-phosphoethanolamines with [^{14}C]methyl iodide in the presence of a crown ether (Kumar et al., 1984). PAF containing a ^{14}C-labelled alkyl moiety can be prepared by incubating a cell suspension culture of rape with a tagged 1-O-alkyl-sn-glycerol (Weber and Mangold, 1985). Plant cell cultures could, of course, also be used for the facile preparation of PAF labelled with ^{31}P. In addition to radioactively labelled compounds, those marked with a stable isotope, such as deuterium, may be of interest in certain areas of research (Wyrick et al., 1985).

Also of interest in this connection are two analogs of the platelet activating factor that are labelled with an azido group in position 2 of the glycerol backbone, VIII, (Ponpipom and Bugianesi, 1984) and with both a photoreactive azido group and radioactive iodine, IX, (Bette-Bobillo et al.,

1985). Both compounds VIII and IX exhibit physiological activities comparable with those of PAF.

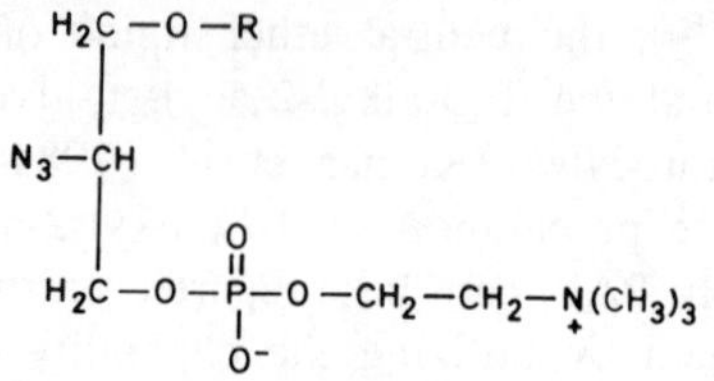

$$\text{VIII} \qquad\qquad \text{IX}$$

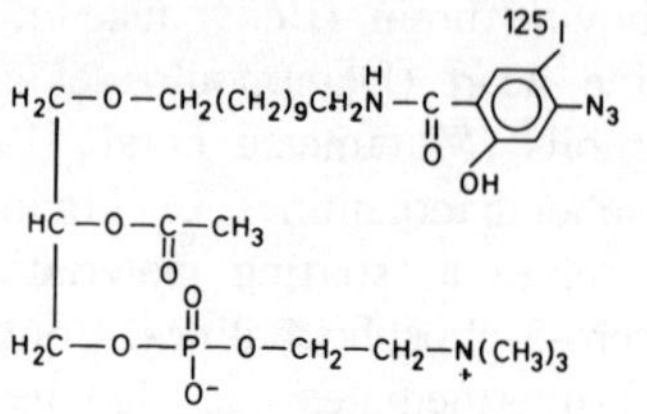

A fluorescent PAF analog in which one of the methyl groups of the choline moiety is replaced by a 9-anthrylmethyl group can be prepared from the choline plasmalogens of bovine heart (Imbs et al., 1985).

ANTAGONISTS

A number of observations indicate that PAF is bound to specific receptor sites of various human and animal cells. Binding experiments with [^{3}H]PAF prove that human and animal platelets are indeed provided with such receptors and biochemical studies show that these are proteins (Braquet and Godfroid, 1986). The binding of PAF to receptors can be inhibited by a great number of substances. In order to find a way to neutralize the adverse effects of PAF, much effort is being devoted to the synthesis of receptor antagonists or their isolation from natural sources. Some of the antagonists known so far are structurally related to PAF, whereas others are not, as shown below.

The structures of two synthetic antagonists, U-66985 and U-66982, show close resemblance to that of PAF inasmuch as these substances have an alkyl moiety and an acetyl group; their polar headgroup, however, differs from choline by having six and ten methylene groups, respectively, between the phosphate and the trimethyl-ammonium group, instead of two. The two compounds have the sn-3 configuration. Diacyl analogs of this type of glycerophospholipids have been known for many years (Diembeck and Eibl, 1979). The synthesis of the afore-mentioned alkylacylglycerophospholipids is described in the patent literature and a publication (Wissner et al., 1986) that appeared after the antagonistic effects of these substances had been reported (Tokumura et al., 1985).

Some N-alkylcarbamoyl glycerolipids inhibit the action of PAF. The synthesis of a series of such compounds is described in a recent publication

Antagonists of 1-*O*-alkyl-2-acetyl-*sn*-glycero-3-phosphocholines

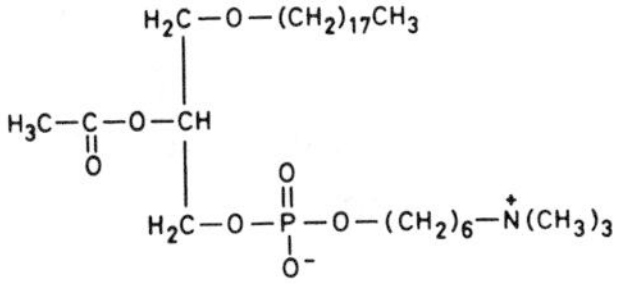

U - 66985

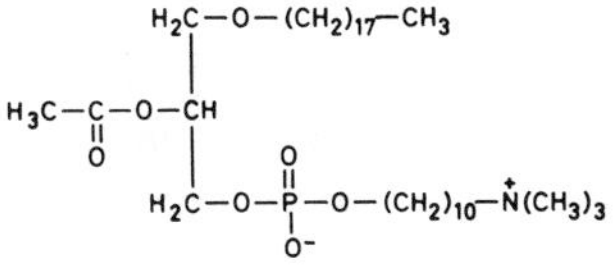

U - 66982

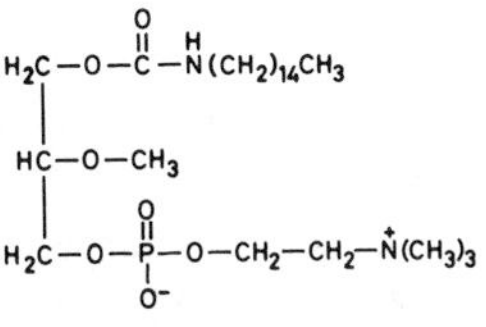

U-68043

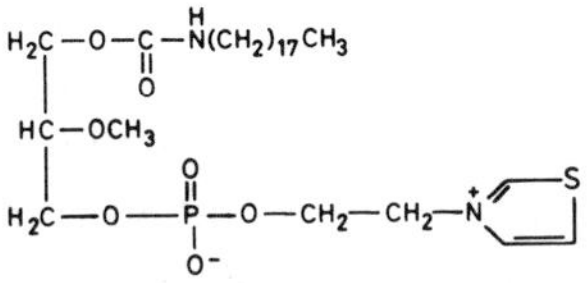

CV-3988

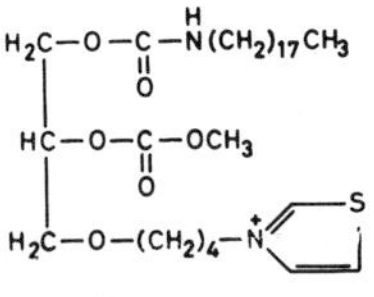

RO 19-3704

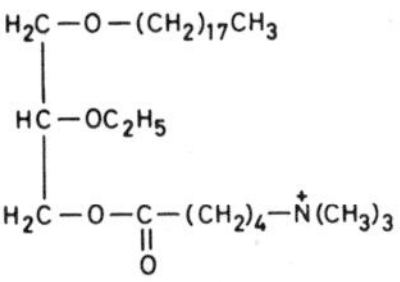

RU - 45, 703

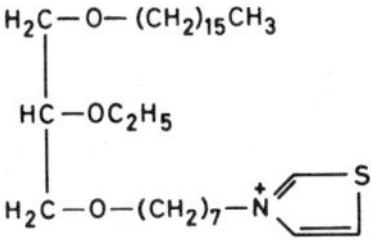

ONO - 6240

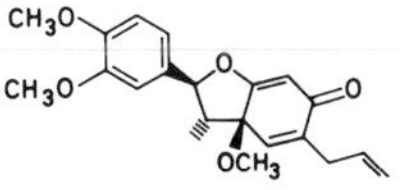

Kadsurenone

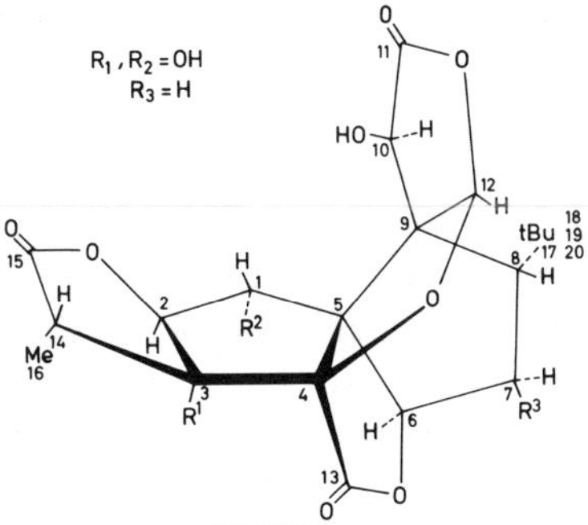

$R_1 , R_2 = OH$
$R_3 = H$

BN 52021

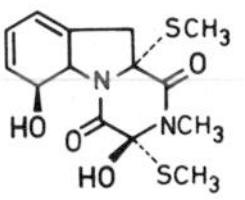

FR-49175

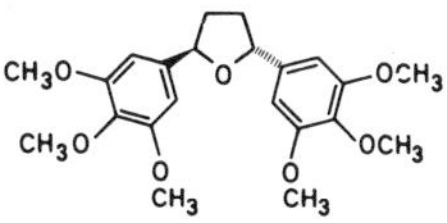

L-652,731

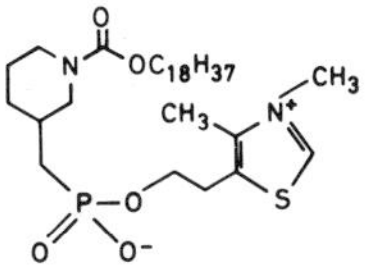

SRI 63-073

(Tsushima et al., 1984). Receptor antagonists with an N-alkylcarbamoyl moiety in position 1 and a methyl group in position 2 of glycerol include U-68043, which contains a phosphorylcholine moiety (Tokumura et al., 1985), and CV-3988, which contains a phosphorylethylthiazolium moiety (Terashita et al., 1983, 1985). The latter compound constitutes the first PAF antagonist described in the literature. RO 19-3704, an N-alkylcarbamoyl glycerolipid without a phosphate group, is a particularly potent antagonist (Barner et al., 1984). Other glycerol-derived antagonists include RU-45 703, a dialkylacylglycerol, and ONO-6240, a trialkylglycerol.

In addition to these synthetic compounds there is a great number of natural products that inhibit the action of PAF on human and animal blood cells. Some of these substances are lignans and neo-lignans, such as kadsurenone (Shen et al., 1985), which occurs in Piper futokadsurae, a plant indigenous to Southern China. Kadsurenone is also available through total chemical synthesis (Ponpipom et al., 1986). Other antagonists are terpenes, such as ginkgolides A, B, C, and M, the most potent being ginkgolide B (BN 52021) (Braquet, 1984). These substances are isolated from ginkgo (Ginkgo biloba) leaves on an industrial scale; syntheses are not available. Microbial products that are known to be PAF antagonists are the gliotoxins, such as FR-49 175, which can be isolated from Penicillium terlikowskii (Okamoto et al., 1986, a, b, c).

A synthetic lignan described recently is several times more potent than kadsurenone, a structurally related natural product (Hwang et al., 1985). An interesting synthetic phospholipid, SRI 63-073, which incorporates structural features of both PAF and thiamine phosphate, also represents a potent PAF antagonist (Winslow and Lee, 1986). There is evidence of the existence of endogenous PAF antagonists in human tissues and body fluids (J. Benveniste, private communication, 1985; P. Braquet, private communication, 1985). It is striking that most receptor antagonists known so far contain aliphatic ether groups or tetrahydrofuran rings.

A comprehensive review article provides further information on the chemical structures and biological effects of receptor antagonists as well as unspecific antagonists of PAF (Braquet et al., 1986), whereas another thorough and imaginative treatise describes the conformational properties of the binding sites in platelet membranes as deduced from the results of structure-activity studies (Braquet and Godfroid, 1986). Specific antibodies against PAF are available as tools for a more detailed characterization of PAF receptors in various cells (Nishihira et al., 1984). A recent publication describes a comparative study of the activity and specificity of inhibition of PAF-induced platelet activation by three different antagonists (Nunez et al., 1986).

CONCLUSION

It is generally believed that PAF is a key mediator of various pathological events, and it is hoped that antagonists will be developed which can be administered orally for the treatment of asthma, thrombosis, and shock.

REFERENCES

Barner, R., Hadvary, P., Burri, K., Hirth, G., Cassal, J.M., and Muller, K. (1984) Eur. Patent 147 768.

Baumann, W.J. and Mangold, H.K. (1964) J. Org. Chem. 29 : 3055-3057.

Benveniste, J., Henson, P.M., and Cochrane, C.G. (1972) J. Exp. Med. 136 : 1356-1377.

Benveniste, J. and Arnoux, B., Eds. (1983) Platelet activating factor and structurally related ether lipids. Elsevier, Amsterdam.

Benveniste, J., Tencé, M., Varenne, P., Bidault, J., Boullet, C., and Polonsky, J. (1979) C.R. Hebd. Seances Acad. Sci., Ser. D. 289 : 1037-1040.

Benveniste, J. and Vargaftig, B.B. (1983) in: Ether lipids: Biochemical and biomedical aspects, pp 355-387, Mangold, H.K. and Paltauf, F., Eds. Academic Press, New York.

Bette-Bobillo, P., Bienvenue, A., Broquet, C., and Maurin, L. (1985) Chem. Phys. Lipids 37 : 215-226.

Bittman, R., Rosenthal, A.F., and Vargas, L.A. (1984) Chem. Phys. Lipids 34 : 201-205.

Braquet, P. (1984) Brit. Patent GB 8, 418, 424.

Braquet, P. and Godfroid, J.J. (1986) in: Platelet activating factor, Snyder, F., Ed. Plenum Press, New York.

Braquet, P., Touqui, L., Shen, T.Y., and Vargaftig, B.B. (1986) J. Med. Chem., in press.

Demopoulos, C.A., Pinckard, R.N., and Hanahan, D.J. (1979) J. Biol. Chem. 254 : 9355-9358.

Diembeck, W. and Eibl, H. (1979) Chem. Phys. Lipids 24 : 237-244.

Eibl, H. (1981) in: Liposomes: From physical structure to therapeutic applications, pp 19-50, Knight, C.G., Ed. Elsevier, Amsterdam.

Eibl, H. (1984) Angew. Chem. Intl. Ed. Engl. 23 : 257-271.

Eibl, H. and Woolley, P. (1986) Chem. Phys. Lipids, in press.

Fujita, K., Nakai, H., Kobayashi, S., Inoue, K., Nojima, S., and Ohno, M. (1982) Tetrahedron Lett. 23 : 3507-3510.

Godfroid, J.J., Heymans, F., Michel, E., Redeuilh, C., Steiner, E., and Benveniste, J. (1980) FEBS Lett. 116 : 161-164.

Heymans, F., Michel, E., Borrel, M.C., Wichrowski, B., Godfroid, J.J., Convert, O., Coëffier, E., Tencé, M., and Benveniste, J. (1981) Biochim. Biophys. Acta 666 : 230-237.

Hirth, G. and Barner, R. (1982) Helv. Chim. Acta 65 : 1059-1084.

Hirth, G., Saroka, H., Bannwarth, W., and Barner, R. (1983) Helv. Chim. Acta 66 : 1210-1240.

Hwang, S.-B., Lam, M.-H., Biffu, T., Beattie, T.R., and Shen, T.Y. (1985) J. Biol. Chem. 260 : 15639-15645.

Imbs, A.B., Smirnova, M.M., Molotkovsky, J.G., and Bergelson, L.D. (1985) Bioorg. Khim. 11 : 1135-1139.

Kumar, R., Weintraub, S.T., McManus, L.M., Pinchard, R.N., and Hanahan, D.J. (1984) J. Lipid Res. 25 : 198-208.

Mangold, H.K. (1983) in: Platelet activating factor and structurally related ether lipids, pp 23-35, Benveniste, J. and Arnoux, B., Eds. Elsevier, Amsterdam.

Mueller, H.W., O'Flaherty, J.T., and Wykle, R.L. (1984) J. Biol. Chem. 259 : 14554-14559.

Muramatsu, T., Totani, N., and Mangold, H.K. (1981) Chem. Phys. Lipids 29 : 121-127.

Nishihira, J., Ishibashi, J., and Imai, Y. (1984) J. Biochem. 95 : 1247-1251.

Nunez, D., Chignard, M., Korth, R., Le Couedic, J.P., Norel, X., Spinnewyn, B., Braquet, P., and Benveniste, J. (1986) Eur. J. Pharmacol. 123 : 197-206.

Ohno, M., Fujita, K., Nakai, H., Kobayashi, S., Inoue, K., and Nojima, S. (1985) Chem. Pharm. Bull. 33 : 572-582.

Okamoto, M., Yoshida, K., Nishikawa, M., Ando, T., Iwami, M., Kohsaka, M., and Aoki, H. (1986a) J. Antibiotics 39 : 198-204.

Okamoto, M., Yoshida, K., Uchida, I., Kohsaka, M., and Aoki, H. (1986b) Chem. Pharm. Bull. 34 : 345-348.

Okamoto, M., Yoshida, K., Uchida, I., Nishikawa, M., Kohsaka, M., and Aoki, H. (1986c) Chem. Pharm. Bull. 34 : 340-344.

Paltauf, F. (1983) in: Ether lipids: Biochemical and biomedical aspects, pp 49-84. Mangold, H.K. and Paltauf, F., Eds. Academic Press, New York.

Pinckard, R. N., McManus, L.M., and Hanahan, D.J. (1982) Adv. Inflammation Res. 4 : 147-180.

Ponpipom, M.M. and Bugianesi, R.L. (1984) Chem. Phys. Lipids 35 : 29-37.

Ponpipom, M.M., Yue, B.Z., Bugianesi, R.L., Brooker, D.R., Chang, M.N., and Shen, T.Y. (1986) Tetrahedron Lett. 27 : 309-312.

Shen, T.Y., Hwang, S.-B., Chang, M.N., Doebber, T.W., Lam, M.-H., Wu, M.S., Wang, X., Han, G.Q., and Li, R.Z. (1985) Proc. Natl. Acad. Sci. USA 82 : 672-676.

Snyder, F. (1985) Med. Res. Rev. 5 : 107-140.

Snyder, F., Ed. (1986) Platelet activating factor. Plenum Press, New York.

Surles, J.R., Wykle, R.L., O'Flaherty, J.T., Salzer, W.L., Thomas, M.J., Snyder, F., and Piantadosi, C. (1985) J. Med. Chem. 28 : 73-78.

Terashita, Z., Imura, Y., Nishikawa, K., and Sumida, S. (1985) Eur.J. Pharmacol. 109 : 257-261.

Terashita, Z., Tsushima, S., Yoshioka, Y., Nomura, H., Inada, Y., and Nishikawa, K. (1983) Life Sci. 32 : 1975-1982.

Tokumura, A., Homma, H., and Hanahan, D.J. (1985) J. Biol. Chem. 260 : 12710-12714.

Tsuri, T. and Kamata, S. (1985) Tetrahedron Lett. 26 : 5195-5198.

Tsushima, S., Yoshioka, Y., Tanida, S., Nomura, H., Nojima, S., and Hozumi, M. (1984) Chem. Pharm. Bull. 32 : 2700-2713.

Weber, N. (1986) Pharmazie in unserer Zeit 15 : 107-112

Weber, N., Benning, H., and Mangold, H.K. (1984) Appl. Microbiol. Biotechnol. 20 : 238-244.

Weber, N. and Mangold, H.K. (1985) J. Lipid Res. 26 : 495-500.

Weintraub, S.T., Ludwig, J.G., Mott, G.E., McManus, L.M., Lear, C., and Pinckard, R.N. (1985) Biochem. Biophys. Res. Commun. 129 : 868-876.

Wells, M.A. and Hanahan, D.J. (1969) Biochemistry 8 : 414-424.

Winslow, C.M. and Lee, M.L., Eds. (1986) New horizons in platelet activating factor research, Churchill Livingston Publ., Edinburgh.

Wissner, A., Schaub, R.E., Sum, P.-E., Kohler, C.A., and Goldstein, B.M. (1986) J. Med. Chem. 29 : 328-333.

Wyrick, S.D., McClanahan, J.S., Wykle, R.L., and O'Flaherty, J.T. (1985) J. Labelled Comp. Radiopharm. 22 : 1169-1174.

ETHER GLYCEROGLYCOLIPIDS WITH PHYSIOLOGICAL ACTIVITY

Nikolaus Weber and Hildegard Benning

Bundesanstalt für Fettforschung, Institut für Biochemie und Technologie, H.P. Kaufmann-Institut, Piusallee 68/76, D-4400 Münster, Federal Republic of Germany

INTRODUCTION

Platelet activating factor (PAF) - a mediator of different physiological effects including platelet aggregation as well as allergical and shock reactions - was found to be a mixture of 1-O-alkyl-2-acetyl-sn-glycero-3-phosphocholines having alkyl groups with 16 and 18 carbon atoms (Benveniste and Vargaftig, 1983). Recently, it was shown that many of the undesirable activities mediated by PAF are suppressed by various receptor antagonists of this ether glycerophospholipid, e.g. (RS)-2-methoxy-3-(carbamoyloxy)-propyl-1-(3-thiazolio)-ethylphosphate (CV-3988; Figure 1) (Winslow and Lee 1986; Braquet and Godfroid, 1986). Furthermore, 1-O-alkyl-2-O-methylglycero-3-phosphocholines are known to be effective cancerostatics (Weltzien and Munder, 1983; Berger and Schmähl, 1986). It is presumed that antineoplastic effects of these substances may be due both to their cytotoxicity and their activation of cells of the immune system, e.g. macrophages (Weltzien and Munder, 1983). 1-O-Octadecyl-2-O-methylglycero-3-phosphocholine (ET-18-OCH$_3$; Figure 1) has already been tested in a clinical phase I pilot study (Berdel et al., 1985).

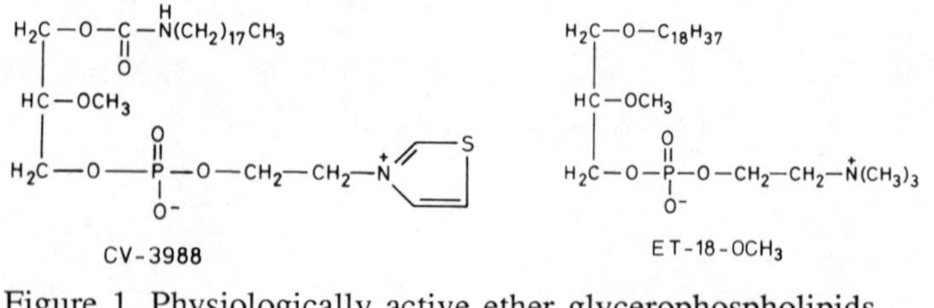

Figure 1. Physiologically active ether glycerophospholipids.

We have synthesized glycolipid analogs of the above mentioned ether glycerophospholipids in which the ionic phosphoester group is replaced by a β-glycosidically bound sugar moiety (Weber and Benning, 1986). These compounds, i.e. 1-O-alkyl-2-O-methyl-3-β-D-glycosyl-glycerols 2, 4 and 6 (Figure 2), may be useful as antagonists of platelet activating factor and/or as antineoplastic drugs. The present communication describes a facile procedure for the synthesis of various ether glyceroglycolipids having a 1-O-alkyl-2-O-methyl-glycerol backbone which is based on methods for the preparation of cerebrosides (Shapiro, 1969) as well as alkyl and steryl β-D-glycosides (Weber, 1977; Weber and Benning, 1982).

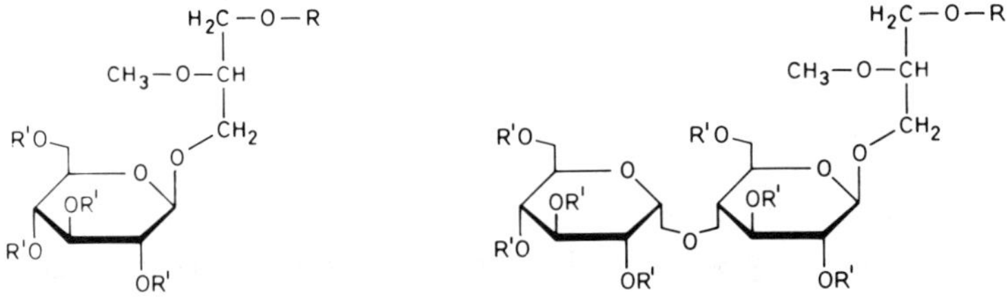

Figure 2. 1-O-Alkyl-2-O-methyl-3-β-D-glycosylglycerols with potential physiological activity: (1) R=$C_{16}H_{33}$; R'=3-β-D-(tetraacetyl)-glucosyl; 1a=rac; 1b=sn-1 (2) R=$C_{16}H_{33}$; R'=3-β-D-glucosyl; 2a=rac; 2b=sn-1 (3) R=$C_{18}H_{37}$; R'=3-β-D-(tetraacetyl)-glucosyl; 3a=rac; 3b=sn-1; 3c=sn-3 (4) R=$C_{18}H_{37}$; R'=3-β-D-glucosyl; 4a=rac; 4b=sn-1; 4c=sn-3 (5) R=$C_{18}H_{37}$; R'=3-β-D-(heptaacetyl)-maltosyl; rac (6) R=$C_{18}H_{37}$; R'=3-β-D-maltosyl; rac

EXPERIMENTAL METHODS

Syntheses of 1-O-alkyl-2-O-methyl-3-β-D-glycosyl-glycerols

The procedure for the preparation of rac-1-O-hexadecyl-2-O-methyl-3-β-D-glucosylglycerols is described in detail; other 1-O-alkyl-2-O-methyl-3-β-D-glycosylglycerols were synthesized similarly.

rac-1-O-Hexadecyl-2-O-methyl-3-(2,3,4,6-tetra-O-acetyl-β-D-glucosyl)-glycerol.

rac-1-O-Hexadecyl-2-O-methylglycerol (Biochemisches Labor R. Berchtold, Bern, Switzerland), 2.0 g (7.1 mmol), acetobromoglucose, 6.0 g (14.5 mmol), and Hg(CN)$_2$ 3.6 g (14.5 mmol), were reacted in 50 ml benzene - nitromethane (1:1, v/v) at 80°C under argon for 24 h. The resulting peracetylated β-glycoside was purified by column chromatography on silica gel 60 (E. Merck, Darmstadt, FRG) as described earlier [7, 9]. Fraction 5 (500 ml hexane - diethyl ether 1:1, v/v) contained 3.6 g of almost pure 1a. In addition, small amounts of this compound together with some sugar derivatives were eluted with 300 ml diethyl ether (fraction 6); yield 4.0 g (85%; Weber and Benning, 1986).

rac-1-O-Hexadecyl-2-O-methyl-3-β-D-glucosylglycerol

Compound 1a, 1.3 g (2 mmol), was subjected to alkaline hydrolysis by stirring for 1 h with 20 ml of 0.25 M methanolic KOH. The hydrolyzate was neutralized by the addition of ethyl formate; the solution was filtered and the solvents evaporated under vacuum. The resulting compound 2a was crystallized from dichloromethane (Weber and Benning, 1986).

TOXICITY IN NMRI-MICE

A single dose (10 mg/animal, i.e. 0.5 g/kg body weight; i.p.) of rac-1-O-hexadecyl-2-O-methyl-3-β-D-glucosylglycerol was lethal within 24-48 hrs, whereas doses of 5 mg per animal given i.p. on five consecutive days were tolerated.

DISCUSSION

The reaction of 1-O-alkyl-2-O-methylglycerols with acetobromosugars in the presence of mercury(II) cyanide leads in high yields (77-86%) to stereochemically uniform peracetylated 1-O-alkyl-2-O-methyl-3-β-D-glycosylglycerols. Alkaline hydrolysis of the latter compounds affords 1-O-alkyl-2-O-methyl-3-β-D-glycosylglycerols. The method is applicable to the preparation of radioactive substances. ^{1}H-NMR spectra of the various ether glyceroglycolipids show characteristic signals at 4.55 - 4.60 ppm ($J_{1,2}$=7.8 Hz, spin-spin interaction between two axial protons) of the anomeric proton at C-1 of the β-glycosidically bound sugar moiety (Prinz et al., 1985). Both the relatively low chemical shift of this proton at C-1 of the sugar moiety and the relatively high constant of spin-spin coupling point to a β-glycosidic bond in the ether glyceroglycolipids.

Figure 3 shows the chemical shifts of the anomeric proton of compounds 3a, 3b, and 3c. Each of the compounds 3b (sn-1-O-octadecyl) and 3c (sn-3-O-octadecyl) shows a doublet of the proton at C-1 of the sugar moiety at 4.56 ppm and 4.58 ppm ($J_{1,2}$=7.9 Hz), respectively, whereas the same proton in the spectrum of the mixture of these two diastereomers (3a) appears as a pseudo-triplet ("t") at 4.57 ppm (J,J'=7.8 Hz) that originates by superposition of the spectra of the diastereomers 3b and 3c. From these data it is evident that racemisation at C-2 of the glycerol moiety does not occur.

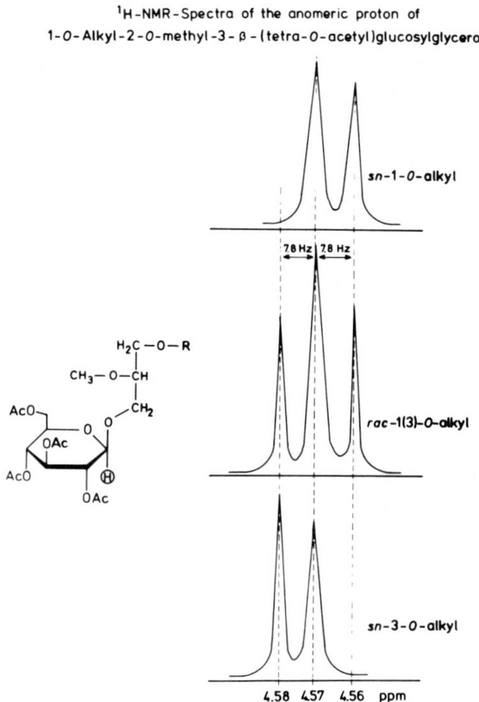

Figure 3. Chemical shifts of the anomeric protons of compounds 3a (rac-1-O-octadecyl), 3b (sn-1-O-octadecyl), and 3c (sn-3-O-octadecyl).

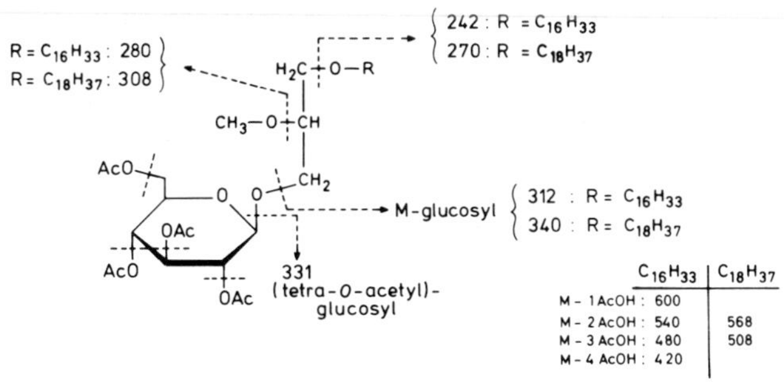

	C₁₆H₃₃	C₁₈H₃₇
M – 1AcOH :	600	
M – 2AcOH :	540	568
M – 3AcOH :	480	508
M – 4AcOH :	420	

Figure 4. Fragmentation pattern of 1-O-alkyl-2-O-methyl-3-(2,3,4,6-tetra-O-acetyl-β-D-glucosyl)glycerols.

Figure 4 shows the fragmentation pattern of peracetylated 1-O-alkyl-2-O-methyl-3-β-D-glucosylglycerols. Compounds 1a and 3a show the diagnostic peak for the 2,3,4,6-tetra-O-acetylglucosyl moiety at m/e 331, and compound 5 the corresponding peak for the 2,3,6,2',3',4',6'-hepta-O-acetylmaltosyl moiety at m/e 619 (see also Weber and Benning, 1986).

The mass spectra of the various ether glyceroglycolipids show the characteristic peaks as given in the literature (Egge, 1983). Efficient and stereoselective syntheses of various other ether glyceroglycolipids have been described recently (Gigg, 1980; Ogawa and Beppu, 1982; Prinz et al., 1985).

The compounds described may be useful as antagonists of the platelet activating factor and as antineoplastic agents. Preliminary results show cytotoxic and/or tumoricidal effects of these ether glyceroglycolipids. Studies concerning both fields are in progress.

ACKNOWLEDGEMENT

We thank Dr. H.P.K. Schiller for recording mass spectra. This work has been supported by Deutsche Forschungsgemeinschaft (DFG), D-5300 Bonn 2 (Grant We 927/2-1)

REFERENCES

Benveniste, J., and Vargaftig, B. B. (1983) in: Ether lipids: Biochemical and biomedical aspects. Mangold, H. K., and Paltauf, F. (Eds.) Academic Press, New York, pp. 355-376.

Berdel, W. E., Fink, U., Weiß, G., Emslander, H. P., Wüst, I., Nisenbaum, J., Thiel, E., Babic, R., Gößner, W., Rastetter, R., and Bloemer, H. (1985) in: Kabara, J. J. (Ed.) The pharmacological effects of lipids II. The American Oil Chemists' Society, Champaign, IL. pp. 281-293.

Berger, M. R., and Schmähl, D. (1986) Cancer Lett. 30:73-78.

Braquet, P., and Godfroid, J. J. (1986) in: Platelet activating factor. Snyder, F. (Ed.) Plenum Press, New York, in press.

Egge, H. (1983) in: Ether lipids. Biochemical and biomedical aspects. Mangold, H. K., and Paltauf, F. (Eds.) Academic Press, New York, pp. 17-47 and 141-159.

Gigg, R. (1980) Chem. Phys. Lipids 26:287-404.

Ogawa, T., and Beppu, K. (1982) Agric. Biol. Chem. 46:255-262.

Prinz, H., Six, L., Rueß, K.-P., and Liefländer, M. (1985) Liebigs Ann. Chem. 1985:217-225.

Shapiro, D. (1969) in: Chemistry of natural products. Lederer, E. (Ed.) Part IX, Hermann Publ., Paris, pp. 92-102.

Weber, N. (1977) Chem. Phys. Lipids 18:145-148.

Weber, N., and Benning, H. (1982) Chem. Phys. Lipids 31:325-329.
Weber, N., and Benning, H. (1986) Chem. Phys. Lipids, in press.
Weltzien, H. U., and Munder, P. G. (1983) in: Ether lipids. Biochemical and biomedical aspects. Mangold, H. K., and Paltauf F. (Eds.) Academic Press, New York, pp. 277-308.
Winslow, C. M., and Lee, M. L. (Eds.) (1986) New horizons in platelet activating factor research. Churchill Livingstone, Edinburgh.

PLATELET-ACTIVATING FACTOR AS A BIOLOGICAL AGONIST

T.J. Rink

Smith Kline & French Research Ltd., The Frythe, Welwyn, Herts. AL6 9AR, U.K.

INTRODUCTION

The main objective here is to outline the reasons why we believe that platelet-activating factor, PAF, is a specific ligand that activates cells by binding to specific membrane receptors. But before doing this I want briefly to consider lipids as biological signals.

Training as a physiologist in the sixties, one thought of the phospholipid bilayer of the plasma membrane as a relatively inert boundary that provided electrical insulation and capacitance, and acted as a selectively impermeant diffusion barrier separating the extracellular fluid from the very different cytoplasm. This lipid membrane served as support for the interesting bits like the voltage-gated sodium and potassium channels, the acetylcholine-gated channel and the sodium pump. We now know that there are tens, or more likely hundreds, of specific membrane proteins subserving many functions including: primary and secondary active solute transport; dissipative solute transport; specific binding for hormones, neuro-transmitters and drugs; coupling proteins and enzymes for signal transducton; cell surface recognition sites; and anchors for internal and external force transmission. Also, we now know that the membrane lipids have a life of their own and a function beyond the support of the proteins. A major role is in the generation of signals, both intracellular second messengers acting on cytoplasmic targets, and also intercellular messengers that act at specific receptors on similar or different cells types.

The first such system to be discovered and analysed was the eicosanoid derivatives of arachidonic acid, including the prostaglandins, thromboxane and the leukotrienes (see e.g. Moncada, 1983). The details of the formation, interconversion, transfer and degradation of these substances is far from resolved and there is still much to be learned of their physiological and patho-physiological functions. Nor have the intricacies of the arachidonate cascade been fully worked out; products of monooxygenase are only now beginning to be identified and their chemical and biological properties investigated.

A recent development of great significance has been the discovery of the signal transduction function of the inositol lipids. An early response of many cells to many agonists is the stimulated hydrolysis of the head group of phosphatidylinositol bisphosphate; the water-soluble inositol trisphosphate is released into the cytosol and the hydrophobic stearoyl-arachidonoyl diacyl-glycerol is left on the inner face of the bilayer. Inositol 1,4,5,-trisphosphate has been shown to cause discharge of Ca^{2+} ions, from Ca^{2+}-sequestering organelles (Berridge and Irvine, 1984), almost certainly by interaction with a specific intracellular membrane receptor. Various other inositol phosphates are formed following cell stimulation but as yet their function is unknown. Diacyl-glycerol is also an important second messenger in its own right with the major function, so far as we know, of activating protein kinase-C (Nishizuka, 1984). It is not our present purpose to discuss either of these major mechanisms in any detail; but a mention is necessary since PAF is another messenger derived from the phospholipids of the plasma membrane, and one that exerts its biological action partly by liberating arachidonate and by hydrolysing inositol lipids.

PAF was discovered as an "activity" released by IgE-stimulated basophils, that caused activation of platelets (Benveniste, Henson and Cochrane, 1972). The platelet served as a bioassay or detection system and not unnaturally the first known action provided the name; PAF is a pleasant enough acronym. It is not obvious, however, that platelet activation is the major, or indeed an important, biological function of this substance. Words are very powerful and strong connotations can be a surprising barrier to objective and innovative thinking. For instance, if you call something vasopressin, or epidermal growth factor, then people may find it hard to remember that production of concentrated urine, or modulation of gastric acid secretion, may actually be more important in physiological function than vascular constriction, or stimulation of epidermal (skin) growth. On the other hand, if you give something a neutral or chemical name fewer people may read your papers, and convention usually dictates that only somebody very eminent can give the substance your name, and probably after you are dead.

The chemical nature of PAF was worked out some seven years after its discovery. It turned out to be a mixture of 1-O-alkyl, 2-acetyl-sn-glycero-3-phosphorylcholine (Demopoulos, Pinckard and Hanahan, 1979; Snyder, 1985). The alkyl group is usually hexadecyl or octadecyl. The biosynthesis of PAF appears to require the deacylation of the alkoxylipid in the 2-position and subsequent acetylation. More detail of the biosynthesis of PAF and its regulation is given in the accompanying chapter in this volume

(Ninio et al.). In order to be biologically active, PAF has to bind with receptors on the surface of cells, as discussed in more detail below. We presume, therefore, that the PAF formed at the inner leaflet of the plasma membrane of activated cells must be rapidly released by permeation of the plasma membrane and discharged into the extracellular fluid. In view of the rapidity (seconds rather than minutes) with which biologically generated PAF can exert its effect on cells, this membrane permeation must be rather rapid. Whether the transfer occurs by simply permeation of the bilayer due to the natural lipophilicity of PAF, or by a specific transporter is not presently known.

Further studies of PAF transport through artificial bilayers and through cell membranes are needed to resolve this question.

BIOLOGICAL ACTIONS OF PAF

As with most biologically active molecules, we know more about the biosynthesis of PAF, and the effects of PAF added to in vitro systems and injected into experimental animals than we do about its actual physiological or pathophysiological role under natural circumstances. Judging from its known sites of production, macrophages, neutrophils, basophils and platelets, we might guess that PAF has functions in mediating inflammatory and allergic reactions. In vitro PAF has many activities which fit with this idea including: stimulation of neutrophils with aggregation, secretion and release of oxygen radicals; contraction of various smooth muscles; endothelium, dependent vasodilatation and also direct vasoconstriction; and the release of glucose from the liver cells (Snyder, 1985; Venuti, 1985). In experimental animals, intravenous PAF causes hypotension, airways constriction and decreased lung compliance, neutropenia, thrombocytopenia (Snyder, 1985; Venuti, 1985) and gastric ulceration (Rosam, Wallace and Whittle, 1986). Intradermal injection of PAF causes both a typical rapid wheal and flare inflammatory response and also a more persistent inflammation reminiscent of that associated with delayed hypersensitivity (Archer et al. 1985). Inhalation of PAF causes a prolonged hypersensitivity of the airways to bronchoconstrictor stimuli (Morley, Page and Sanjar, 1985). It seems a reasonable hypothesis that PAF is an important mediator in a number of types of inflammatory disease, perhaps in various kinds of asthma and in various dermatologic conditions. The increasing availability of potent and effective PAF antagonists should allow the dissection of the role of this mediator in animal models, and the availability of such agents for human testing will help unravel the part played by PAF in different human diseases. Since there are many different mediators of inflammation one would not expect PAF antagonists to have significant therapeutic impact in all conditions; but it is a reasonable hope that some inflammatory diseases will be particularly dependent on mediation by PAF, and susceptible to this form of treatment.

PAF RECEPTORS

General acceptance of the idea that many biologically active molecules exert their effect by binding to highly specific receptors in the surface membrane is relatively recent. A brief survey of the reasons why this is believed for PAF will also provide a general overview of the background to this view of cell stimulation. Major points of evidence in support of this idea can be summarised as follows: PAF is potent; its effects are stereospecific; homologous desensitisation is observed; the chemical structures of PAF analogues required for biological activity are relatively constrained; several classes of specific antagonists have been identified; specific high-affinity binding sites have been identified; typical receptor-mediated signal transduction processes follow application of PAF; and a PAF binding protein has been extracted from plasma membranes. The determination of the amino acid sequence of the PAF receptor and the reconstitution of the isolated receptor into an artificial PAF-receptor-transduction system has yet to be achieved.

POTENCY

The more potent the cell activator, the more difficult most people find it to ascribe its actions to some "non-specific" effect. In vitro PAF is active in the nanomolar range, EC_{50} values typically being in the range of 1-10nM. This is very much in the middle of the range of potencies for biologically active molecules believed to act at specific receptors. Many biologically active peptides are active in the picomolar range while neurotransmitter substances such as acetylcholine and noradrenaline typically have potencies in the micromolar range. In vivo, doses less than 1μg per kg have substantial effects.

STEREOSPECIFICITY

PAF has one chiral carbon atom, in the 2-position of the glycerol backbone. L-PAF is the naturally occurring stereoisomer and not surprisingly this turns out to be the biologically effective one. D-PAF is at least 2 orders of magnitude less effective as a biological agonist and in displacing radiolabelled PAF from membrane binding sites (Snyder, 1985; Tuffin et al. 1985). Stereospecificity is exceedingly common in biological ligand binding and in enzyme activity. It is quite difficult to imagine how a "non-specific" effect on cell membranes could be stereospecific. To my mind, the combination of high potency and stereospecificity is already very strong evidence for a specific receptor molecule.

HOMOLOGOUS DESENSITISATION

This term refers to the observation that exposure of a cell or tissue to a specific agonist can result in the tissue becoming refractory to that agonist and that agonist only. PAF displays this property quite strongly in some cells. For instance in rabbit platelets exposure of the cell to concentrations of PAF that are below the threshold for observable responses renders the platelets refractory to subsequent stimulation by an optimal concentration of PAF, but they still respond to other agonists such as thrombin and collagen (Hallam, Scrutton and Wallis, 1983).

Desensitisation can be quite rapid and is thought to be partly responsible for the fact that responses to a maximal concentration of PAF may be only transient, and that when the response has subsided further application of the same, or even a larger, concentration of PAF is virtually without effect. This is seen in experiments where application of PAF rapidly increases the cytosolic free calcium in human platelets loaded with the fluorescent calcium indicator quin-2 (Hallam, Sanchez and Rink, 1984). The mechanism of rapid homologous desensitisation of this type is not well understood. It presumably reflects some structural modification of the receptor that modifies its ability to interact with the various coupling systems by which transduction occurs.

AGONIST STRUCTURE ACTIVITY

The synthesis and testing of structural analogues of the biologically effective molecules can help define which parts of the molecule are most critical for its biological activity. The combination of a detailed structure-activity analysis with molecular modelling can generate plausible hypotheses about the nature of the biological binding site of the agonist. Because of its potential therapeutic importance, as well as its intrinsic interest, a great deal of work has been done on structural modifications of PAF (Snyder, 1985) and a detailed discussion is beyond the scope of this article. The major findings can be summarised as follows. Attempts to replace the O-alkyl function result in a substantial loss of activity. The length of the alkyl chain is not critical. The hexadecyl substituent gives a more potent compound than the octadecyl and chain lengths below 14 carbon atoms give rapidly reducing activity. A number of small substituents in the sn-2 position can give active molecules; longer chain lengths can give inactive molecules. The picture is complicated by the fact that some compounds with a long chain in the two-position and a shorter substituent in the one-position may still show a considerable activity. Another important feature, not surprisingly, is the positive charge in the choline substituent.

SPECIFIC ANTAGONISTS

Several types of PAF antagonists have been identified: both natural products extracted from plants and various synthetic analogues of either PAF or the natural products (Snyder, 1985; Venuti, 1985). Because of the possible therapeutic importance of such agents these are the subject of intensive investigation by many pharmaceutical companies. Details of the medicinal chemistry of some of these compounds are found in the chapter by Mangold. Two examples are shown in Fig. 1, which also gives the structure of PAF itself. In one case the analogy between the natural agonist and the antagonist is quite apparent; the reason for the high affinity of kadsurenone for the PAF receptor is not obvious. It is not, however, apparently beyond the skills or machinations of the molecular modeller to find a conformation which produces a satisfactory overlay with PAF and a reasonable hypothesis for a fit to a PAF receptor. The ability of molecules structurally or conformationally related to a biological agonist specifically to

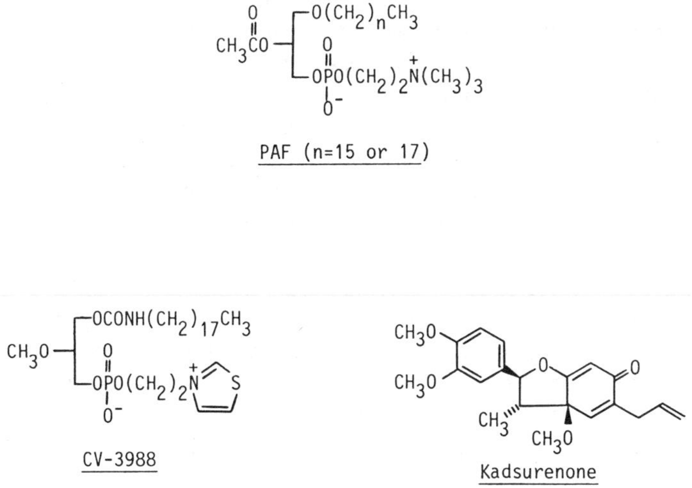

PAF (n=15 or 17)

CV-3988

Kadsurenone

Figure 1. For details see text.

antagonise its actions on a cell or tissue while leaving that cell or tissue responsive to other stimuli is strong evidence for a specific receptor. Finding that different antagonists have very different potencies in antagonising different responses to a single agonist is a strong hint that there may be sub-types of the receptor, with different structures and different modes of binding. Only limited evidence is so far available for receptor sub-types with PAF; Lambrecht and Parnham (1986) have reported that measure of pA_2 for kadsurenone antagonism of PAF indicate an apparent affinity of the antagonist 91 times higher for guinea pig macrophage PAF receptors than for neutrophil receptors.

SPECIFIC BINDING

The idea of biologically active molecules stimulating cells by acting at receptors implies the existence of specific binding. Measurement of specific binding to receptors requires a high affinity ligand, and with PAF the natural agonist itself has sufficient affinity for measurement of binding. A number of studies have reported specific high affinity binding to both intact cells and plasma membranes. Most work has been done with human platelet membranes. "Specific" binding is usually defined as the component of binding of radio-labelled ligand, rapidly displaceable by an excess of "cold" ligand. With this approach several groups have reported high-affinity binding of PAF to intact platelets and neutrophils (e.g. Shaw and Henson, 1980; Valone et al. 1982; Hwang, Lam and Pong, 1986; Tuffin et al. 1985). Tuffin et al. (1985) analysed their results as indicating two high affinity binding sites and a large, non-saturable, non-specific binding component. One site had a $\underline{K}_d$ value of 0.26nM and approximately 250 sites per platelet; the other had a $\underline{K}_d$ of 9.2nM and 1600 sites per platelet. It is not clear if one or both of these sites is responsible for the receptor-mediated responses. A detailed analysis of dose-response curves, under as closely as possible the conditions used for binding, could help resolve this issue. Bound (^{3}H)-PAF was displaced by cold L-PAF but much less so by D-PAF and barely at all by lyso-PAF. PAF binding to isolated plasma membranes has also been studied. For instance Hwang, Lam and Pong (1986) demonstrated high affinity binding of (^{3}H)-PAF to membranes prepared from rabbit platelets. Depending on the conditions (see below) the affinity varied between about 0.3 and 2.0nM; i.e. in the range of the high affinity binding to intact platelets. A typical feature of agonist-receptor binding, found in this study of PAF, is inhibition by increased [Na^+] and enhancement by increased [Mg^{2+}].

GTP-BINDING PROTEIN

The linkage of receptor occupancy to the enzymic generation of second messengers often, perhaps always, occurs via GTP-binding proteins, or G-proteins (see, e.g., Houslay, 1984). These are membrane-associated hetero-oligomers typically composed of α, β and γ sub-units. The most fully analysed are "N_s" and "N_1" which mediate stimulation and inhibition of adenylate cyclase. A number of tests for the involvement of G-proteins have become almost routine and most been been done for PAF. G-proteins have GTPase activity, stimulated by receptor occupation. PAF has been shown to stimulate GTPase activity in rabbit and human platelet membranes. Because of the interaction between G-proteins and receptors, the presence of GTP reduces the affinity of ligand binding; this has been demonstrated for PAF binding to rabbit platelet membranes. Certain non-hydrolysable GTP analogues, e.g. GTP-γS, can promote G-protein-mediated enzyme activation, e.g. activation of cyclase via N_s or of phospholipase C via another G-protein. This effect can synergise with the stimulation produced by the ligand receptor interaction; such a synergy has been demonstrated for PAF-activated diacylglycerol formation and dense-granule secretion in permeabilised platelets (Haslam, Williams and Davidson, 1985). Certain bacterial toxins appear to exert their biological effects by catalysing the ADP-ribosylation of G-proteins. Cholera toxin thereby stimulates N_s, while pertussis toxin inhibits N_1. Pertussis toxin also seems to interfere with the linkage between receptor occupation and phospholipase C or Ca^{2+} mobilisation in some cells (see e.g. Joseph, 1985; Naccache et al. 1985). However, pertussis toxin does not influence receptor coupling to phospholipase C in all cell types. The failure of this toxin or cholera toxin to affect PAF-stimulated GTPase in human platelet membranes suggests the presence of a G-protein distinct from N_s and N_1 (Houslay, Bojanic and Wilson, 1986).

Finally, a note of caution: in isolated platelet membranes PAF can reduce the activity of adenylate cyclase, presumably via N_1. In intact platelets, however, PAF does not appear to have this effect (Haslam et al. 1985). We should view work on isolated membranes for what it is: a study of a cell fragment whose relation to what goes on in intact cells must be established by expriments, not guess-work.

SIGNAL TRANSDUCTION

PAF appears to act mainly by promoting an influx of Ca^{2+}, mobilising intracellular Ca^{2+} and causing the hydrolysis of phosphatidylinositol bisphosphate. These responses have been most clearly shown in platelets (Hallam et al. 1984; Lapetina and Siegel, 1983), but other cells that respond to PAF including macrophages (e.g. Conrad and Rink, 1986) and neutrophils (Naccache et al. 1985) also seem to use these transduction pathways. Kinetic aspects of Ca^{2+} mobilisation by PAF in human platelets are described in this book, in the article by Stewart Sage. These pathways of signal transduction are like those used by many excitatory agonists that clearly act at specific surface receptors, e.g. thrombin and vasopressin (at V_1 receptors).

RECEPTOR ISOLATION, SEQUENCING, CLONING AND RECONSTITUTION

These are the final steps in defining a specific receptor protein. A 180,000 dalton PAF-binding protein has been isolated from platelet membranes (Valone, 1984; Nishihira et al. 1985). This may be the PAF receptor, or a part of it, but as the authors point out it could also be part of a transport protein or a PAF-processing enzyme. Evidence in favour of this protein being the PAF receptor could be provided by looking at the binding of agonist analogues of PAF, and the binding of antagonists to see if affinities and structure-affinity relationships fit with those of the functional responses. If this protein is the PAF receptor a natural progression will be to analyse the primary structure of this protein, and by the wizardry of modern molecular genetics derive a gene, and expression system, so that large amounts of the receptor can be made and further examined. This would open up the possibility of studying the receptor in a simplified system, reconstituted in phospholipid vesicles of defined composition along with, for instance, purified G-proteins and adenylate cyclase. The many techniques of protein biochemistry and biophysics could be applied to working out the secondary and tertiary structure. Then computer modelling and site-directed mutagenesis can help define the molecular structure-function relations of the agonist and antagonist interactions and the transduction process. Even with today's sophisticated technologies success in such ventures may not be assured, and many man-years of effort in many disciplines will be needed. But with the nicotinic acetylcholine receptor much of the above has been done (e.g. Colquhuon, 1986), so if the manpower is applied it should be do-able for the PAF receptor.

ACKNOWLEDGEMENTS

I thank Peter MacIntosh for his help in preparing this article, and David Tuffin for letting me see preprints of his papers.

REFERENCES

Archer, C.B., Page, C.P., Morley, J. and MacDonald, D.M. (1985) Br. J. Dermatol. 113, Suppl. 28, 133-135

Benveniste, J., Henson, P.J. and Cochrane, C.G. (1972) J. Exp. Med. 136, 1356-1377

Berridge, M.J. and Irvine, R.F. (1984) Nature 312, 315-321

Colquhoun, D. (1986) Nature 321, 382-383

Conrad, G.W. and Rink, T.J. (1986) J. Cell Biol. In press

Demopoulos, C.A., Pinckard, R.N. and Hanahan, D.J. (1979) J. Biol. Chem. 254, 9355-9358

Hallam, T.J., Scrutton, M.C. and Wallis, R.B. (1983) FEBS Lett. 162, 142-146

Hallam, T.J., Sanchez, A. and Rink, T.J. (1984) Biochem. J. 218, 819-827

Haslam, R.J., Williams, K.A. and Davidson, M.M.L. (1985) In: "Mechanisms of Stimulus-Response Coupling in Platelets" Eds. Westwick, J., Scully, M.F., MacIntyre, D.E. and Kakkar, V.V. pp. 265-280 Plenum, New York

Houslay, M.D. (1984) Trends in Biochem. Sci. 9, 39-40

Houslay, M.D., Bojanic, D. and Wilson, A. (1986) Biochem. J. 234, 737-740

Hwang, S.B., Lam, M.H. and Pong, S.S. (1986) J. Biol. Chem. 261, 532-537

Joseph, S. (1985) Trends in Biochem. Sci. 10, 297-298

Lambrecht, G. and Parnham M.J. (1986) Br. J. Pharmacol. 87, 287-298

Lapetina, E.G. and Siegel, F.L. (1983) J. Biol. Chem. 258, 7241-7244.

Moncada, S. (Ed.) (1983) Prostacyclin, Thromboxane and Leukotrienes. Vol. 39. Brit. Med. Bull. pp. 209-300

Morley, J., Page, C.P. and Sanjar, S. (1985) Lancet 2, 451

Naccache, P.H., Molski, M.M., Volpi, M., Becker, E.L. and Shaafi, R.I. (1985) Biochem. Biophys. Res. Commun. 130, 677-684

Nishihira, Y., Ishibashi, T., Imai, Y., Muramatsu, T. (1985) Tohoku J. Exp. Med. 147; 145-152

Nishizuka, Y. (1984) Nature 308, 693-698

Rosam, A.C., Wallace, J.L. and Whittle, B.J. (1986) Nature 319, 54-56

Shaw, J.O. and Henson, P.M. (1980) Am. J. Path. 98, 792-810

Snyder, F. (1985) Med. Res. Reviews 5, 107-140

Tuffin, D.P., Davey, P., Dyer, R.L., Lunt, D.O. and Wade, P.J. (1985) In: "Mechanisms of Stimulus-Response Coupling in Platelets". Eds. Westwick, J., Scully, M.F., MacIntyre, D.E. and Kakkar, V.V. pp. 83-96 Plenum, New York

Valone, S.H. (1984) Immunology 52, 169-174

Valone, S.H., Coles, E., Reinhold, V.R. and Goetzl, E.J. (1982) J. Immunol. 129, 1637-1641

Venuti, M.C. (1985) Annual Reports in Medicinal Chemistry, 20, 193-202

REGULATIONS INVOLVED IN PAF-ACETHER (PLATELET-ACTIVATING FACTOR) BIOSYNTHESIS

E. Ninio, J.M. Mencia-Huerta and J. Benveniste

INSERM U 200, Université Paris-Sud, 32 rue des Carnets, 92140 Clamart, France

INTRODUCTION

The discovery of platelet-activating factor is fairly recent. Benveniste (1972) observed that basophils isolated from immunized rabbits and challenged with specific antigen released a substance that was able to aggregate and to liberate histamine and serotonin from washed rabbit platelets. Several years later, Benveniste (1977) reported on the structure of the mediator. On the basis of lipid extraction and treatment with various phospholipases it appeared to be an analogue of phosphatidylcholine, certainly bearing an ether bound at sn-1 position of glycerol. Two years later, the full structure of the molecule was elucidated independently by two groups (Benveniste, 1979 and Demopoulos, 1979) (Fig.1). Since then it has been named paf-acether to underline the major pecularities of its structure i.e. the ether bound at the sn-1 position of glycerol and the acetate moiety at the sn-2 position.

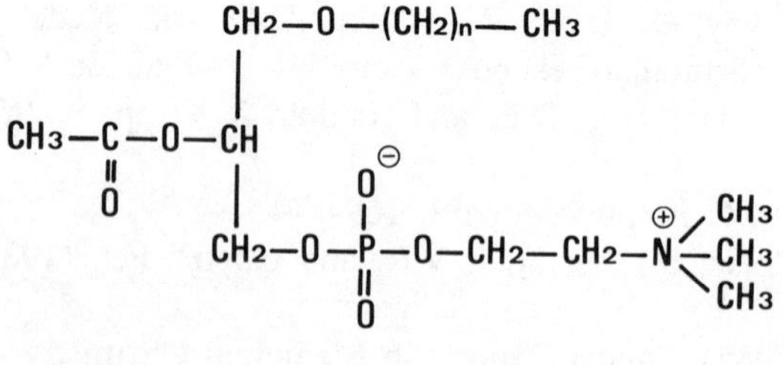

$$CH_2-O-(CH_2)_n-CH_3$$

$$CH_3-C-O-CH$$

$$CH_2-O-P-O-CH_2-CH_2-N^{\oplus}(CH_3)_3$$

$$n = 15,17\ldots$$

Figure 1 Paf-acether structure: 1-O-alkyl-2-acetyl-sn-glycero-3-phosphocholine (from Benveniste, 1979 and Demopoulos, 1979).

SYNTHESIS AND ASSAY

Total synthesis of paf-acether was obtained by Godfroid (1980) and research concerning the biosynthesis of this mediator, as well as on its cellular origin, started in several laboratories. A very convenient bioassay was developed to characterize and to quantitate paf-acether formed and released by various cell types and organs. It is based on the observation that the aggregation of washed rabbit platelets by paf-acether is independent of arachidonic acid cyclooxygenase metabolites and ADP (Cazenave, 1979). In practice this means that platelets rendered refractory to arachidonic acid by treatment with aspirin would, in the presence of an ADP scavenger complex such as creatine phosphate/creatine phosphokinase, aggregate specifically in response to paf-acether. In this assay, synthetic paf-acether serves to calibrate the platelet preparation. It is noteworthy that the sensitivity of this method is very high, since one can detect and precisely quantitate picomoles of paf-acether. Another bioassay was developed based on the ^{14}C-serotonine release from washed rabbit platelets pre-labelled with this radioactive amine. This method is less sensitive and in fact more time consuming than the previous one. Recently, some sophisticated chemical methods have been described to quantitate and to characterize paf-acether. Mass spectrometry studies performed by several groups allowed the determination of the composition of the alkyl chain at the <u>sn</u>-1 position of the glycerol (Hanahan, 1980 and Satouchi, 1983). Concomitantly with paf-acether the <u>sn</u>-1 acyl analog was shown to be released from stimulated human neutrophils (Oda, 1985). Although these methods are very useful they cannot substitute for the very rapid platelet aggregation assay. Efforts are now focused towards the development of specific anti-paf-acether antibodies and of a radioimmunoassay.

FORMATION OF PAF-ACETHER

Studies conducted in our and several other laboratories have shown that various pro-inflammatory cells were able to form paf-acether when stimulated with appropriate stimuli. Amongst them are murine peritoneal and alveolar macrophages, rat and human monocytes, platelets, neutrophils, eosinophils, endothelial cells and alveolar macrophages and recently, murine bone marrow-derived mast cells (BMMC). Paf-acether is not extracted from unstimulated cells demonstrating that it is not a preformed mediator. Its formation is strictly dependent on physiological temperature and pH, the presence of Ca^{2+} and the presence of membrane stimuli like calcium ionophore A 23187, zymosan opsonized or not, or specific antigen.

The concept of the initial action of a phospholipase A_2 on ether-linked phosphatidylcholine came from the following observations. Hog leukocytes incubated overnight at alkaline pH formed large amounts of the biologically inactive lyso analogue of paf-acether (1-$\underline{O}$-alkyl-$\underline{sn}$-glycero-3-phosphocholine) (Polonsky, 1980). This compound could be transformed into paf-acether by chemical acetylation (Polonsky, 1980). Later, several cell types including murine macrophages (Roubin, 1982), human leukocytes (Jouvin-Marche, 1984) or platelets (Benveniste, 1982) were shown to form lyso paf-acether during stimulation with their specific secretagogues. The inhibition of phospholipase A_2 activity in macrophages and platelets by EDTA, bromophenacyl bromide, 874 CB (Clin Midy) or mepacrine led to subsequent abrogation of paf-acether formation during cell stimulation (Mencia-Huerta, 1981 ; Benveniste, 1982). In the case of murine peritoneal macrophages, we showed (Mencia-Huerta, 1981) that a 10 min pretreatment of the cells with bromophenacyl bromide (0.1 mM) inhibited paf-acether formation upon stimulation with zymosan, and that addition of lyso paf-acether and acetyl-CoA to such cell populations induced, even in the absence of stimulus, the synthesis of paf-acether. These experiments suggested that paf-acether was synthesized in a two-step process involving (i) activation of a phospholipase A_2 that generates lyso paf-acether and (ii) subsequent acetylation of the latter compound by an acetyltransferase (Fig.2).

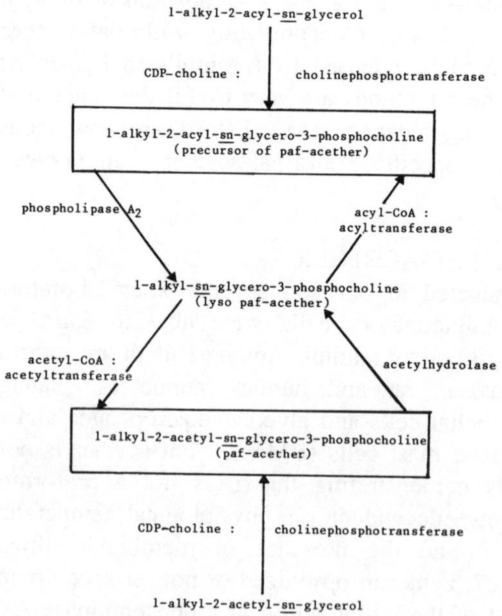

Figure 2 Paf-acether metabolism (adapted from Snyder, 1985).

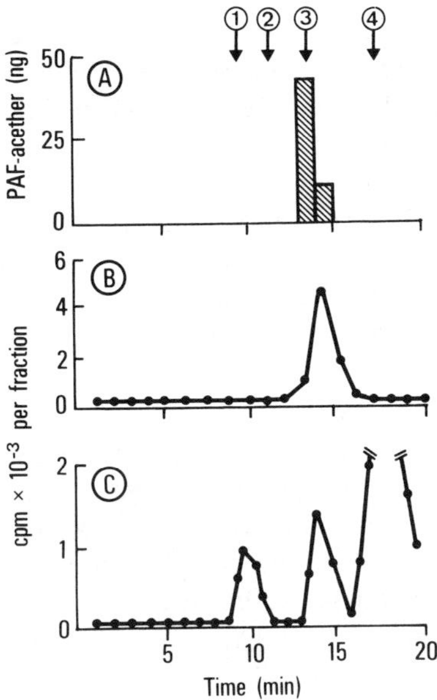

Figure 3 Typical high performance liquid chromatography elution pattern of paf-acether obtained from zymosan-stimulated rat peritoneal cells in the presence of tritiated substrates. (A) Biological activity; (B) Radioactivity in the presence of ^{3}H-acetyl-CoA; (C) Radioactivity in the presence of ^{3}H-lyso paf-acether and unlabelled acetyl-CoA. Paf-acether was extracted with 80 % ethanol in water. After centrifugation and evaporation of the supernatant in an air stream, paf-acether was recovered in dichloromethane/methanol (1/1 v/v). Column, Microporasil; solvent, dichloromethane, methanol, water (60/50/5) ; flow rate, 1.7 ml/min. The arrows indicate the rentention times of (1) phosphatidylcholine, (2) sphingomyelin, (3) hog leukocyte, semi-synthetic and synthetic paf-acether, and (4) lysophosphatidylcholine (from Mencia-Huerta, 1982, with permission from the American Association of Immunologists).

Several experiments supported the implication of an enzymatic acetylation step in paf-acether biosynthesis. An acetyltransferase (EC : 2.3.1.67) capable of forming paf-acether from synthetic lyso paf-acether and acetyl-CoA was described in microsomes derived from several rat tissues (Wykle, 1980) and from murine macrophages (Ninio, 1982). As well, exposure of platelets to the ionophore A 23187 in the presence of ^{3}H-acetate yielded radiolabelled paf-acether (Chap, 1981). Finally, rat peritoneal adherent cells stimulated with zymosan in the presence of ^{3}H-acetyl-CoA generated ^{3}H-paf-acether (Fig.3) (Mencia-Huerta, 1982) and similar results were obtained later using rabbit neutrophils (Mueller, 1983).

REGULATION OF SYNTHESIS

The question of the regulation of paf-acether biosynthesis was soon raised since this mediator is synthetized only during cell stimulation of defined cell types. The most important observation linked to paf-acether formation was the activation of the acetyltransferase upon cell stimulation. Activation of several cell types results in a very early several-fold increase in acetyltransferase level prior to paf-acether synthesis. This phenomenon was observed in human neutrophils stimulated with opsonized zymosan

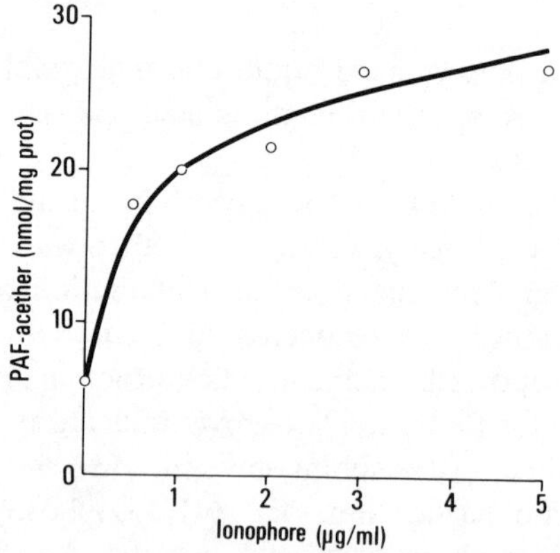

Figure 4 Ionophore A 23187-induced acetyltransferase activity in intact rat peritoneal cells. Intact cells were incubated for 5 min at 37°C in the presence of various concentrations of ionophore A 23187 or vehicle (0.1 % dimethyl sulfoxide), washed in saline and then lysed by sonication. Acetyltransferase activity was measured by incubating for 10 min at 37°C the cell lysate in the presence of 40 µM lyso paf-acether and 200 µM ^{3}H-acetyl-CoA (0.2 µCi/100 nmol).

(Alonso, 1982) or the ionophore A 23187 (Lee, 1982 and Jouvin-Marche, 1984), murine peritoneal and alveolar macrophages (Ninio, 1983 ; Albert, 1983) human and rabbit platelets stimulated with thrombin (Coëffier, 1986) and human eosinophils stimulated with the ionophore A 23187 or specific secretagogues (Lee, 1982 and 1984). In Fig. 4 we show the enhancement of acetyltransferase activity in murine peritoneal cell lysates obtained after stimulation of intact cells with various concentrations of the ionophore A 23187. The time-course studies showed that the activity had already doubled after 30 sec of incubation with the ionophore, suggesting that a post-translational protein modification controls the level of acetyltransferase. Kinetic studies using preparations from stimulated and unstimulated cells indicated that the enzymes were very similar and the observed difference in the activity was probably due to the increase in the number of active molecules present in the lysate from stimulated cells. Actually, no difference in the $\underline{K}_m$ of the enzyme for acetyl-CoA could be observed before as opposed to after stimulation.

We confirmed a role for acetyltransferase and its activation in paf-acether formation in the BMMC i.e. a model of immediate hypersensitivity (Ninio, 1986 submitted). BMMC were passively sensitized with an optimal dose of dinitrophenyl (DNP)-specific monoclonal IgE. Upon challenge with DNP coupled to bovine serum albumin, they released paf-acether and expressed higher levels of acetyltransferase than unstimulated cells. The activity of acetyltransferase doubled within 30 sec upon antigen challenge, and in keeping with the results previously obtained (Mencia-Huerta, 1983), the formation of paf-acether was maximal by 2-3 min. Thus, the acetyltransferase stimulation preceded the formation and the release of paf-acether by BMMC.

The post-translational concept (Ninio, 1983) was further studied in BMMC. Recently, this hypothesis was supported by Lenihan and Lee (1984) who showed that a phosphorylation-dephosphorylation process might control the level of acetyltransferase in rat spleen microsomes. However, this model does not reflect the physiological or pathological synthesis of paf-acether, so we made an attempt to determine whether such a phosphorylation/ dephosphorylation mechanism was operational in antigen-stimulated BMMC. Indeed, we showed that the activity of the enzyme present in lysates from unstimulated cells doubled after preincubation with ATP and Mg^{2+} whereas adenylyl imidodiphosphate, a non-phosphorylating analogue of ATP, was without effect (Ninio, 1986 submitted). In contrast, ATP and Mg^{2+} were ineffective in lysates from antigen-challenged cells that exhibited already high levels of acetyltransferase activity, suggesting that phosphorylation of the enzyme took place at the time of cell stimulation. Thus we report on both the activation of a key enzyme for paf-acether synthesis and on the mechanism of this activation in an antigen-stimulated pure mast cell population. A link between bridging of IgE receptors and an enzyme activity critical to the formation of a lipid mediator is thereby evidenced.

Recently, Benhamou (1986) demonstrated that treatment of BMMC with dexamethasone (1 μM, 24 hr) inhibited the immunologic release of paf-acether and the preformed granule marker, β-hexosaminidase, in a dose- and time-dependent fashion. In contrast, no inhibition of mediator release was observed when dexamethasone-treated cells were stimulated with the ionophore A 23187 (1 μM). Dexamethasone treatment induced a dose- and time-dependent inhibition of ^{125}I-IgE binding to the cells. Scatchard analysis of ^{125}I-IgE binding to dexamethasone-treated BMMC revealed a 51 % decrease in the IgE Fc receptor number without alteration of the $\underline{K}_d$. Cytofluorometer analysis showed that every cell in the dexamethasone-treated cell population exhibited a decreased IgE binding. The antigen-induced increase in acetyltransferase activity, used as an index of cellular activation, was inhibited by 40% in dexamethasone-treated cells as compared to untreated ones (Table I). However when the half-sensitization procedure was employed to mimick the action of dexamethasone, only an 11 % decrease in acetyltransferase activation was observed as compared with a 29 % decrease in paf-acether release. These results suggest that dexamethasone inhibits in part the immunologic release of paf-acether by decreasing the number of IgE Fc receptors available for sensitization. Thus, the modulation of IgE Fc receptor number plays a role in the anti-inflammatory properties of glucocorticosteroids. These results are not in contradiction with the role of glucocorticosteroids in inducing lipocortin synthesis (Di Rosa, 1984). It is possible that the inhibitory effect of lipocortin on phospholipase A_2 could modify the phospholipid environment of the IgE Fc receptor thus leading to a decreased expression on the cell surface.

The acetyltransferase stimulation appears to be a key event in paf-acether synthesis in several cell types and the lack of paf-acether formation could be related to the absence of, or a very low level of, acetyltransferase activity in thioglycollate-elicited mouse macrophages (Roubin, 1982), human lymphocytes (Jouvin-Marche, 1984) and rat kidney tubular cells (Pirotzky, 1984). However, rat kidney medullary cells that exhibit very low levels of acetyltransferase are capable of forming paf-acether upon ionophore A 23187 stimulation (Pirotzky, 1984). The involvement of an alternative route for paf-acether biosynthesis via the alkylacetylglycerol : CDP-choline : cholinephosphotransferase (EC : 2.7.8.16) (Renooij and Snyder, 1981) was proposed. In addition, work by Woodard (1984) has shown that the activity of the latter enzyme was 10 times higher in the medullary part of the kidney

TABLE I

**EFFECT OF DEXAMETHASONE TREATMENT AND HALF-SENSITIZATION PROCEDURES
ON ACETYLTRANSFERASE ACTIVITY AND ON PAF-ACETHER RELEASE[a]**

Preincubation with dexamethasone	–	+	–
Sensitization period (min) with IgE	60	60	5
	Controls	Percent decrease or inhibition	
IgE-binding ($\underline{n}$ = 3)	79.0 ± 12.3 (ng/1 x 10^6 cells)	50.9 ± 16.7	40.7 ± 18.6
Paf-acether release ($\underline{n}$ = 3)	25.9 ± 9.2 (ng/1 x 10^6 cells)	39.8 ± 8.2	29.2 ± 3.5
Acetyltransferase activity ($\underline{n}$ = 7)	467.0 ± 226.8 (pmol paf-acether/ min/1 x 10^6 cells)	40.3 ± 14.8	10.5 ± 3.7

[a]BMMC were incubated with 1 μM dexamethasone or ethanol for 24 hr. Replicates of 1 x 10^6 BMMC were sensitized with IgE or ^{125}I-IgE (25 μg/ml) for the indicated periods. Binding of ^{125}I-IgE and measurement of acetyltransferase activity and paf-acether release were performed as described in Materials and Methods. The results are expressed as the mean ± 1 SD of the indicated number of experiments. (From Benhamou, 1986, with permission from the American Association of Immunologists.)

than in the cortex, suggesting that the CDP-choline choline-phosphotransferase is the major pathway of paf-acether synthesis in rat medullary cells. Recently, two enzymes implicated in the biosynthesis of alkylacetylglycerol, namely alkylglycerophosphate : acetyltransferase and alkylacetylglycerophosphate : phosphatase, were described by Lee (1986).

Besides the activation of acetyltransferase, the generation and/or removal of lyso paf-acether may control in part the level of paf-acether production. In human neutrophils, the generation of lyso paf-acether preceeds paf-acether formation (Jouvin-Marche, 1984). Addition of exogenous lyso paf-acether to stimulated neutrophils doubled the amount of paf-acether formed (Jouvin-Marche, 1984) (Table II). Interesting was also the lack of effect of acetyl-CoA alone on paf-acether biosynthesis in this cell type. From these data a cooperation between platelets, which generate large amounts of lyso paf-acether, and neutrophils which are able to transform it into paf-acether was postulated by Coëffier (1984). Indeed, when mixing neutrophils stimulated with opsonized zymosan together with platelets stimulated with thrombin, paf-acether formation doubled with respect to the mere summation. Two explanations are likely: (i) lyso paf-acether released by platelets is transformed into paf-acether by neutrophils, or (ii) stimulated platelets produce a mediator capable of stimulating one of the biosynthetic steps of paf-acether formation, i.e. phospholipase A_2 or acetyltransferase. In favour of the second hypothesis is the recent report by Billah (1985) on the modulatory effects of 5-lipoxygenase products on the activity of neutrophil phospholipase A_2.

Another aspect of paf-acether regulation was studied in activated mouse macrophages (Roubin, 1986). These cells, which were obtained from mice injected with bacteria or immunostimulants derived from bacteria, synthesized in response to a zymosan challenge 2-3 times less paf-acether than resident macrophages although the level of phospholipid precursors for paf-acether synthesis and acetyltransferase levels were similar in both populations. The biosynthesis of paf-acether could be restored in activated macrophages by addition of acetyl-CoA. In contrast, resident macrophages cultured for at least 24 hr lose their ability to form paf-acether and addition of acetyl-CoA to the monolayers does not restore the mediator production (Haye-Legrand, 1986). When the catabolic activity, i.e. paf-acether : acetylhydrolase, was studied it became evident that the enhanced paf-acether degradation by the latter enzyme was responsible for the lower output of the mediator in cultured macrophages.

TABLE II

RELEASE OF PAF-ACETHER FROM HUMAN NEUTROPHILS
IN THE PRESENCE OF ACETYLTRANSFERASE SUBSTRATES[a]

Stimulating Agent

	None	Io (1 μg)	ZC (1 mg)
None	0	16.9 ± 1.3	12.3 ± 1.1
Acetyl-CoA (100 μM)	0	20.5 ± 2.4	15.1 ± 0.9
lyso paf-acether (100 nM)	0	29.3 ± 1.3	25.8 ± 1.4
LPC[b] (100 nM)	0	16.4 ± 1.4	12.6 ± 1.3
Acetyl-CoA (100 μM) + lyso paf-acether (100 nM)	0	31.9 ± 1.6	29.4 ± 1.5
Acetyl-CoA (100 μM) + LPC (100 nM)	0	19.3 ± 1.3	24.4 ± 1.6

Io Ionophore A 23187
ZC opsonized zymosan

[a] 1×10^6 cells in 1 ml HEPES buffer were incubated for
1 hr at 37°C. The results in pmol paf-acether are the
means ± 1 SD of five experiments.

[b] LPC, lysophosphatidylcholine.
(from Jouvin-Marche, 1984, with permission from the
American Association of Immunologists)

CATABOLISM OF PAF-ACETHER

The catabolic route for paf-acether can be summarized as follows :
acetylhydrolase cleaves the acetyl moiety of paf-acether leading to the
formation of lyso paf-acether which is in turn reacylated into alkyl-
acylglycerophosphocholine (Fig.2). In rabbit platelets (Malone, 1985) and
human neutrophils (Chilton, 1983) the reacylation reaction of lyso paf-
acether is very efficient. By contrast, rat alveolar macrophages (Robinson,
1985) and capillary endothelial cells (Tan and Snyder, 1985) appear to
possess a low acyltransferase activity since a considerable amount of lyso
paf-acether is not metabolized.

Presently we know that different agonists trigger more or less
efficiently the release of paf-acether from cells into culture medium (Roubin,
1982) and that the intracellular paf-acether is not randomly distributed. In

human neutrophils stimulated by the ionophore A 23187 or opsonized zymosan, paf-acether was found in granules or in phagosomes, respectively (Riches, 1985). The acetyltransferase in ionophore A 23187-stimulated and unstimulated human neutrophils is located in the internal membranes (Ribbes, 1985). These results imply the presence of a paf-acether transport system within the cell, the regulation of which could also control the release of the mediator.

The regulation of paf-acether biosynthesis occurs at several levels the most important being certainly the activation of acetyltransferase. However, it is possible that a given cell type is able to use the alternative pathway through CDP-choline : cholinephosphotransferase under certain circumstances. The precise knowledge of the mechanism of the activation of the acetyltransferase should allow one to develop new drugs acting at this level and that are required to block the synthesis of this potent pro-inflammatory mediator.

REFERENCES

Albert D.H. and Snyder F. (1983) J. Biol. Chem. 258 : 97-102.

Alonso F., Gil M.G., Sanchez-Crespo M. and Mato J.M. (1982) J. Biol. Chem. 257 : 3376-3378.

Benhamou M., Ninio E., Salem P., Hiéblot C., Bessou G., Pitton C., Liu F.-T. and Mencia-Huerta J.M. (1986) J. Immunol. 136 : 1386-1392.

Benveniste J., Chignard M., Le Couedic J.P., Vargaftig B.B. (1982) Thromb. Res. 25 : 375-386.

Benveniste J., Henson P.M. and Cochrane C.G. (1972) J. Exp. Med. 136 : 1356-1377.

Benveniste J., Le Couedic J.P., Polonsky J., Tencé M. (1977) Nature. 269 : 170-171.

Benveniste J., Tencé M., Varenne P., Bidault J., Boullet C. and Polonsky J. (1979) C.R. Acad. Sc. Paris, 289D : 1037-1040.

Billah M.M., Bryant R.W., Siegel M. (1985) J. Biol. Chem. 260 : 6899-6906.

Cazenave J.P., Benveniste J. and Mustard J.F. (1979) Lab. Invest. 41 : 275-285.

Chap H., Mauco G., Simon M.F., Benveniste J. and Douste-Blazy L. (1981) Nature 289 : 312-314.

Chilton F.H., O'Flaherty J.T., Ellis J.M., Swendsen C.L. and Wykle R.L. (1983) J. Biol. Chem. 258 : 6357-6361.

Coëffier E., Chignard M., Delautier D. and Benveniste J. (1984) Fed. Proc. 43 : 781 (Abs).

Coëffier E., Ninio E., Le Couedic J.P. and Chignard M. (1986) Br. J. Haematol. 62 : 641-651.

Demopoulos C.A., Pinckard R.N. and Hanahan D.J. (1979) J. Biol. Chem. 254 : 9355-9358.

Di Rosa M., Flower R.J., Hirata F., Parente L. and Russo-Marie F. (1984) Prostaglandins 28 : 441-442.

Godfroid J.J., Heymans F., Michel E., Redeuilh C., Steiner E., Benveniste J. (1980) FEBS Lett. 116 : 161-164.

Hanahan D.J., Demopoulos C.A., Liehr J. and Pinckard R.N. (1980) J. Biol. Chem. 255 : 5514-5516.

Haye-Legrand I., Dulioust A., Vivier E., Meslier N., Benveniste J. and Roubin R. (1986) Biochim. Biophys. Acta, in press.

Jouvin-Marche E., Ninio E., Beaurain G., Tencé M., Niaudet P. and Benveniste J. (1984) J. Immunol. 133 : 892-898.

Lee T-c., Lenihan D.J., Malone B., Roddy L.L. and Wasserman S.I. (1984) J. Biol. Chem. 259 : 5526-5530.

Lee T-c., Malone B. and Snyder F. (1986) J. Biol. Chem. 261 : 5373-5377.

Lee T-c., Malone B., Wasserman S.I., Fitzgerald V. and Snyder F. (1982) Biochem. Biophys. Res. Comm. 105 : 3303-3308.

Lenihan D.J. and Lee T-c. (1984) Biochem. Biophys. Res. Comm. 120 : 834-839.

Malone B., Lee T-c. and Snyder F. (1985) J. Biol. Chem. 260 : 1531-1534.

Mencia-Huerta J.M., Lewis R.A., Razin E. and Austen F. (1983) J. Immunol. 131 : 2958-2964.

Mencia-Huerta J.M., Ninio E., Roubin R. and Benveniste J. (1981) Agents and Action 11 : 556-558.

Mencia-Huerta J.M., Roubin R., Morgat J. and Benveniste J. (1982) J. Immunol. 129 : 804-808.

Mueller H.W., O'Flaherty T.J. and Wykle R.L. (1983) J. Biol. Chem. 258 : 6211-6218.

Ninio E., Mencia-Huerta J.M. and Benveniste J. (1983) Biochim. Biophys. Acta 751 : 298-304.

Ninio E., Mencia-Huerta J.M., Heymans F. and Benveniste J. (1982) Biochim. Biophys. Acta 710 : 23-27.

Oda, M., Satouchi, K., Yasunaga, K. and Saito, K. (1985) J. Immunol. 134: 1090-1093.

Pirotzky E., Ninio E., Bidault J., Pfister A. and Benveniste J. (1984) Lab. Invest. 51 : 567-572.

Polonsky J., Tencé M., Varenne P., Das B.C., Lunel J. and Benveniste J. (1980) Proc. Natl. Acad. Sci. USA 77 : 7019-7023.

Renooij W. and Snyder F. (1981) Biochim. Biophys. Acta 663 : 545-556.

Ribbes, G., Ninio, E., Fontan, P., Record, M., Chap, H., Benveniste, J. and Douste-Blazy, L. (1985) FEBS Letters 191: 195-199.

Riches D.W.H., Young S.K., Seccombe J.F., Lynch J.M. and Henson P.M. (1985) Fed. Proc. 44 : 737 (Abs).

Robinson M., Blank M.L. and Snyder F. (1985) J. Biol. Chem. 260 : 7889-7895.

Roubin, R., Dulioust A., Haye-Legrand I., Ninio E. and Benveniste J. (1986) J. Immunol. 136 : 1796-1802.

Roubin, R. Mencia-Huerta J.M., Landes A. and Benveniste J. (1982) J. Immunol. 129 : 809-813.

Satouchi, K., Oda M., Yasunaga K. and Saito K. (1983) J. Biochem. 94 : 2067-2070.

Snyder F. (1985) Med. Res. Rev. 5 : 107-140.

Tan E.L. and Snyder F. (1985) Thromb. Res. 38 : 713-717.

Woodard D.S., Lee T-c. and Snyder F. (1984) Fed. Proc. 43 : 1655 (Abs).

Wykle R.L., Malone B. and Snyder F. (1980) J. Biol. Chem. 225 : 10256-10260.

EFFECT OF IONIC SUBSTITUTION ON HUMAN PLATELET RESPONSES TO PAF

Stewart O. Sage

The Physiological Laboratory, Downing Street, Cambridge, CB2 3EG, U.K.

INTRODUCTION

Platelets, like most cells, maintain large transmembrane gradients of Na^+ and K^+ and have a negative membrane potential near -70 mV, as estimated with fluorescent carbocyanine dyes (MacIntyre and Rink, 1982; Pipili,1985) and radiolabelled, membrane permeant ions (Wencel-Drake and Feinberg, 1985). The role of this membrane potential in stimulus-response coupling is still unclear. Thrombin, ADP and PAF have been reported to change platelet membrane potential (Horne and Simons, 1978; Greenberg-Sepersky and Simons, 1984; Pipili, 1985), but these changes do not correlate well with platelet activation and are typically only 4 to 10 mV (Wencel-Drake and Feinberg, 1985; Pipili, 1985). It is known that high $[K^+]$ solutions depolarise but do not activate platelets (MacIntyre and Rink ,1982; Doyle and Ruegg, 1985; Hallam and Rink, 1985a) and responses to thrombin persist after replacement of Na^+ by choline or N-methyl-D-glucamine (Connolly and Limbird, 1983; Pipili, 1985). These observations argue against the existence of voltage-dependent Ca^{2+}-channels in platelet membranes.

In this study, the effects of substitution of external Na^+ by choline and K^+ on responses to PAF have been further investigated in human platelets loaded with fluorescent Ca^{2+}-indicator dyes. The effects of these conditions on both the magnitude of the agonist-induced rise in $[Ca^{2+}]_i$ in quin2-loaded cells and the kinetics of this rise in fura-2-loaded cells are reported.

METHODS

Human platelets were isolated from freshly drawn blood and loaded with either approximately 1 mM quin2 or 30 μM fura-2 by incubation with the appropriate acetoxymethyl ester at 20 μM or 2 μM respectively (Rink et al., 1983; Tsien et al., 1985). The standard medium for resuspending

"

platelets was: 145 mM NaCl, 5 mM KCl, 1 mM $MgCl_2$, 10 mM HEPES, 10 mM glucose, pH 7.4 at $37^{\circ}C$. For low-[Na] solutions choline chloride replaced NaCl, for high-[K] solutions KCl replaced NaCl. Hirudin (0.05 U/ml) and Apyrase (20 µg/ml) were added to all platelet suspensions to prevent activation by residual traces of thrombin and ADP, respectively. The platelet-rich plasma was pretreated with 100 µM aspirin to prevent the generation of prostaglandin endoperoxides and thromboxanes. This procedure allows the study of primary responses to agonists, uncomplicated by secondary effects of cyclooxygenase products. Prior to experimental procedures the external $[Ca^{2+}]$ was adjusted by the addition of $CaCl_2$ or EGTA (Fluka) as required and the cells were then equilibrated at $37^{\circ}C$ for about 5 min.

Quin2 fluorescence was measured in a stirred, thermostated cuvette in a Perkin-Elmer MPF44A spectrophotometer with excitation 339 nm and emission 500 nm as previously described (Tsien et al., 1982; Rink et al., 1982). The procedure for calibration of the fluorescence signal is detailed elsewhere (Tsien et al., 1982; Rink et al., 1982). PAF (Calbiochem) was added from aqueous stock, A23187 (Calbiochem) was added from DMSO stock. The maximum concentration of DMSO was 0.1% (v/v) which was itself without effect.

Kinetic measurements were made on fura-2-loaded cells in a Hi-Tech Scientific SFA-II Rapid Kinetic Accessory mounted in the fluorimeter as above. Fura-2-loaded cells were injected through one port and an agonist solution in the same medium through the other (Rink and Sage, 1985). The amplified signal from the photomultiplier was fed directly to the analogue input port of a BBC microcomputer. The input was scanned at 10 msec intervals after triggering of the microswitch on the stopped-flow apparatus. Twelve sequential scans were summed and averaged. The delay in fluorescence increase and the time to reach a peak were measured by hand from printouts of the averaged trace.

RESULTS AND DISCUSSION

Fig. 1 shows the effects of ionic substitution on the calcium transients produced by 20 ng/ml PAF in quin2-loaded cells. Fig. 1A shows responses in 1 mM external Ca^{2+}. Substitution of Na^+ by choline has no effect on the PAF induced $[Ca^{2+}]_i$ rise. The change in $[Ca^{2+}]_i$ was 791 $\pm$ 36 nM (S.E.M.,n=10) in controlled medium, and 793 $\pm$ 43 nM (S.E.M.,n=10) in choline medium. The $[Ca^{2+}]_i$ rise in high $[K^+]$ medium is slowed and considerably reduced. The rise in $[Ca^{2+}]_i$ was only 121 $\pm$ 8 nM (S.E.M.,n=11) compared to 726 $\pm$ 61 nM (S.E.M.,n=11) in paired controls.

Fig. 1B shows the effects of ionic substitution in the absence of external Ca^{2+}. Na^+ replacement with choline or K^+ has no effect on the PAF-induced $[Ca^{2+}]_i$ rise. The Ca^{2+} rise in choline medium was 56 $\pm$ 7 nM (S.E.M.,n=10) compared with 64 $\pm$ 12 nM (S.E.M.,n=10) in paired controls. In high $[K^+]$ medium the $[Ca^{2+}]_i$ rise was 48 $\pm$ 2 nM (S.E.M.,n=12) compared with 42 $\pm$ 2 nM (S.E.M.,n=12) in paired controls. In both choline and high $[K^+]$ the difference from controls is not statistically significant.

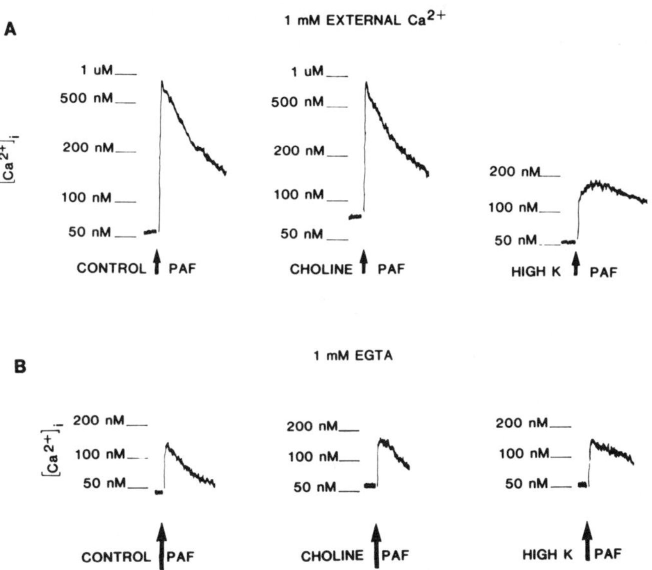

Figure 1. Effect of substituting Na^+ for choline or K^+ on reponses to 20 ng/ml PAF: A, in the presence of 1 mM external Ca^{2+} and B, in the presence of 1 mM EGTA.

The Ca^{2+}-ionophore A23187 (200 nM) raised $[Ca^{2+}]_i$ to levels that saturated the quin2 response in control, choline and high $[K^+]$ media (not shown).

Fig. 2A shows $[Ca^{2+}]_i$ rises produced by PAF at a final concentration of 200 ng/ml in fura-2-loaded cells. The trace shows the initial fluorescence level of the newly injected unstimulated cells followed by the rise to the stimulated level. Activated cells already occupying the chamber are displaced by the incoming cells before the microswitch is triggered; the fall in fluorescence which occurs is therefore not shown in the trace. Mixing time in the apparatus (assessed by quin2-acid dilution) was 40 msec; this mixing time precedes microswitch triggering. The $[Ca^{2+}]_i$ rise in choline medium did not differ from the control. The delay in the rise in fluorescence was the same under both conditions, 310 ± 10 msec (S.E.M.,$\underline{n}$=9). The time to peak fluorescence was 930 ± 30 msec (S.E.M.,$\underline{n}$=9) in choline medium and 930 ± 20 msec (S.E.M.,$\underline{n}$=9) in control medium. Substitution of Na^+ by K^+ increased slightly both the delay in the rise in $[Ca^{2+}]_i$ and the time to peak fluorescence. The delay was 350 ± 10 ms (S.E.M.,$\underline{n}$=12) in high K^+ compared to 280 ± 10 ms (S.E.M.,$\underline{n}$=14) in controls. The time to peak in high K^+ was 1170 ± 20 ms (S.E.M.,$\underline{n}$=12) compared to 960 ± 10 ms (S.E.M.,$\underline{n}$=14) in controls.

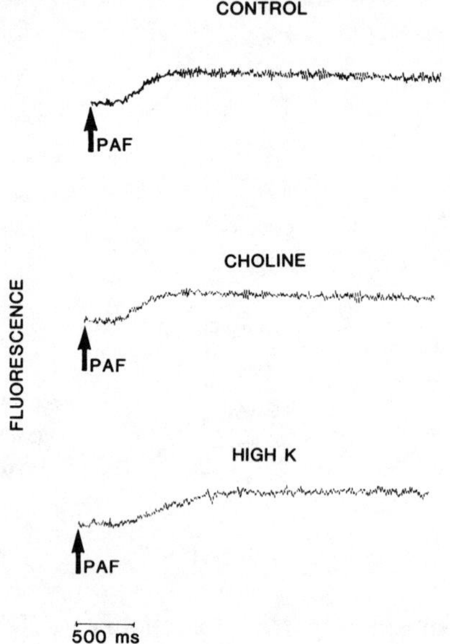

Figure 2. Effect of substituting Na^+ for choline or K^+ on fura-2 fluorescence changes evoked by 200 ng/ml PAF in the presence of 1 mM external Ca^{2+}. For explanation of traces see text.

The results presented in Figs. 1 and 2 show that replacement of external Na^+ with the non-permeant cation choline affects neither the extent nor the kinetics of the rise in $[Ca^{2+}]_i$ induced by PAF. The larger rise in $[Ca^{2+}]_i$ seen in the presence of external Ca^{2+} compared to that seen in its absence is attributed to Ca^{2+} influx from the medium additional to release from internal stores, which is recorded alone in the presence of EGTA (Rink et al., 1982; Hallam et al., 1984; Hallam and Rink, 1985b).

A number of investigations using potential-sensitive dyes have demonstrated small agonist-induced depolarisations of platelet membranes (Horne and Simons, 1978; Greenberg-Sepersky and Simons, 1984; Pipili, 1985). Replacement of external Na^+ with a non-permeant cation abolishes or reverses the depolarisation, suggesting that it is Na^+-dependent (Greenberg-Sepersky and Simons, 1984; Pipili, 1985). Since both the extent and kinetics of the rises in $[Ca^{2+}]_i$ induced by PAF were unaffected by Na^+ substitution with choline, Na^+-dependent depolarisation is presumably not important in the Ca^{2+} mobilisation, whether it arises from influx or internal stores. These new results support previous conclusions (MacIntyre and Rink, 1982; Doyle and Ruegg, 1985; Hallam and Rink, 1985b; Pipili, 1985) that Ca^{2+} influx does not occur through voltage-dependent Ca^{2+}-channels.

Figs. 1 and 2 show that replacement of external Na^+ with K^+ has a marked effect on the increase in $[Ca^{2+}]_i$. K^+ substitution considerably reduces Ca^{2+} influx induced by PAF, whilst internal release is unaffected. Similar results have been shown for the responses of platelets to ADP (Hallam and Rink, 1985a) and of lymphocytes to the lectin phytohemagglutinin (Gelfand et al., 1984). This reduced signal was not due to some effect of high $[K^+]$ on the dye, either intracellularly or during calibration, since the Ca^{2+} ionophore A23187 was able to raise $[Ca^{2+}]_i$ to the quin2 fluorescence maximum. One possible explanation for the observed reduction in Ca^{2+} influx is a decreased driving force for Ca^{2+} entry due to the K^+-induced depolarisation (Gelfand et al, 1984; Hallam and Rink, 1985a). However, Ca^{2+} enters platelets down both electrical and chemical gradients. Even if the potential gradient were completely eliminated by depolarisation, a ten thousand-fold difference in $[Ca^{2+}]$ would still exist across the membrane and it seems unlikely that loss of the potential component of the gradient would have the severe effect on Ca^{2+} influx reported here.

Direct inhibition of PAF-binding in high $[K^+]$ medium seems unlikely since internal release was unaffected. Since influx but not internal release is suppressed, the most likely explanation of the reduced rises in $[Ca^{2+}]_i$ seen in high $[K^+]$ solutions in the presence of external Ca^{2+} seems to be that high external $[K^+]$ affects the transduction process involved in influx but not internal release. This could be at the level of the plasma membrane Ca^{2+} transporter, or its coupling to the receptor.

ACKNOWLEDGEMENT

S.O.S. held an MRC Scholarship.

REFERENCES

Connolly, T.M. and L.E. Limbird (1983). J. Biol. Chem. 258: 3907-12.

Doyle, V.M. and U.T. Ruegg (1985). Biochem. Biophys. Res. Comm.127: 161-167.

Gelfand, E.W., R.Y. Cheung, and S. Grinstein (1984). J. Cell. Physiol. 121: 533-539.

Greenberg-Sepersky, S.M. and E.R. Simons (1984). J. Biol. Chem. 259: 1502-1508.

Hallam, T.J. and T.J. Rink (1985a). J. Physiol. 368: 131-146.

Hallam, T.J. and T.J. Rink (1985b). FEBS Lett. 186: 175-179.

Hallam, T.J., A. Sanchez, and T.J. Rink (1984). in: Prostaglandins and Membrane Ion Transport, eds. P. Braquet et al. (Raven Press, New York) pp. 157-163.

Horne, W.C. and E.R. Simons (1978). Blood 51: 741-749.

MacIntyre, D.E. and T.J. Rink (1982). Thromb.Haemostas.(Stuttgart) 47: 22-26.

Pipili, E. (1985). Thromb Haemostas (Stuttgart) 54: 645-649.

Rink, T.J. and T.J. Hallam (1984). Trends in Biochemical Sciences 9: 215-219.

Rink, T.J., A. Sanchez, and T.J. Hallam (1983). Nature 305: 317-319.

Rink, T.J. and S.O. Sage (1985). J. Physiol. 369: 115P.

Rink, T.J., S.W. Smith, and R.Y. Tsien (1982). FEBS Lett. 149: 21-26.

Tsien, R.Y., T. Pozzan, and T.J. Rink (1982). J. Cell. Biol. 94: 325- 334.

Tsien, R.Y., T.J. Rink and M. Poenie (1985). Cell Calcium 6: 145-157.

Wencel-Drake, J.D. and H. Feinberg (1985). Thromb. Haemostas. 53: 75-79.

ENOLATE REACTIONS OF TRIGLYCERIDES AND RELATED COMPOUNDS

Margareta Herslöf and Salo Gronowitz

Organic Chemistry 1, Chemical Center, Box 124, S-221 00 Lund, Sweden

INTRODUCTION

In saturated fatty acids and esters, the α-position is the most reactive site in the hydrocarbon chain. Despite this, until recent years only few direct substitutions have had preparative value. Among these are halogenation (Harwood, 1962) and sulfonation (Gilbert, 1965) in acidic media, and base-catalyzed Claisen condensation (Hauser and Hudson, 1942). Free radical addition of acids to olefins (Ault et al., 1965; Maerker et al., 1966; Ault et al., 1967) has been used to prepare α-alkylated fatty acids, although the alkylation of malonic esters (Cason et al., 1953) has been more frequently employed. Other indirect methods leading to α-substituted derivatives have involved formation through displacement of an α-halogen substituent. The problems with abstracting a proton from the slightly acidic α-methylene group were not solved until the development of strong non-nucleophilic bases in 1967 (Creger, 1967).

Since then, a number of papers that describe procedures to generate enolate anions from aliphatic acids and esters has been published (Pryde, 1979; Seebach and Geiss, 1976; Petragnani and Yonashiro, 1982). The α-anions are versatile intermediates, since several derivatives are accessible by reaction with suitable electrophiles; for review cf. Pryde (1979).

Little research has been performed on the preparation of α-substituted triacyl glycerols (Aydin, 1977), especially on reactions starting from triacyl glycerols. Convenient synthetic methods for modified triglycerides would be of interest both for a study of biologically active compounds such as platelet activating factors and for the preparation of compounds having technical uses (Berg, 1976; Calogero, 1979; Gama and Narazaki, 1979; Inagaki et al., 1978; Murakami et al., 1978; Watanabe et al., 1983). No vegetable oils containing α-alkyl branched triacyl glycerols are known, but such lipids have been identified in the preen glands of birds (Garton, 1985; Nuhn et al.,

1985). The aim of this work has been to study the synthesis of α-substituted triacyl glycerols, with special interest in reactions usable with mono- and triglycerides as starting materials. The substituents, mainly alkyl groups, were chosen to give triacyl glycerols with new properties, like lower melting points and greater stability toward hydrolysis and oxidation.

In principle it is possible to obtain α-substituted triacyl glycerols via three different approaches: a) In the first route the substituent is introduced into the fatty acid or its protected form. The triacyl glycerols can then be synthesized by esterification of glycerol or interesterification. b) In the second alternative the chemical modification is performed on the protected monoacyl glycerol and the triacyl glycerols are obtained after deprotection and subsequent acylation. c) Finally, it is also possible to make the α-substitution directly on the triacyl glycerol. This is an attractive possibility, since it might be extended to α-substitution of vegetable oils.

SYNTHESIS OF α-SUBSTITUTED LIPID DERIVATIVES

FATTY ACIDS AS STARTING MATERIALS

α-Alkyl substituted fatty acids have been prepared via alkylation of the corresponding 2-oxazolines, as described by Meyers (Meyers and Temple, 1970; Meyers et al., 1974).

The preparation of the 2-alkyl-4,5-dihydro-4,4-dimethyloxazoles was carried out either directly from the fatty acid and 2-amino-2-methyl-1-propanol, or stepwise from the acid chloride via the amido alcohol. The overall yield was 60-70% in both methods. The oxazolines from hexanoic, dodecanoic and octadecanoic acid were treated with n-butyllithium and alkyl iodides at -70°C, to give the α-alkyl branched oxazolines as shown in Scheme 1 (Herslöf and Gronowitz, 1983).

Most of them were prepared in two ways: either a long chain oxazoline was alkylated with a short alkyl iodide, or a short chain oxazoline was alkylated with a long chain alkyl iodide. The α-alkylated oxazolines were isolated in yields ranging from 65-82%, irrespective of the route chosen. The branched oxazolines were hydrolysed to the corresponding acids. Due to steric hindrance the α-alkyl oxazolines were more resistant to hydrolysis than the straight chain analogues. Prolonging the reaction time from 20 minutes to 3 days gave the α-alkyl fatty acids in yields varying from 60 to 85%. From the fatty acids the monoglycerides were prepared according to a standard procedure (Jensen and Pitas, 1976). The reaction path is shown in Scheme 2. The overall yields of α-alkyl branched monoacyl glycerols starting from straight chain acids were 20-25% (Herslöf et al., 1983). The greatest advantage of the α-alkylation via the oxazolines

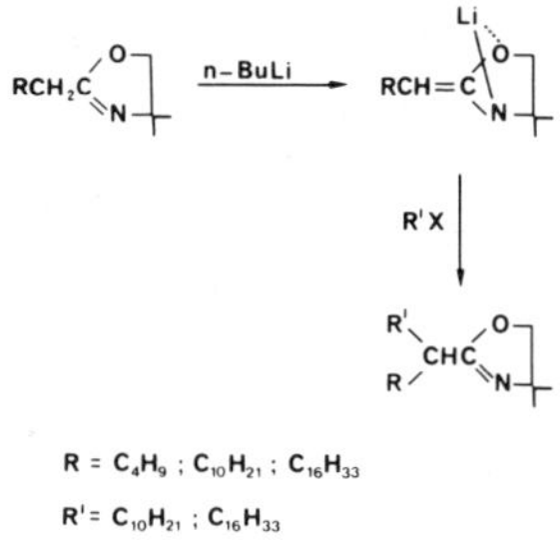

SCHEME 1

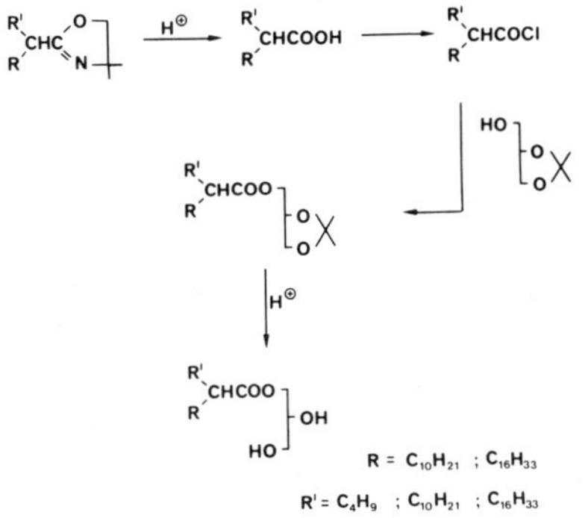

SCHEME 2

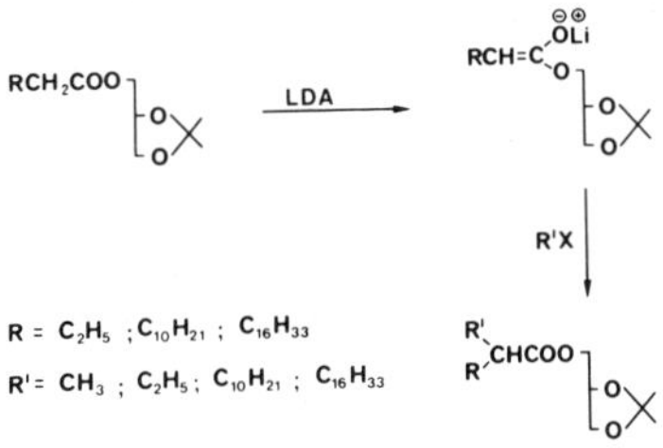

SCHEME 3

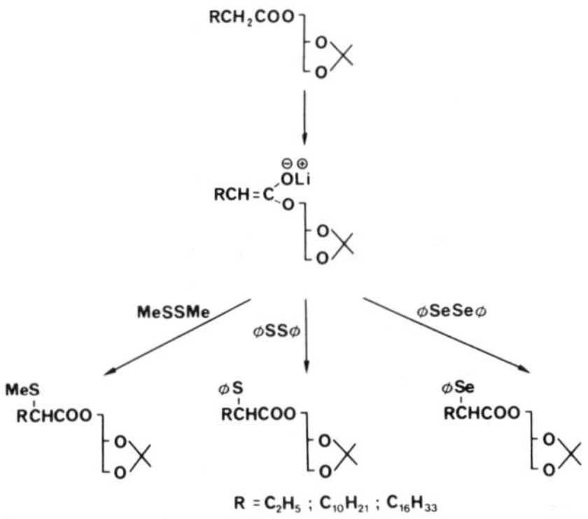

SCHEME 4

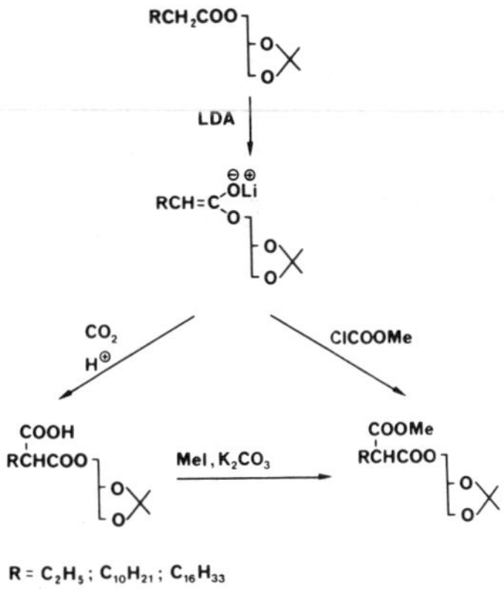

SCHEME 5

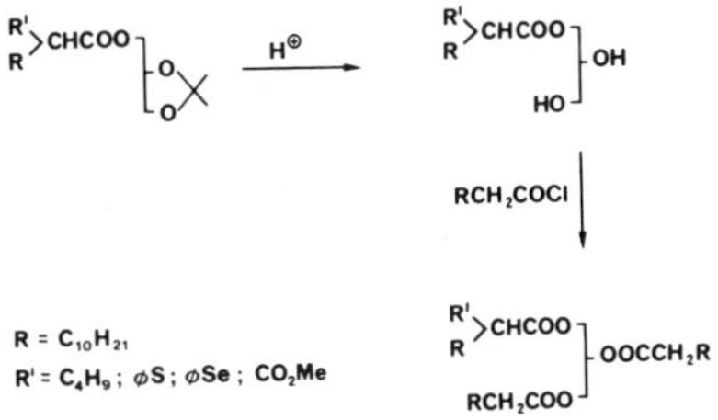

SCHEME 6

compared to other alkylation methods (Pryde, 1979) is the easy purification of the branched oxazolines by column chromatography.

PROTECTED MONOACYL GLYCEROLS AS STARTING MATERIALS

Acylisopropylidene glycerols have been used as starting materials for the preparation of enolate anions. Butanoyl-, dodecanoyl- and octadecanoylisopropylidene glycerols were used as starting materials, in order to represent short, medium and long chain monoacyl glycerols. The anions of these compounds were reacted with alkyl halides (Rathke and Lindert, 1971; Cregge et al., 1973) of different chain lengths (Herslöf and Gronowitz, 1985), disulfides (Trost and Salzmann, 1973; Trost, 1978; Seebach and Teschner, 1976; Trost et al., 1976) and diselenides (Brocksom et al., 1974; Reich et al., 1975; Sharpless et al., 1973). Carboxylation in the α-position was performed by reaction of the anions with carbon dioxide (Reiffers et al., 1971) or methyl chloroformate (Krapcho et al., 1974; Herslöf and Gronowitz, 1985). The reactions are summarized in Schemes 3, 4 and 5.

ALKYL HALIDES AS ELECTROPHILES

The enolate anions of butanoyl- and dodecanoylisopropylidene glycerol were prepared at -70°C by the addition of lithium diisopropylamide to the acylisopropylidene glycerols in THF. The solutions of the anions were transferred with cooling (-70°C) to solutions of the alkyl halides in THF-DMSO at room temperature. The yields of alkylated products after reaction with methyl, butyl, decyl and hexadecyl iodide were between 60 and 90%. With octadecanoylisopropylidene glycerol as starting material, no alkylated products were formed under identical conditions. This was probably due to the low solubility of octadecanoylisopropylidene glycerol in THF at -70°C. When the temperature for anion formation was raised to between -40 and -20°C, alkylated products were formed in 50-75% yields after treatment with alkyl iodides in THF-DMSO at room temperature. The reactions are shown in Scheme 3.

The resulting alkylated acylisopropylidene glycerols were isolated by column chromatography or distillation. Acidic cleavage of the ketal group, as shown in Scheme 3, gave the α-alkyl branched monoacyl glycerols in an overall yield of about 35%, based on the straight-chain free fatty acids. This may be compared with the 20-25% yields that were obtained when the α-alkylated monoacyl glycerols were prepared according to the multistep route via alkylation of oxazolines, as shown in Schemes 2 and 3. The reaction of the acylisopropylidene glycerols offers another advantage in that many different alkylated monoacyl glycerols can be obtained from the same starting material. It is also possible to introduce other groups into the molecules, here exemplified by the reaction with disulfides, diselenides and carboxylating agents (Herslöf and Gronowitz, 1985).

DISULFIDES AND DISELENIDES AS ELECTROPHILES

In the sulfenylation reactions of the enolate anions (Scheme 4) the product yields were influenced by the reactivity of the disulfide and by the chain length of the acylisopropylidene glycerol. With dimethyl disulfide as electrophile, about 50% of the α-methyl-thiobutanoylisopropylidene glycerol was formed according to GLC. With the anions of dodecanoyl- and octadecanoylisopropylidene glycerols, only between 20 and 30% of the α-methylthio-substituted acylisopropylidene glycerols were obtained. Unlike the reactions with alkyl halides, some starting material was also found (15%). This might be formed by desulfenylation of the reaction product back to the starting acylisopropylidene glycerol, a reaction reported by Trost and Salzmann (1973). In the reaction of the enolate of butanoyl-isopropylidene glycerol with dimethyl disulfide, some of the side products are probably condensation products with or without thiomethyl groups. The more reactive diphenyl disulfide gave higher yields of the α-sulfenylated products. According to GLC, the yields were 68-69% after reaction with the three different acylisopropylidene glycerols. The isolated yields were between 40 and 50%, with the lowest yield for 2-phenyl-thiooctadecanoylisopropylidene glycerol. Reaction of the acylisopropylidene glycerols with diphenyl diselenide gave results similar to those in the reaction with diphenyl disulfide (Herslöf and Gronowitz, 1985).

CARBON DIOXIDE AND METHYL CHLOROFORMATE AS ELECTROPHILES

Polar groups were introduced into the acylisopropylidene glycerols by reaction of the enolate anions with carbon dioxide and methyl chloroformate. The resulting α-carboxyacylisopropylidene glycerols were methylated with methyl iodide in acetone in the presence of potassium carbonate (Moore et al., 1979). The yields of the crude α-carboxy-acylisopropylidene glycerols were between 62 and 85%, and the overall yields after the transformation to methyl esters were 40-55%. Higher yields (40-71%) were obtained by the reaction of the anions with methyl chloroformate, but the purification of these compounds was more complicated than that of those with a free carboxyl group. The reactions are summarized in Scheme 5 (Herslöf et al., 1985).

SYNTHESIS OF TRIACYL GLYCEROLS

From the α-substituted acylisopropylidene glycerols described above, the corresponding triacyl glycerols, of potential interest as lubricating agents, could be obtained. This was exemplified by the preparation of some triacyl glycerols from the α-substituted dodecanoyl derivatives. The reactions were performed according to standard procedures described in the literature as summarized in Scheme 6.

By the use of mono-α-butyldodecanoyl glycerol, the tridodecanoyl glycerol containing one butyl group was prepared. The tridodecanoyl glycerols with two and three butyl groups were obtained by reaction of monododecanoyl glycerol or glycerol with α-butyldodecanoyl chloride as shown in Scheme 7.

The triacyl glycerol containing three α-decyldodecanoyl groups was also prepared (Bohlin et al., 1986). The α-alkyltriacyl glycerols had lower melting points than the corresponding straight-chain analogues, as expected.

While the syntheses of the α-phenylthio- and α-phenylselenotri-dodecanoyl glycerols were straightforward, the preparation of the α-carbomethoxytridodecanoyl- glycerol was more complicated. After acylation of the monoacyl glycerol with dodecanoyl chloride, three products were formed in the proportions 1.2 : 1.7 : 1.0 according to GLC. These compounds could not be separated by column chromatography on silica gel or by preparative TLC.

TRIACYL GLYCEROLS AS STARTING MATERIALS

Application of the method described above to the synthesis of triacyl glycerols was unsuccessful. Compared to simple alkyl esters, the triacyl glycerols are likely to have a greater tendency to undergo self condensation, since there are always possibilities for internal reactions. To circumvent these problems, it was desirable to trap the enolate anions by the use of a reactive electrophile, before condensation took place. The electrophile of choice was trimethylsilyl chloride (TMSCl), which is very reactive at low temperatures and is also compatible with lithium diisopropylamide (LDA), and thus can be present when the anions are formed (Corey and Gross, 1984). The $\underline{O}$-trimethylsilyl ketene acetals formed by the reaction of ester enolates with TMSCl are rather stable compounds that can be handled at room temperature, as long as acids are avoided (Rathke and Sullivan, 1973; Ainsworth et al., 1972). According to the literature, they are also versatile, since several derivatives can be obtained by further reactions without the use of strongly basic media. It is for instance possible to perform α-t-alkylations (Reetz et al., 1983; Reetz and Schwellnus, 1978; Lion and Dubois, 1981), α-acylations (Rathke and Sullivan, 1973), aldol additions (Yamamoto et al.,

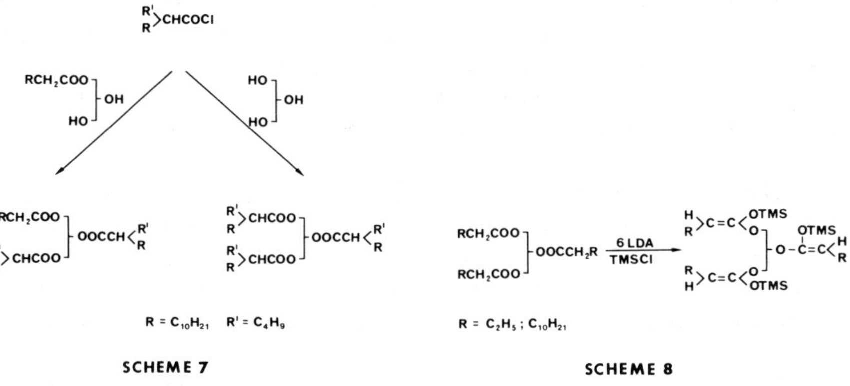

SCHEME 7

SCHEME 8

SCHEME 9

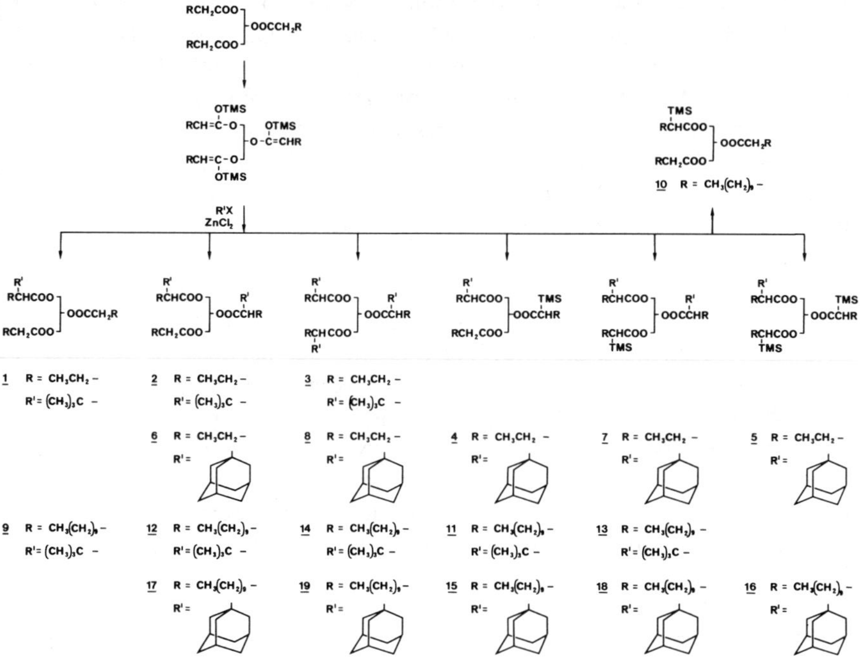

1984; Chan et al., 1979), Michael additions (Rajan-Babu, 1984; Miyashita et al., 1984) and oxidations (Rubottom et al., 1983). Alkenyl and alkyl groups can be introduced in the α-position through alkylation with the phenylsulfinyl chloride adduct of alkenes (Patel and Paterson, 1983), or through alkylation with chloromethyl phenyl sulfide (Paterson and Fleming, 1979).

The trimethylsilyl ketene acetals of tributanoyl glycerol and tridodecanoyl glycerol were formed by the addition of the triacyl glycerols to cooled solutions of LDA and TMSCl in THF. By the use of two equivalents of LDA per acyl group, silyl ketene acetals containing three trimethylsilyloxy groups were formed, as shown in Scheme 8 (Herslöf et al., 1986).

With tridodecanoyl glycerol some unreacted starting material was also obtained. Analyses of the product mixtures by ^{1}H NMR spectroscopy indicated that no trimethylsilyl ketene acetals with only one or two trimethylsilyloxy groups were formed under the conditions used. The ^{1}H NMR spectra also showed that the E forms of the ketene acetals were probably obtained, since the vinyl protons appeared at δ=3.64 ppm (Chan et al., 1979; Wilcox and Babston, 1984; Dubois et al., 1984). This is the expected isomer when the reaction is kinetically controlled (Corey and Gross, 1984). With three equivalents of base, the amounts of unreacted triacyl glycerols were increased, but mixed ketene acetal-acyl glycerols were still not formed. The addition of the cooled (-70°C) base to cooled (-70°C) mixtures of the triacyl glycerols and TMSCl also gave mainly ketene acetals containing three trimethylsilyloxy groups and starting material, probably together with some mixed ketene acetal-acyl glycerol. This behaviour suggests that the triacyl glycerols reacted in all three positions immediately when they were added to the LDA-TMSCl mixtures, and that some of the base was consumed in side reactions before the addition of the triacyl glycerols was completed, leaving some unreacted starting material. By adding the base to the triacyl glycerol and TMSCl the probability for the formation of ketene acetals with one and two TMS groups was increased, since the triacyl glycerol was present in excess during the reaction. Hydrolysis of the trimethylsilyl ketene acetals regenerated the starting material, together with small amounts of C-silylated compounds.

The trimethylsilyl ketene acetals from tributanoyl glycerol and tridodecanoyl glycerol were reacted with t-butyl chloride and 1-adamantyl bromide in the presence of zinc chloride in methylene chloride, according to the method described by Reetz and his coworkers (Reetz et al., 1983; Reetz and Schwellnus, 1978). The reactions and the products formed are shown in Scheme 9. The reaction of the trimethylsilyl ketene acetal of tributanoyl glycerol with t-butyl chloride was studied in some detail. It was then found that three new compounds were always formed. They were identified by GLC-MS (CI) to be the tributanoyl glycerols containing one, two and three α-t-butyl groups (compounds 1-3 in Scheme 9). Some starting material was also obtained. These results were always obtained, even though the starting ketene acetal contained three trimethylsilyloxy groups according to ^{1}H NMR spectroscopy. The highest yields of α-alkylated products were obtained when the trimethylsilyl ketene acetal, formed by the addition of the tributanoyl glycerol to a cooled solution of TMSCl and LDA (2 equiv./acyl group), was reacted with t-butyl chloride (1 equiv./acyl group) and zinc chloride in methylene chloride at room temperature. The reaction was complete in one hour. From the crude mixtures thus obtained, the triacyl glycerol fraction was isolated in about 50% yield. The specific triacyl glycerols were isolated by repeated flash chromatography and were characterized by ^{1}H NMR spectroscopy, MS (CI) and elemental analyses. It should be noted (Scheme 9) that only the kinds and numbers of the substituents were determined, and not in which of the acyl chains (1, 2 or 3) the substituents actually were situated. The purified triacyl glycerols might thus be mixtures of several positional isomers. In addition, new asymmetric centres were created during the α-alkylation, generating mixtures of diastereomers.

The reaction conditions described above (except that the reaction time was two hours) were used to alkylate tributanoyl glycerol with 1-adamantyl bromide, leading to the formation of five new products (4-8 in Scheme 9). They were isolated in the form of a triglyceride fraction, and each triacyl glycerol was purified by semi-preparative HPLC and characterized by[1]H NMR spectroscopy, MS (CI) and elemental analyses. It was found that in addition to the α-adamantyltriacyl glycerols 6 and 8, the triacyl glycerols containing both α-adamantyl and α-trimethylsilyl groups were formed. Products with alkyl groups and α-trimethylsilyl groups were also formed in the reactions of tridodecanoyl glycerol with t-butyl chloride and 1-adamantyl bromide (10,11,13 and 15,16,18 in Scheme 9 respectively). All compounds were isolated and identified by ^{1}H NMR spectroscopy and MS (CI). The yields of the triacyl glycerol fractions were between 40 and 80%.

The formation of the α-trimethylsilyltriacyl glycerols was quite unexpected. Since the hydrolysis of the silyl ketene acetals gave only minor amounts of C-silylated products, it seemed reasonable to believe that the α-TMS groups were introduced through a rearrangement during the alkylation. It was indeed found that treatment of the trimethylsilyl ketene acetal of tributanoyl glycerol with zinc chloride in methylene chloride at room temperature in the absence of alkylating agent led to the formation

(GLC-MS (CI)) of the α-trimethylsilyl-butanoyl glycerols shown in Scheme 10. The α-TMS derivatives were beginning to form after 15 minutes, and after 2 hours the composition of the mixture did not change. This reaction thus appeared to be somewhat slower than the alkylation of the same ketene acetal with t-butyl chloride. This would explain why α-TMS derivatives are not found in the t-butylation of the tributanoyl glycerol.

Rearrangements to the α-TMS esters instead of alkylations have been reported only when the TMS ketene acetals of acetates were used as starting materials (Reetz et al., 1983; Lutsenko et al., 1966). Others have reported that treatment of ketene acetals with Lewis acids in the absence of alkylating agents results in the formation of β-keto esters or/and succinate products (Inaba and Ojima, 1977; Totten et al., 1985; Jacobsen et al., 1985).

Desilylation of the α-trimethylsilyltributanoyl glycerols with potassium fluoride regenerated the starting material (Scheme 10). When the crude t-butylated tributanoyl glycerol mixture was desilylated under identical conditions, the product composition did not change, further indicating that no α-silylated products were formed in this particular reaction.

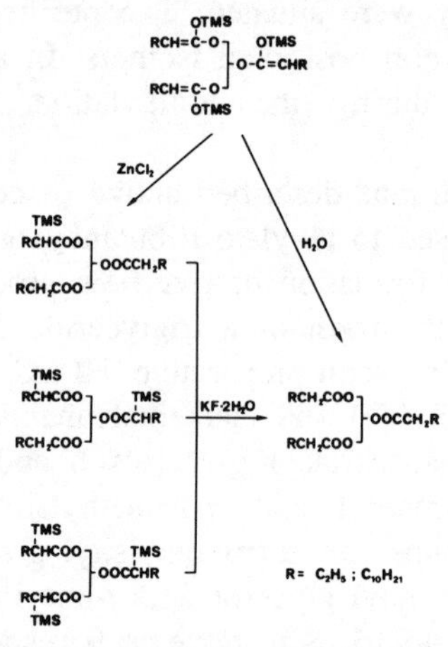

SCHEME 10

The reactions of the ketene acetals of tributanoyl glycerol and tridodecanoyl glycerol with adamantyl bromide, and of the ketene acetal of tridodecanoyl glycerol with t-butyl chloride, needed two hours for completion. The rate of formation of the α-silylated derivatives through rearrangement could thus be comparable. The α-trimethylsilylated compounds in the product mixtures could be desilylated leaving the compounds containing the t-butyl and adamantyl groups. In this way, the purification of the reacted triacyl glycerols was facilitated.

REFERENCES

Ainsworth, C., Chem, F. and Kuo, Y-N., J. Organometal. Chem. 46, 59 (1972).

Ault, W.C., Eisner, A., Bilyk, A. and Dooley, C.J., J. Am. Oil Chemists' Soc. 44, 506 (1967).

Ault, W.C., Micich, T.J., Stirton, A.J. and Bistline Jr., R.J., J. Am. Oil Chemists' Soc. 42, 233 (1965).

Aydin, A., Breusch, F.L. and Ulusoy, E., Chemica Acta Turcica 5, 93 (1977).

Berg, A., Dragoco Rep. 23, 159 (1976).

Bohlin, L., Hernqvist, L. and Herslöf, M., Fette Seifen Anstrichmittel, in press.

Brocksom, T.J., Petragnani, N. and Rodrigues, R., J. Org. Chem. 39, 2114 (1974).

Calogero, A.V., Cosmetics, Toiletries 94, 77 (1979).

Cason, J., Allinger, N.L. and Williams, D.E., J. Org. Chem. 18, 842 (1953).

Chan, T.H., Aida, T., Lau, P.W.K., Gorys, V. and Harpp, D.N., Tetrahedron Lett. 42, 4029 (1979).

Corey, E.J. and Gross, A.W., Tetrahedron Lett. 25, 495 (1984).

Creger, P.L., J. Am. Chem. Soc. 89, 2500 (1967).

Cregge, R.J., Herrmann, J.L., Lee, C.S., Richman, J.E. and Schlessinger, R.H., Tetrahedron Lett. 26, 2425 (1973).

Dubois, J-E., Axiotis, G. and Bertounesque, E., Tetrahedron Lett. 25, 4655 (1984).

Gama, Y. and Narazaki, H., Jpn. Kokai Tokkyo Koho, 79,106,425 (1979), Chemical Abstracts 92: 41351h.

Garton, G.A., Chem. Ind. 295 (1985).

Gilbert, E.E., "Sulfonation and Related Reactions", Interscience Publ., John Wiley and Sons Inc., New York, N.Y., 1965.

Harwood, H.J., Chem. Rev. 62, 99 (1962).

Hauser, C.R. and Hudson Jr., B.E., Org. Reactions 1, 266 (1942).

Herslöf, M. and Gronowitz, S., Chemica Scripta 22, 230 (1983).

Herslöf, M. and Gronowitz, S., J. Am. Oil Chemists' Soc. 62, 1013 (1985).

Herslöf, M. and Gronowitz, S., Chemica Scripta 25, 257 (1985).

Herslöf, M. and Gronowitz, S., Chemica Scripta, in press.

Inaba, S. and Ojima, I., Tetrahedron Lett. 23, 2009 (1977).

Inagaki, T., Nakagawa, H. and Isa, H., Jpn. Kokai Tokkyo Koho, 78,40,215 (1978), Chemical Abstracts 90: 88308y.

Jacobsen, E.N., Totten , G.E., Wenke, G., Karydas, A.C. and Rhodes, Y.E., Synth. Commun. 15, 301 (1985).

Jensen, R.G. and Pitas, R.E., Adv. Lipid Res. 14, 213 (1976).

Krapcho, A.P., Jahngen, E.G.E.,Jr. and Kashdan, D.S., Tetrahedron Letters 2721 (1974).

Lion, C. and Dubois, J.-E., Tetrahedron 37, 319 (1981).

Lutsenko, I.F., Baukov, Yu.I., Burlachenko, G.S. and Khasapov, B.N., J. Organometal. Chem. 5, 20 (1966).

Maerker, G., Kenney, H.E., Bilyk, A. and Ault, W.C., J. Am. Oil Chemists' Soc. 43, 104 (1966).

Meyers, A.I. and Temple, D.L., J. Am. Chem. Soc. 92, 6644 (1970).

Meyers, A.I., Temple, D.L., Nolen, R.L. and Mihelich, E.D., J. Org. Chem. 39, 2778 (1974).

Miyashita, M., Yanami, T., Kumazawa, T. and Yoshikoshi, A., J. Am. Chem. Soc. 106, 2149 (1984).

Moore, G.G., Foglia, T.A. and McGahan, T.J., J. Org. Chem. 44, 2425 (1979).

Murakami, T., Kawashima, M., Matsutani, N., Watanabe, N., Takai, M. and Onoda, K., Jpn.Kokai Tokkyo Koho, 78,108,084 (1978), Chemical Abstracts 90: 24270x.

Nuhn, P., Gutheil, M. and Dobner, B., Fette Seifen Anstrichmittel 87, 135 (1985).

Patel, S.K. and Paterson, I., Tetrahedron Lett. 24, 1315 (1983).

Paterson, I. and Fleming, I., Tetrahedron Lett. 11, 993 (1979).

Petragnani, N. and Yonashiro, M., Synthesis 521 (1982).

Pryde, E.H., "Fatty Acids", Am. Oil Chemists' Soc., Campaign, Illinois, 1979.

Rajan-Babu, T.V., J. Org. Chem. 49, 2083 (1984).

Rathke, M.W. and Lindert, A., J. Am. Chem. Soc. 93, 2318 (1971).

Rathke, M.W. and Sullivan, D.F., Synth. Commun. 3, 67 (1973).

Rathke, M.W. and Sullivan, D.F., Tetrahedron Lett. 15, 1297 (1973).

Reetz, M.T. and Schwellnus, K., Tetrahedron Lett. 17, 1455 (1978).

Reetz, M.T., Schwellnus, K., Hubner, F., Massa, W. and Schmidt, R.E., Chem. Ber. 116, 3708 (1983).

Reich, H.J., Renga, J.M. and Reich, I.L., J. Am. Chem. Soc. 97, 5434 (1975).

Reiffers, S., Weinberg, H. and Strating, J., Tetrahedron Letters 3001 (1971).

Rubottom, G.M., Gruber, J.M., Marrero, R., Juve Jr., H.D. and Kim, C.W., J. Org. Chem. 48, 4940 (1983).

Seebach, D. and Geiss, K.H., J. Organometal. Chem. Libr. 1, 1 (1976).

Seebach, D. and Teschner, M., Chem. Ber. 109, 1601 (1976).

Sharpless, K.B., Lauer, R.F. and Teranishi, A.Y., J. Am. Chem. Soc. 95, 6137 (1973).

Totten, G.E., Wenke, G. and Rhodes, Y.E., Synth. Commun. 15, 291 (1985).

Trost, B.M., Chem. Rev. 78, 363 (1978).

Trost, B.M. and Salzmann, T.N., J. Am. Chem. Soc. 95, 6840 (1973).

Trost, B.M. and Salzmann, T.N., J. Org. Chem. 40, 148 (1975).

Trost, B.M., Salzmann, T.N. and Hiroi, K., J. Am. Chem. Soc. 98, 4887 (1976).

Watanabe, S., Fujita, T., Suga, K. and Sugahara, K., J. Am. Oil Chemists' Soc. 60, 116 (1983).

Wilcox, C.S. and Babston, R.E., J. Org. Chem. 49, 1451 (1984).

Yamamoto, Y., Maruyama, K. and Matsumoto, K., Tetrahedron Lett. 25, 1075 (1984).

<u>Section</u> <u>2</u>

EICOSANOIDS

STRUCTURAL ASPECTS OF THE PROSTAGLANDIN ENDOPEROXIDES, THROMBOXANES AND PROSTACYCLINS

R.L. Jones

Department of Pharmacology, University of Edinburgh, Edinburgh, EH8 9JZ

INTRODUCTION

Over the past 20 years major advances have been made in our understanding of the role of prostaglandins in cellular function. In this review article I will deal with three classes of compound: the prostaglandin endoperoxides, the thromboxanes and the prostacyclins. In each case the natural agent is a chemically unstable, highly potent biological agonist and this has created several challenges to which both chemists and biologists have responded vigorously. These challenges include (a) the elucidation of the structure of the natural agent and its biosynthetic pathway, (b) chemical synthesis of the agent and comparison with biologically derived material, (c) estimation of the active species or its metabolites in biological fluids, (d) synthesis of stable mimetics, and (e) synthesis of biosynthesis inhibitors and receptor antagonists.

PROSTAGLANDIN ENDOPEROXIDES

The enzyme which converts arachidonic acid (20 : 4 ω6) to the prostaglandin endoperoxide PGH_2 is called prostaglandin endoperoxide synthetase (Fig.1). It is widely if not ubiquitously distributed in mammalian tissues and a particularly rich source is the seminal vesicles of the sheep. In 1964, both Bergström and his colleagues in Sweden (Bergström et al., 1964) and van Dorp and his colleagues in the Netherlands (van Dorp et al., 1964) first described the conversion of radioactive arachidonic acid into labelled PGE_2 by homogenates of sheep seminal vesicles and provided preliminary evidence for an endoperoxide intermediate. Some nine years later both groups described the isolation of pure PGG_2 and PGH_2 (Hamberg and Samuelsson, 1973; Nugteren and Hazelhof, 1973). Detailed studies of the purified enzyme complex from sheep and bovine seminal vesicles have indicated two separate catalytic activities, a cyclo-oxygenase which uses molecular oxygen to produce PGG_2 and a hydroperoxidase which reduces the 15- hydroperoxy group of PGG_2 to produce PGH_2 (van der Ouderaa and Buytenhek, 1982; Yamamato, 1982).

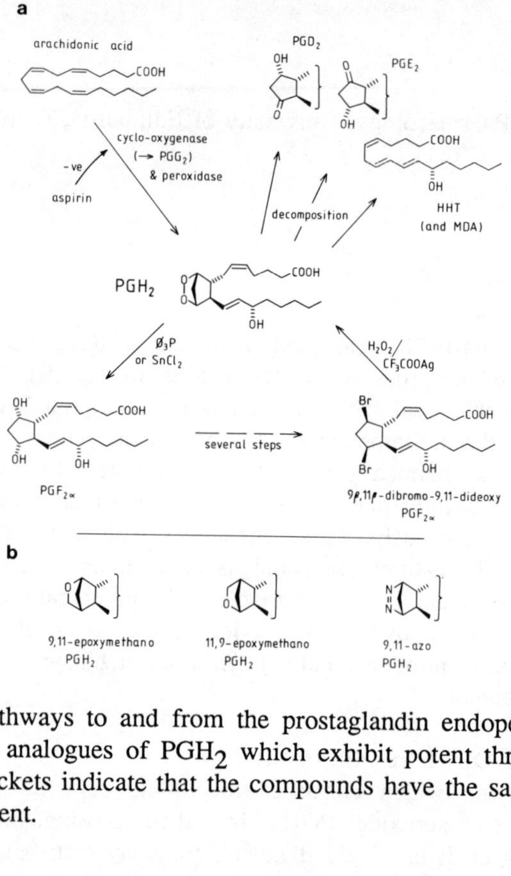

Figure 1. (a) Pathways to and from the prostaglandin endoperoxide PGH_2. (b) Some stable analogues of PGH_2 which exhibit potent thromboxane-like activity. The brackets indicate that the compounds have the same side-chains as the natural agent.

Non-steroidal anti-inflammatory drugs inhibit the cyclo-oxygenase moiety and this contributes to their therapeutic properties (Vane, 1971). In the case of aspirin, its acetyl group is transferred to a serine hydroxyl at the active site of the enzyme, producing an irreversible type of inhibition (van der Ouderaa et al., 1980; Roth et al., 1980). More potent inhibitors, such as indomethacin and meclofenamic acid, do not covalently bind to cyclo-oxygenase. In the case of ibuprofen the inhibition is stereospecific, the S-enantiomer being 160 times more active than the R-enantiomer on the enzyme in vitro. However inversion of configuration of the weakly active R-enantiomer occurs <u>in</u> <u>vivo</u> and the potency difference becomes much smaller (Adams et al., 1976).

Vesicular gland preparations continue to be of value. They are used to prepare PGG_2 and PGH_2 for biological study (Gorman et al., 1977; Ubatuba and Moncada, 1977) and high specific activity tritiated prostaglandins for use in radio-immunoassays. However, chemical methods of synthesis have been developed as well. In the earliest study (Johnson et al., 1977b) PGH_2 methyl ester was synthesized in 3% yield by treatment of $9\beta,11\beta$-dibromo-9,11-dideoxy $PGF_{2\alpha}$ methyl ester with potassium superoxide. Later workers (Porter et al., 1979b) improved the method by using 90% H_2O_2/silver trifluoroacetate to give about 20% yields of PGH_2 itself. The same group have elegantly prepared PGG_2 by reacting the H_2O_2/silver ion system with 15β-chloro-$9\beta,11\beta$-dibromo-9,11-dideoxy $PGF_{2\alpha}$ to simultaneously introduce the 9,11- and 15-peroxide groups (Porter et al.1979a).

The half-life of PGH_2 in neutral aqueous solution at 37°C is 3-4 min, the main product being PGE_2. In the presence of bovine or guinea-pig serum albumen PGD_2 predominates, whereas with egg albumen $PGF_{2\alpha}$ and 12<u>L</u>-hydroxy- 5,8,10-heptadecatrienoic acid (HHT) were the major

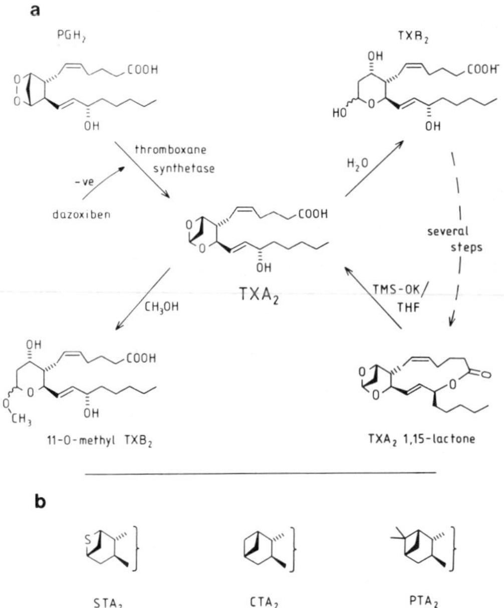

Figure 2. (a) Pathways to and from thromboxane A_2 (TXA_2). (b) Some stable mimetics of TXA_2.

components identified (Hamberg and Fredholm, 1976) (Fig.1). These decomposition profiles are quite dissimilar to that seen with the parent compound 2,3-dioxabicyclo(2.2.1)heptane (Salomon and Salomon, 1977, 1978). The instability of PGH_2 creates difficulties for the pharmacologist attempting to study its biological activity. The first problem is quantitation of PGH_2 in freshly prepared aqueous stock solutions. Treatment with stannous chloride or triphenylphosphine rapidly produces $PGF_{2\alpha}$ (Hamberg et al., 1974) which can be estimated by radio-immunoassay or gas chromatography mass spectrometry (with 3,3,4,4-tetradeutero $PGF_{2\alpha}$ as an internal standard): not all workers have indicated how their endoperoxide solutions were assayed. The second problem is that the decomposition products (e.g. PGD_2 and PGE_2) may exert biological actions of their own.

The activities which can be reasonably attributed to PGH_2 itself comprise vasoconstriction, bronchoconstriction and stimulation of platelet aggregation. However, it always appears to be less active than thromboxane A_2 and there is little evidence for the existence of distinct prostaglandin endoperoxide receptors. Nevertheless a number of stable analogues of PGH_2 have been synthesized which show potent thromboxane/prostaglandin endoperoxide-like activity. These include 11,9-epoxymethano PGH_2, 9,11-epoxymethano PGH_2 (Yankee et al., 1976) and 9,11-azo PGH_2 (Corey et al., 1975, 1976) (Fig.1). 9,11-Ethano PGH_2 showed partial agonist activity in several thromboxane-sensitive systems (Jones et al., 1982) and was the starting point for the synthesis of specific thromboxane receptor antagonists (see later).

THROMBOXANES

In 1975, Samuelsson and his colleagues reported that human platelets produced a novel product TXB_2 from PGH_2 (Hamberg et al., 1975). They proposed that TXB_2 arose from hydrolysis of a unique and unstable cyclic acetal TXA_2 (Fig.2). A major part of their evidence for the existence of TXA_2 came from trapping experiments in which excess amounts of nucleophiles (e.g. methanol or azide) were added to compete with water for the ring opening reaction. When methanol was used the 11-O-methyl derivative of TXB_2 was produced and this was identified by g.c. - m.s. TXA_2 was found to be a highly potent stimulant of human platelet aggregation in vitro. For some years previously Vane and colleagues had been studying the release of biologically active substances from guinea-pig lungs during anaphylactic shock (Piper and Vane, 1969). One agent, whose production was inhibited by aspirin and indomethacin, contracted the rabbit aorta and was named RCS (rabbit-aorta contracting substance). It had properties similar to the prostaglandin endoperoxides (Gryglewski and Vane,

1973). However, it was subsequently demonstrated that the major component of RCS was more unstable than the prostaglandin endoperoxides (Svensson et al., 1975) and it soon became obvious that it was the putative TXA_2 (t $1/2 = 30s$ under physiological conditions).

Recently TXA_2 has been synthesized from TXB_2 and its chemical nature established (Bhagwat et al., 1985). One crucial aspect of the synthesis is the temporary schielding of the acidic carboxyl group by formation of a 1,15-lactone. Subsequent oxetane formation yields the 1,15-lactone of TXA_2 which is a stable precursor of TXA_2 (Fig.2). Saponification of the lactone proceeds without affecting the cyclic acetal.

The conversion of PGH_2 to TXA_2 in biological tissues is catalyzed by thromboxane synthetase (Needleman et al., 1976). The enzyme has been separated from PGH synthetase found in bovine platelets (Yoshimoto et al., 1977). It is weakly inhibited by imidazole (Moncada et al., 1977; Needleman et al., 1977), whereas certain 1-substituted derivatives (e.g. dazoxiben) show much greater inhibitory activity (Tai and Yuan, 1978; Yoshimoto et al., 1978; Cross et al., 1981). It is only in platelet suspensions from a small proportion of human volunteers that thromboxane synthetase inhibitors produce a significant inhibition of aggregation induced by arachidonic acid (Heptinstall et al., 1980; Bertelé et al., 1984). In contrast thromboxane receptor antagonists are effective in all samples (Jones et al., 1985). These results indicate the importance of the prostaglandin endoperoxides in this test system.

Several stable analogues of TXA_2 have been prepared of which the 9,11-thia-11a-carba analogue STA_2 (Fig.2) appears to be most useful as a mimetic (Katsura et al., 1983). The dicarba analogue of TXA_2, CTA_2, exhibits some interesting biological properties. Like TXA_2 it shows high contractile activity on vascular smooth muscle, but is inhibitory in platelet aggregation tests (Nicolaou et al., 1980; Lefer et al., 1980; Ohuchida et al., 1983). Using material synthesized by the May & Baker Group (Ansell et al., 1982), we have found that CTA_2 does indeed show thromboxane-like activity on human platelets, but true activity in this respect is masked through its ability to raise cyclic AMP levels (Armstrong et al., 1985). This observation merits further investigation in terms of the ability of thromboxane mimetics to interact with either PGD or PGI receptors mediating inhibition of the human platelet function (see Armstrong et al., 1986). Interestingly the May and Baker Group have observed thromboxane-like activity with the 8-epi,12-epi diastereoisomer of CTA_2 (M.P.L. Caton, personal communication). Here again further work is required particularly in relation to assumptions made in the reported work regarding assignment of structure to separated diastereoisomers. Finally in the case of PTA_2 (Fig.2), which can be prepared from the chiral natural products (-)-myrtenol (Nicolaou et al., 1979) or (-)-nopol (Ansell et al., 1984; Wilson et al., 1982), partial agonist activity on a range of thromboxane-sensitive preparations including human platelets was seen (Armstrong et al., 1985).

The characterization of thromboxane receptors is gathering momentum owing to the development of specific competitive antagonists. A major proportion of these compounds have a natural α-chain, a stable ring system closely related to either PGH_2 or TXA_2, and a radically altered ω-chain. Typical examples are EP 092 (Armstrong et al., 1985) and ONO 11120 (Katsura et al., 1983) (Fig.3). Other compounds are obviously prostanoid in nature, for example AH 23848 (Brittain et al., 1985) whereas the achiral linear molecule BM 13177 bears little resemblance to either TXA_2 or PGH_2 (Patscheke and Stegmeier, 1984). In general pharmacological measurements on isolated smooth muscle preparations and platelet suspensions indicate a competitive mode of action for these antagonists. This is supported by data from ligand binding studies on human platelets. The following radioactive ligands have been used with some success - $[^3H]$-9,11-epoxymethano PGH_2 (Armstrong et al., 1983) and $[^{125}I]$-PTA-OH (Mais et al., 1985; Narumiya et al., 1986). In each case about 2000 specific binding sites per platelet were found. Improved binding systems may allow a comparison of thromboxane receptors in platelets with those in smooth muscle.

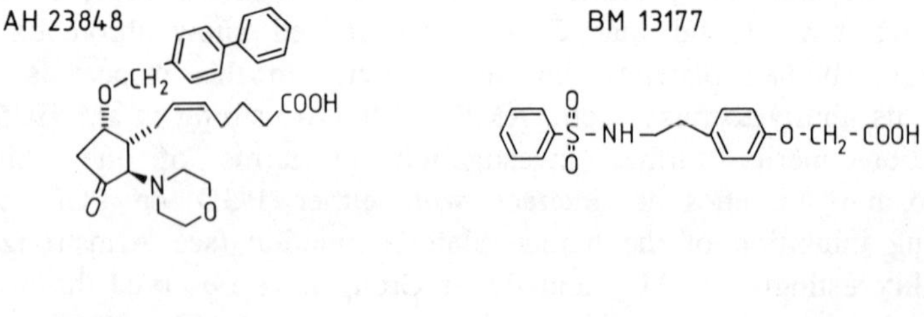

Figure 3. Structures of some thromboxane receptor antagonists. The radioligand $[^{125}I]$PTA-OH is the p-hydroxy-m-iodo derivative of ONO 11120.

PROSTACYCLINS

In general, many of the biological actions of prostacyclin (PGI_2) are opposite in nature to those of TXA_2. It is a potent inhibitor of platelet aggregation and a dilator in many peripheral vascular beds (Moncada et al., 1976a). It is formed from PGH_2 by prostacyclin synthetase and vascular endothelium is a rich source of the enzyme. Lipid peroxides inhibit the enzyme (Moncada et al., 1976b).

A joint effort between the prostaglandin research groups of Wellcome, U.K. and Upjohn, U.S.A. led to the structural proof of the PGI_2 structure and its hydrolysis product 6-oxo-$PGF_{1\alpha}$ (Johnson et al., 1976, 1977a). Treatment of $PGF_{2\alpha}$ methyl ester with basic iodine solution gives two diastereoisomeric iodoethers (Fig.4). Reaction of either isomer with DBN gives PGI_2 methyl ester, which is saponified and freeze-dried to give the sodium salt. The latter shows good stability if kept dry at -20°C; stock solutions are buffered to pH 9 to reduce hydrolysis.

PGI_2 may have a local role in preventing the spread of a thrombus

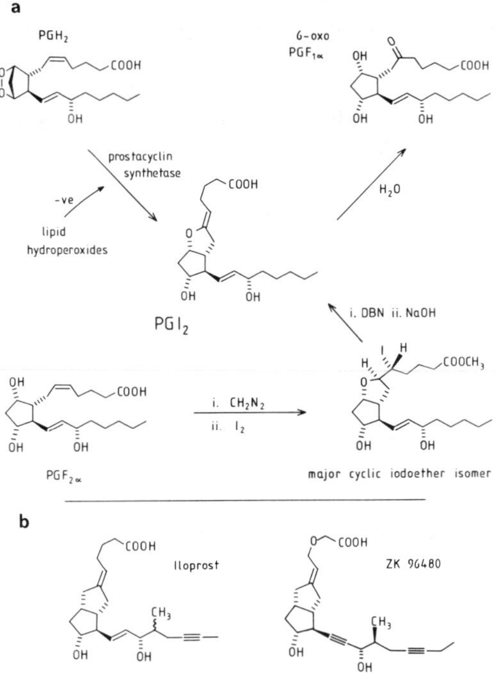

Figure 4. (a) Pathways to and from prostacyclin (PGI_2). (b) Two stable mimetics of PGI_2.

beyond that required for blood vessel sealing. Its discovery has opened up the possibility of suppressing platelet reactivity in certain cardiovascular disease states. For this purpose many groups have attempted to overcome the instability of PGI_2 ($\underline{t}$ 1/2 = 2-5 min at neutrality depending on buffer composition) by making changes to the vinyl ether unit. Three changes which have been attempted are reduction of the double bond, substitution of electron - withdrawing groups on or close to the vinyl ether, and substitution of the oxygen of the vinyl ether with sulphur, nitrogen or carbon. In general the PGI_1 analogues show only weak prostacyclin-like activity (Axen et al., 1978; Nicolaou et al., 1978), probably resulting from a loss of the unique stereochemistry at the junction of the α-chain and oxygen-containing ring. Substitution of two fluorine atoms at C10 produces at least a 100-fold increase in stability with little loss of bioactivity (Fried et al., 1980). Both the 7-oxo (Tömösközi et al., 1982) and 4-oxo (Galombos et al., 1983; Ono et al., 1983) derivatives of PGI_2 have platelet inhibitory and vasodilator activities. Replacement of the vinyl ether oxygen with carbon has been a very common strategy and a compound which has received much attention is Iloprost (Fig.4) (Skubulla and Vorbrüggen, 1981). As with the natural compound the 5$\underline{Z}$ isomer has much higher activity than the 5$\underline{E}$ isomer on human platelets (Schrör et al., 1982). Several carbocyclic analogues including Iloprost show considerable PGE-like contractile activity on smooth muscle (Dong et al., 1986). A recent addition to the carbacyclin armoury, ZK 96480 (Skubulla et al., 1986) (Fig.4), shows very high prostacyclin-like activity (Stürzebecher et al., 1986), but is devoid of PGE-like activity (Dong et al., 1986). The compound is of interest in terms of molecular modelling by computer because of its enhanced rigidity. Detailed discussion of the biological actions of prostacyclin and its analogues may be found in the review by Whittle and Moncada (Whittle and Moncada, 1984).

CONCLUDING REMARKS

Many research groups are still synthesizing novel variants of the prostaglandin endoperoxides, thromboxane A_2 and prostacyclin and it seems likely that interesting new biological activities will emerge. For more detailed information on the chemistry of the prostanoids the reader is referred to Volume 14 of the series Advances in Prostaglandin, Thromboxane and Leukotriene Research, edited by J.E. Pike and D.R. Morton, Jnr. (Raven Press, New York, 1985).

REFERENCES

Adams, S.S., Bresloff, P. and Mason, C.G. (1976). J. Pharm. Pharmac. 28: 256-257.

Ansell, M.F., Caton, M.P.L. and Stuttle, K.A.J. (1982). Tetrahedron Lett. 23: 1955-1956.

Ansell, M.F., Caton, M.P.L. and Stuttle, K.A.J. (1984). J. Chem. Soc. (Perkin I), 1069-1077.

Armstrong, R.A., Jones, R.L. and Wilson, N.H. (1983). Br. J. Pharmac. 79: 953-964.

Armstrong, R.A., Jones, R.L., Peesapati, V., Will, S.G. and Wilson, N.H. (1985). Br. J.Pharmac. 84: 595-607.

Armstrong, R.A., Jones, R.L., MacDermot, J. and Wilson, N.H. (1986). Br.J.Pharmac. 87: 543-551.

Axen, U., Lincoln, F.H., Thompson, J.L., Honohan, T. and Nishizawa, E.E. (1978). In: Prostaglandins in Cardiovascular and Renal Function. Edited by A. Scriabine, A.M. Lefer and F.A. Kuehl. pp. 3-8. MTP Press, Ltd., England.

Bhagwat, S.S., Hamann, P.R., Still, C., Bunting, S. and Fitzpatrick, F.A. (1985). Nature 315: 511-513.

Bergström, S., Danielsson, H. and Samuelsson, B. (1964). Biochim. Biophys. Acta 90: 207-210.

Bertelé, V., Falanga, A., Tomasiak, M., Cerletti, C. and de Gaetano, G. (1984). Thromb. Haemostas. 51: 125-128.

Brittain, R.T., Boutal, L., Carter, M.C., Coleman, R.A., Collington, E.W., Geisow, H.P., Hallett,P., Hornby, E.J., Humphrey, P.P.A., Jack, D., Kennedy, I., Lumley, P., McCabe, P.J., Skidmore, I.F., Thomas, M. and Wallis, C.J. (1985). Circulation 72: 1208-1218.

Corey, E.J., Narasaka, K. and Shibasaki, M. (1976). J. Am. Chem. Soc. 98: 6417-6418.

Corey E.J., Nicolaou, K.C., Machida, Y., Malmsten, C.L. and Samuelsson, B. (1975). Proc. Natl. Acad. Sci. U.S.A. 72: 3355-3358.

Cross, P.E., Dickinson, R.P., Parry, M.J. and Randall, M.J. (1981). Agents Actions 11: 274-280.

Dong, Y., Jones, R.L. and Wilson, N.H. (1986). Br.J.Pharmac. 87: 97-107.

Fried, J., Mitra, D.K., Najarajan, M. and Mehrotra, M.M. (1980). J. Med. Chem. 23: 234-237.

Galombos, G., Simonidesz, V., Ivanics, J., Horváth, K. and Kovács, G. (1983). Tetrahedron Lett. 24: 1281-1284.

Gorman, R.R., Sun, F.F., Miller, O.V. and Johnson, R.A. (1977). Prostaglandins 13: 1043-1053.

Gryglewski, R. and Vane, J.R. (1973). Br. J. Pharmac. 46: 449-457.

Hamberg, M. and Fredholm, B.B. (1976). Biochim. Biophys. Acta 431: 189-193.

Hamberg, M. and Samuelsson, B. (1973). Proc. Natl. Acad. Sci. U.S.A. 70: 899-903.

Hamberg, M., Svensson, J. & Samuelsson, B. (1975). Proc. Natl. Acad. Sci. U.S.A. 72: 2994-2998.

Hamberg, M., Svensson, J., Wakabayashi, T. and Samuelsson, B. (1974). Proc. Natl. Acad. Sci. U.S.A. 71: 345-349.

Heptinstall, S., Bevan, J., Cockbill, S.R., Hanley, S.P. and Parry, M.J. (1980). Thrombosis Res. 20: 219-230.

Johnson, R.A., Lincoln, F.H., Thompson, J.L., Nidy, E.G., Mizsak, S.A. and Axen, U. (1977a). J. Am. Chem. Soc. 99: 4182-4184.

Johnson, R.A., Morton, D.R., Kinner, J.H., Gorman, R.R., McGuire, J.C., Sun, F.F., Whittaker, N., Bunting, S., Salmon, J.A., Moncada, S. and Vane, J.R. (1976). Prostaglandins 12: 915-928.

Johnson, R.A., Nidy, E.G., Baczynskyj, L. and Gormon, R.R. (1977b). J. Am. Chem. Soc. 99: 7738-7740.

Jones, R.L., Peesapati, V. and Wilson, N.H. (1982). Br. J. Pharmac. 76: 423-438.

Jones, R.L., Wilson, N.H. and Armstrong, R.A. (1985). In: Mechanism of stimulus-response coupling in platelets. Edited by J. Westwick, M.F. Scully, D.E. MacIntyre and V.V.Kakkar. p.p. 67-81. Plenum Press, New York.

Katsura, M., Miyamoto, T., Hamanaka, N., Kondo, K., Terada, T., Ohyaki, Y., Kawasaki, A. and Tsuboshima, M. (1983). In: Adv. Prostaglandin, Thromboxane and Leukotriene Res. Vol.11, edited by B. Samuelsson, R. Paoletti and P. Ramwell. pp. 351-357, Raven Press, New York.

Lefer, A.M., Smith, E.F., Araki, H., Smith, J.B., Aharony, D., Claremon, D.A., Magolda, R.L. and Nicolaou, K.C. (1980). Proc. Natl. Acad. Sci. U.S.A. 77: 1706-1710.

Mais, D.E., Burch, R.M., Saussy, D.L., Kochel, P.J. and Halushka, P.V. (1985). J. Pharmac. Exp. Ther. 235: 729-734.

Moncada, S., Bunting, S., Mullane, K., Thorogood, P. and Vane, J.R. (1977). Prostaglandins 13: 611-618.

Moncada, S., Gryglewski, R., Bunting, S. and Vane, J.R. (1976a). Nature 262: 663-665.

Moncada, S., Gryglewski, R., Bunting, S. and Vane, J.R. (1976b). Prostaglandins 12: 715-737.

Narumiya, S., Okuma, M. and Ushikubi, F. (1986). Br.J.Pharmac. 88: 323-331.

Needleman, P., Moncada, S., Bunting, S., Vane, J.R., Hamberg, M. and Samuelsson, B. (1976). Nature 261: 558-600.

Needleman, P., Raz, A., Ferrendelli, J.A. and Minkes, M. (1977). Proc.Natl. Acad. Sci. U.S.A. 74: 1716-1720.

Nicolaou, K.C., Barnette, W.E. and Magolda, R.L. (1978). In: Prostaglandins in Cardiovascular and Renal Function. Edited by A. Scriabine, A.M. Lefer and F.A. Kuehl. pp. 9-28. MTP Press Ltd., England.

Nicolaou, K.C., Magolda, R.L. and Claremon, D.A. (1980). J. Am. Chem. Soc. 102: 1404-1409.

Nicolaou, K.C., Magolda, R.L., Smith, J.B., Aharony, D., Smith, E.F. and Lefer, A.M. (1979). Proc. Natl. Acad. Sci. U.S.A. 76: 2566-2570.

Nugteren, D.H. and Hazelhof, E. (1973). Biochim. Biophys. Acta 326: 448-461.

Ohuchida, S., Hamanaka, N. and Hayashi, M. (1983). Tetrahedron 39: 4257-4261.

Ono, T., Sibasaka, I., Nokami, J. and Wakabayashi, S. (1983). Chem.Lett. 1249-1250.

Patscheke, H. and Stegmeier, K. (1984). Thrombosis Res. 33: 277-288.

Piper, P.J. and Vane, J.R. (1969). Nature 223: 29-35.

Porter, N.A., Byers, J.D., Ali, A.E. and Eling, T.E. (1979a). J. Amer. Chem. Soc. 102: 1183-1184.

Porter, N.A., Byers, J.D., Holden, K.M. and Menzel, D.B. (1979b). J. Amer. Chem. Soc. 101: 4319-4322.

Roth, G.R., Siok, C.J. and Ozols, J. (1980). J. Biol. Chem. 255: 1301- 1304.

Salomon, R.G. and Salomon, M.F. (1977). J. Am. Chem. Soc. 99: 3501-3503.

Salomon, R.G. and Salomon, M.F. (1978). J. Am. Chem. Soc. 100: 660-662.

Schrör, K., Darius, H., Ohlendorf, R., Matzky, R. and Klaus, W. (1982). J. Cardiovasc. Pharmac. 4: 554-561.

Skubulla, W. and Vorbrüggen, H. (1981). Angew. Chem. Int. Ed. Engl. 20: 1046-1048.

Skubulla, W., Schillinger, E., Stürzebecher, C-St. and Vorbrüggen, H. (1986). J. Med. Chem. 29: 313-315.

Stürzebecher, S., Haberey, M., Mueller, B., Schillinger, E., Schroeder, G., Skubulla, W., Stock, G., Vorbrüggen, H. and Witt, W. (1986). Prostaglandins 31: 95-109.

Svensson, J., Hamberg, M. and Samuelsson, B. (1975). Acta Physiol. Scand. 94: 222-228.

Tai, H.-H. and Yuan, B. (1978). Biochem. Biophys. Res. Commun. 80: 236- 242.

Tömösközi, I., Kanai, K., Györy, P. and Kovács, G. (1982). Tetrahedron Lett. 23: 1091-1094.

Ubatuba, F.B. and Moncada, S. (1977). Prostaglandins 13: 1055-1066.

Van der Ouderaa, F.J.G. and Buytenhek, M. (1982). In: Methods in Enzymology, Vol.86, pp. 60-68. Eds. W.E. Lands and W.L. Smith, Academic Press, New York.

van der Ouderaa, F.J., Buytenhek, M., Nugteren, D.H. and van Dorp, D.A. (1980). Acetylation of prostaglandin endoperoxide synthetase with acetylsalicylic acid. Eur. J. Biochem. 109: 1-8.

van Dorp, D.A., Beethuis, R.K., Nugteren, D.H. and Vonkeman, H. (1964). Nature 203: 839-841.

Vane, J.R. (1971). Nature (New Biology) 231: 232-235.

Whittle, B.J.R. and Moncada, S. (1984). In: Progress in Medicinal Chemistry, Vol.21. Eds. G.P. Ellis and G.B. West. pp. 237-279. Elsevier, Oxford.

Wilson, N.H., Peesapati, V., Jones, R.L. and Hamilton, K. (1982). J. Med. Chem. 25: 495-500.

Yamamoto, S. (1982). In: Methods in Enzymology, Vol.86, pp. 55-60. Eds. W.E. Lands and W.L. Smith. Academic Press, New York.

Yankee, E.W., Ayer, D.E., Bundy, G.L., Lincoln, F.H., Miller, W.L., Jnr., and Youngdale, G.A. (1976). In: Advances in Prostaglandin and Thromboxane Research, Vol.1. Edited by B. Samuelsson and R. Paoletti. pp. 195-203. Raven Press, New York.

Yoshimoto, Y., Yamamoto, S. and Hayaishi, O. (1978). Prostaglandins 16: 529-540.

Yoshimoto, T., Yamamoto, S., Okuma, M. and Hayaishi, O. (1977). J. Biol. Chem. 252: 5871-5874.

SYNTHESIS OF NOVEL CARBOCYCLIC PROSTACYCLIN ANALOGUES

M.P.L. Caton, E.C.J. Coffee, C.J. Hardy, T.W. Hart and H.J. Mason

The Research Laboratories, May & Baker Ltd., Dagenham, Essex, RM1O 7XS

INTRODUCTION

Prostacyclin (PGI_2) (Fig. 1) is a potent antithrombotic substance formed from the arachidonic acid cascade along with the other prostanoids and the leukotrienes. The discovery of prostacyclin in 1976 (Moncada, Gryglewski, Bunting and Vane, 1976) opened up a new chapter in prostanoid research because of its important role in the physiology of the cardiovascular system and potential application in therapy. Prostacyclin biosynthesis is particularly significant in the vascular vessel walls where its antithrombotic and vasodilator properties are important in preventing the formation of intravascular clots, without interfering with the formation of platelet plugs at bleeding points involving thromboxane A_2. The pharmacological actions of prostacyclin include: inhibition of intravascular blood platelet aggregation; vasodilatation; hypotension; increase of renal blood flow; promotion of renin release; inhibition of gastric acid secretion.

Extensive biological studies and clinical evaluation which have been reviewed in detail elsewhere (Lewis and O'Grady, 1981; Vane and Bergström, 1979; Moncada and Vane, 1980; Szczeklik and Gryglewski, 1985) led to the marketing of prostacyclin as Flolan by Wellcome, and Cycloprostin (epoprostenol) by Upjohn for the preservation of platelet numbers and function during and after pulmonary bypass, for the prevention of platelet aggregation in charcoal columns during the haemoperfusion of patients with fulminant hepatic failure, and as an alternative to heparin during renal dialysis, especially where there is a high risk of bleeding problems due to heparin.

Further clinical trials of prostacyclin and its analogues have suggested other applications, e.g., in angina, peripheral vascular disease and myocardial infarction (Belch et al., 1983; Bugiardini et al., 1984; Chiesa et al., 1985;

Henricksson et al., 1985). However, although some of these trials have shown promising results, there is as yet no clear indication that prostacyclins will find the large scale clinical use which was perhaps optimistically anticipated from the early studies. Side effects are a problem - for example prostacyclin infusion is accompanied by severe headache - and for selective action there is a need to dissociate the antithrombotic and antihypertensive properties which run closely parallel throughout most compounds of this type.

SYNTHESIS

Prostacyclin has been synthesised by a number of methods, notably by the elimination of hydrogen halide from a haloether derived from the readily available prostaglandin $F_{2\alpha}$. A major drawback to the clinical use of prostacyclin is its chemical instability which stems from the lability of the enol ether grouping. Much effort has therefore been directed to the synthesis of stable analogues which retain a sufficient level of useful biological properties and are also, hopefully, more selective in pharmacological action. Chemical stability has been achieved in three principal ways:- (1) Removal of the double bond at C_5. (2) Substitution of a suitably placed electron-withdrawing group at C_4, C_5, C_7 or C_{10} to reduce the electron density on the enol ether and thus render the double bond less susceptible to electrophilic hydrolysis. (3) Replacement of the enol ether oxygen by another atom, e.g., S, N or C to render the C_5 olefin less nucleophilic and therefore less acid labile.

Many analogues have been synthesised and prostacyclin chemistry has been the subject of several reviews (Johnson, 1985; Aristoff, 1985 and Taylor, 1985). The compound where the ring oxygen atom is replaced by a methylene grouping, which is perhaps the most widely known prostacyclin analogue, has been named carbacyclin (Fig. 1). The pharmacological profile of carbacyclin is very similar to that of prostacyclin itself, although it is generally about one tenth as potent.

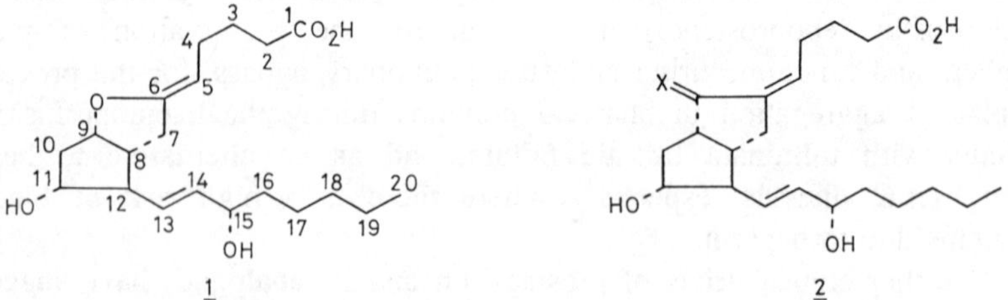

Figure 1

The objective of the present work was to synthesise carbacyclin analogues bearing a ketonic or hydroxy substituent on the ring carbon which had replaced the enol ether oxygen in PGI_2, thus giving a compound having the stability of the carbacyclin ring together with an oxygen function in a region of the molecule where that feature may be important for receptor binding and hence biological activity.

The bicyclo[3,3,0]oct-2-en-6-one <u>6</u>, which bears a ketonic function in the position required in the target molecule, was selected as a key intermediate for this synthesis (Fig. 2). It was envisaged that the hydroxyoctenol chain (β-chain) together with the ring hydroxyl group could be inserted via a suitable addition to the ring olefin and the carboxyalkyl chain (α-chain) attached by a condensation onto the active methylene adjacent to the ketone.

The alcohol <u>5</u> corresponding to <u>6</u> was available, as a racemic mixture, by cyclisation of cycloocta-1,5-diene <u>3</u> to the diacetoxybicyclo[3,3,0]-octene <u>4</u>, followed by pyrolytic cleavage of a molecule of acetic acid and hydrolysis of the remaining acetoxy function (Chernyuk, Melnikova and Pivnitskii, 1981). Oxidation with pyridinium chlorochromate then gave <u>6</u> which was converted to the ethylene glycol ketal <u>7</u>.

Functionalisation of the double bond was achieved by conversion with <u>N,N</u>-dibromo-dimethylhydantoin to the bromohydrin mixture <u>8a</u> and <u>8b</u> (70% yield). As in an analogous reaction in the bicycloheptene series reported by Grudzinski and Roberts (1975), the bromination was stereospecific with respect to the other ring, this being attributed to the formation of an exo-bromonium ion which then underwent nucleophilic attack at the less hindered face. Reaction of the bromohydrin <u>8</u> with sodium hydroxide in methanol afforded the epoxide <u>9</u> in 60% yield which underwent coupling with the chiral intermediate (h, Fig. 2) by the methods of Yamaguchi, Nobayashi and Hirao (1984) to give a 76% yield of a 2:1 mixture of the two adducts 10a and 10b, the required isomer being isolated by chromatography.

Hydrolysis of the protective groupings and conversion of the resulting ketone diol to the disilyl derivative <u>11</u> was followed by aldol condensation (70% yield) with 4-methoxycarbonyl butanal to insert the α-chain. Dehydration of the mesylate of the product <u>12</u> by base treatment (DBU) afforded the <u>trans</u> enone <u>13</u> (61% yield from 12), having the opposite double bond stereochemistry to that required for analogy with prostacyclin. However, irradiation of <u>13</u> with UV, after protecting group hydrolysis brought about partial isomerisation to the required <u>cis</u> enone <u>15</u> from which the latter was isolated by chromatography.

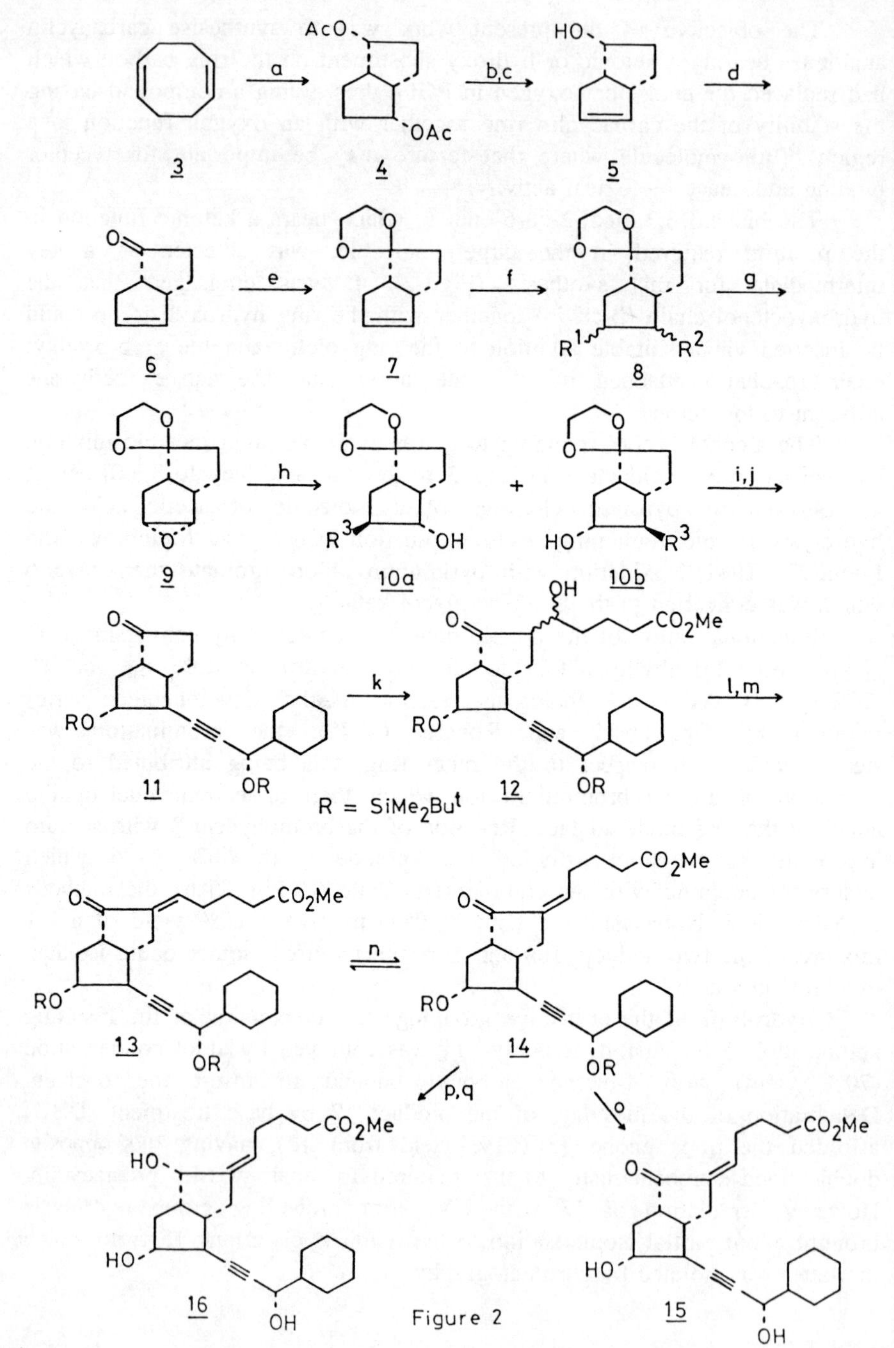

R = SiMe₂Buᵗ

Figure 2

Legend to Fig. 2

a $Pb(OAc)_4$, $LiCl$, $PdCl_2$, $AcOH$

b Pyrolysis (600°)

c aq. $NaOH$

d Pyridinium chlorochromate

e Ethylene glycol

f $\underline{N},\underline{N}$-Dibromodimethylhydantoin

g $NaOH$, $MeOH$

h $Li^+[BF_3C{\equiv}C-\overset{\vdots}{C}H-C_6H_{11}]^-$
   $\quad\quad\quad\quad\quad\overset{\vdots}{O}SiMe_2Bu^t$

i aq. $AcOH$

j Bu^tMe_2SiCl

k $OHC(CH_2)_3CO_2Me$, $LiN(SiMe_3)_2$

l $MeSO_2Cl$, Et_3N, DBU

m aq. $AcOH$

n UV (254 nm)

o $Bu_4^nN^+\ F^-$

p $NaBH_4$

q $Bu_4^nN^+\ F^-$

	$\underline{8a}$	$\underline{8b}$
R^1	---OH	—Br
R^2	—Br	---OH

$R^3 = -C{\equiv}C-\overset{\vdots}{C}H-\text{C}_6\text{H}_{11}$
$\quad\quad\quad\quad\quad\overset{\vdots}{O}SiMe_2Bu^t$

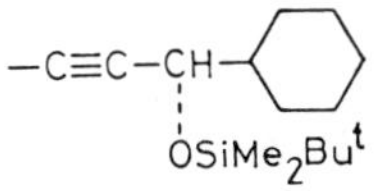

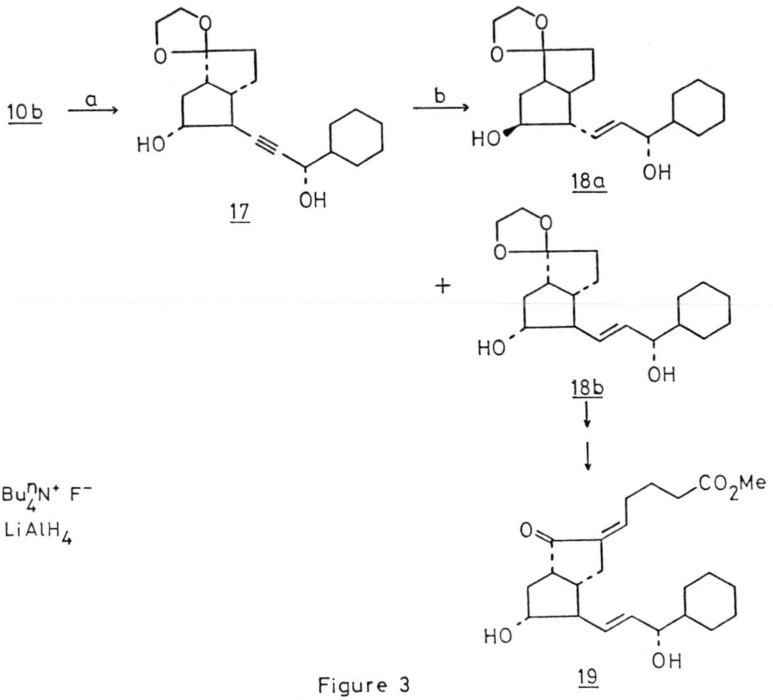

$\underline{10b} \xrightarrow{\ a\ } \underline{17} \xrightarrow{\ b\ } \underline{18a} + \underline{18b} \longrightarrow \underline{19}$

a $Bu_4^nN^+\ F^-$
b $LiAlH_4$

Figure 3

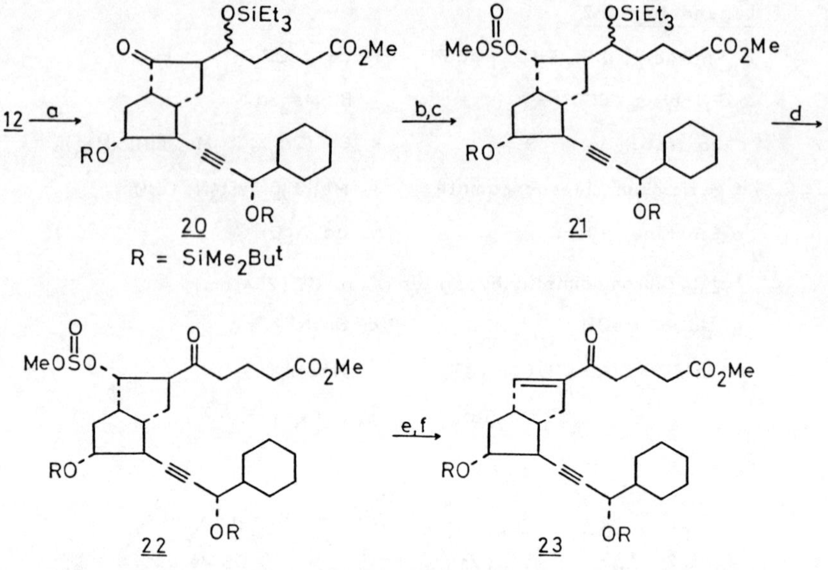

a   Et$_3$SiCl   d   Jones reagent

b   NaBH$_4$ e   DBU

c   MeSO$_2$Cl, Et$_3$N   f   HF, aq. MeCN

Figure 4

TABLE 1

Effect of carbocyclic prostacyclin analogues on blood pressure in anaesthetised rats.

Compound	Dose causing a 25% fall in mean blood pressure µg/kg i.v.
15	1.8
16	8.2
23	6.5
Carbacyclin	3.9
Carbacyclin Me ester	3.6
Prostacyclin Na salt	0.28
Prostacyclin Me ester	0.42

It was subsequently found preferable to carry out the olefinic isomerisation prior to removal of the silyl groups as this afforded a more satisfactory separation of the isomers. This gave a 60:40 <u>trans</u>:<u>cis</u> mixture from which a 36% yield of <u>cis</u> isomer was isolated. Borohydride reduction of the intermediate ketone <u>14</u> followed by hydrolysis afforded the corresponding hydroxy compound <u>16</u>.

The products <u>15</u> and <u>16</u> and intermediates from the coupling stage <u>12</u> onwards were mixtures of two diastereoisomers since they were derived from the racemic intermediate <u>4</u>. These mixtures were chromatographically homogeneous and attempts to achieve a separation were unsuccessful. However, the corresponding olefin, obtained by lithium aluminium hydride reduction of the dihydroxy ketal <u>17</u> from hydrolysis of intermediate <u>10b</u>, was separable into diastereoisomeric components <u>18a</u> and <u>18b</u> (Fig. 3). By a similar reaction sequence to that used to take <u>10b</u> forward, <u>18b</u> afforded the diastereoisomerically pure product <u>19</u>.

Another novel prostacyclin analogue <u>23</u>, isomeric with <u>15</u>, was synthesised from the alcohol <u>12</u> as shown in Fig. 4. The silyl ether <u>20</u> of <u>12</u> was reduced with borohydride and the mesylate <u>21</u> of the product treated with Jones reagent to effect desilylation and oxidation to the ketone <u>22</u>. Base treatment then brought about hydrolysis of the mesylate and dehydration of the β-hydroxyketone gave <u>23</u> after desilylation.

The prostacyclin analogues in these series were synthesised with the cyclohexyl acetylenic rather than the octenyl ω-chain of prostacyclin as these features afford resistance to metabolic inactivation, e.g., by dehydrogenation of the 15-hydroxyl group which is characteristic of the natural prostanoids. The analogues showed biological activity comparable to that of the prostacyclin. The effects of compounds 15, 16 and 23 on rat blood pressure in comparison with prostacyclin and carbacyclin are recorded in Table 1.

ACKNOWLEDGEMENTS
The authors thank Dr A. Ashford and Mr R. Jordan for the pharmacological evaluation.

REFERENCES
Aristoff P.A. (1985) Advances in Prostaglandin, Thromboxane and Leukotriene Research. Vol. 14, Ed. J.E. Pike and D.R. Morton Jr. Raven Press, New York, 309-392

Belch, J.J.F., Drury, J.K., Capell, H., Forbes, C.D., Newman, P., McKenzie, F., Leiberman, P. and Prentice, C.R.M., (1983) Lancet 1: 313-315

Belch, J.J.F., McArdle, B., Pollock, J.G., Forbes, C.D., McKay, A., Leiberman, P., Lowe, G.D.O. and Prentice, C.R.M. (1983) Lancet 1: 315-317

Bugiardini, R., Gridelli, C., Galvani, M., Ferrini, D., Puddu, P. and Lenzi, S. (1984) Circulation 70 (suppl.), II: 44.

Chernyuk, K.Y., Melnikova, V.I. and Pivnitski, K.K. (1981) Biorg.Khim. 7: 1866-1876

Chiesa, R., Vicari, A., Mari, G., Galimberti, M., Di Carlo, V. and Pozza, G. (1985) The Lancet 95-96

Grudzinski, Z. and Roberts, S.M. (1975) J. Chem. Soc. Perkin Trans. 1: 1767-1773

Henriksson, P., Edhag, O. and Wenmalm, A. (1985) Br. Heart J., 53: 173-179

Johnson, R.A. (1985) Advances in Prostaglandin, Thromboxane and Leukotriene Research, Vol. 14. Ed. J.E. Pike and D.R. Morton Jr., Raven Press, New York 131-154

Lewis, P.J. and O'Grady, J. (1981) Clinical Pharmacology of Prostacyclin, Raven Press, New York

Moncada, M., Gryglewski, R., Bunting, S. and Vane, J.R., (1976) Nature (London) 263: 663-665

Moncada, S. and Vane, J.R. (1980) Clinical Pharmacology and Therapeutics, Macmillan, London.Ed. P. Turner, 15-32

Szczeklik, A. and Gryglewski, R.J. (1985) Advances in Prostaglandin, Thromboxane and Leukotriene Research, Vol. 13, eds. G.V.R. Born, J.C. McGiff, G.G. Neri Serneri and R. Paoletti, Raven Press, New York, pp. 345-354

Taylor, R.J.K. (1985) New Synthetic Routes to Prostaglandins and Thromboxanes. Ed. S.M. Roberts and F. Scheinmann, Academic Press, London and New York, pp. 135-158 and 191-241

Vane, J.R. and Bergström, S. (1979) Prostacyclin,Raven Press, New York

Yamaguchi, H., Nobayashi, Y. and Hirao, I. (1984) Tetrahedron,40, 4261-4266 and references cited therein

ANALYSIS OF LIPOXYGENASE PRODUCTS BY THERMOSPRAY LIQUID CHROMATOGRAPHY-MASS SPECTROMETRY

Graham W.Taylor, David Watson, Colin T. Dollery and Ramsay Richmond

Department of Clinical Pharmacology, Royal Postgraduate Medical School, Hammersmith Hospital, Ducane Road, London W12 OHS

INTRODUCTION

Arachidonic acid (5,8,11,14-$\underline{Z},\underline{Z},\underline{Z},\underline{Z}$-eicosatetraenoic acid) is mobilised from phospholipid stores by action of phospholipases, and is metabolised by two main oxidative pathways to a series of biologically important lipids: the prostanoids and the leukotrienes. These species are believed to play an important role in inflammation, and much effort has been expended in their structural analysis, and in the development of suitable quantitative assays for determining <u>in vivo</u> concentrations. Both biological and immunological assays are in current use for many of the eicosanoids; without prior h.p.l.c. purification, they generally lack specificity. Mass spectrometric techniques on the other hand, especially when directly coupled to gas or liquid chromatographs, can offer unrivalled specificity for the analysis of a wide range of species. One such application - that of thermospray liquid chromatography-mass spectrometry - will be discussed in relation to the analysis of some lipoxygenase metabolites of arachidonic acid.

MASS SPECTROMETRIC ANALYSIS

Mass spectrometry (m.s.) has played an important part in eicosanoid studies; indeed, the technique has proved invaluable in the structural elucidation of virtually every member of the prostanoid and leukotriene families (e.g. Hamberg and Samuelsson, 1966; Borgeat and Samuelsson, 1979; Morris, Taylor, Piper and Tippins, 1980). Further, quantitation down to picogramme levels is now routine using mass spectrometric assays (Barrow, Heavey, Ennis, Chappell, Blair and Dollery, 1984). Although mass spectrometric techniques can offer the dual advantages of sensitivity and specificity, there are some drawbacks, notably regarding the purity and physico-chemical state of the sample. There are essentially two basic protocols in current use for mass spectrometric analysis of the eicosanoids: chromatographic purification followed by direct probe analysis (as was used in the structure elucidation of leukotriene D_4: Morris et al. 1980), and secondly, direct coupling of the chromatographic system to the mass spectrometer. In the latter case, the most common form of analysis is by gas chromatography-mass spectrometry, which has been widely applied in the eicosanoid area for many years. The major drawback of g.c.m.s. is the absolute requirement for sample volatility and thermal stability. Many polar or thermally labile species (such as the prostaglandins) must be converted to suitable derivatives prior to g.c.m.s. analysis: for example, ketones are methoximated, hydroxyls are trimethylsilylated and carboxylic acid functions are esterified. Although this approach has been successfully applied for the qualitative and quantitative analysis of most prostanoids, the peptido-leukotrienes are not amenable to g.c.m.s. Direct probe analysis has been successful, both by electron impact m.s. of derivatives and also with soft ionisation techniques such as fast atom bombardment (f.a.b.) m.s. (Morris, Dell, Judkins, McDowell, Panico and Taylor, 1981), although these approaches necessitated h.p.l.c. purification prior to analysis. This clearly limits the sensitivity of the technique; handling losses occur each time the sample is dried down and resuspended (i.e. between h.p.l.c. and m.s.). Direct coupling of the h.p.l.c. to the mass spectrometer should overcome many of the problems associated with g.c.m.s. or h.p.l.c.-probe analysis. Polar species may be chromatographed at ambient temperature, minimising thermal breakdown, derivatisation is unnecessary, eliminating unwanted side reactions, and handling steps are reduced. H.p.l.c.-m.s. interfaces have been available for some time; the most widely used being the "moving belt" and the "chemical ionisation spray" interfaces. Neither has been widely applied to eicosanoid analysis for a variety of reasons. Recently, a new "thermospray" h.p.l.c.-m.s. interface has been introduced, which promises to be a useful tool in eicosanoid studies.

THERMOSPRAY LIQUID CHROMATOGRAPHY – MASS SPECTROMETRY

Thermospray l.c.m.s. has been commercially available for two years, during which time it has been applied to the analysis of a wide variety of substances including drugs and their metabolites (Liberato, Fenselau, Vestal and Yergey, 1983), pharmaceuticals (Catlow, 1985) and steroid conjugates (Watson, Taylor and Murray, 1985; Watson, Taylor and Murray, 1986). Thermospray l.c.m.s. is based on an interface developed by Vestal (Blakey and Vestal, 1984; Vestal, 1984), where the h.p.l.c. eluant is passed into the mass spectrometer ion source through a heated narrow jet (Figure 1). The

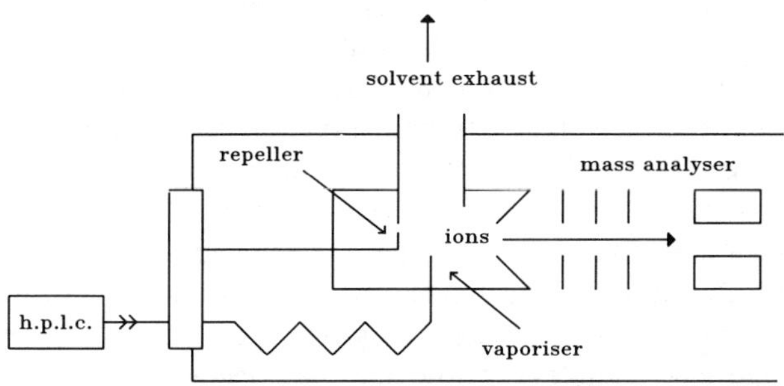

Figure 1. Diagram of the thermospray ion source. H.p.l.c. eluant, at up to 2 ml/min of polar solvents, enters the source and is vaporised. This results in ionisation and formation of molecular ion species $((M+H)^+, (M-H)^-)$ which are mass analysed to generate the thermospray mass spectrum.

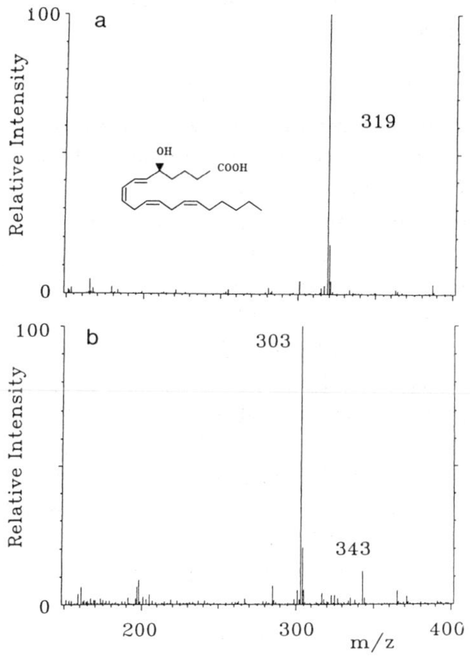

Figure 2(a) Negative ion thermospray mass spectrum of 5-HETE. The deprotonated molecular ion $(M-H)^-$ is observed at m/z 319. (b) Positive ion thermospray mass spectrum of 5-HETE showing the $(M-H_2O+H)^+$ ion at m/z 303. The protonated molecular ion is not observed.

solvent volatilises as a supersonic spray, resulting in solvent-mediated ionisation of the sample. Soft ionisation occurs; molecular ion species are generated $((M+H)^+, (M-H)^-)$ with little fragmentation. The technique is thus suitable for molecular weight determination (cf. f.a.b.m.s.), and, as the ion current is carried mainly by the molecular ion, shows promise for quantitation. Clearly, with mass spectra obtained directly post h.p.l.c., handling losses are minimised. Full mass spectra may be generated on 200-1000 ng, with the limit of detection using selected ion monitoring techniques at about 1-5 ng on column. Spectra may also be obtained via an l.c. bypass loop (on purified samples) allowing the operating parameters to be optimised. In our hands, the negative ion thermospray mass spectra are most reproducible, and although positive spectra may be obtained from the eicosanoids, higher background ionisation, and fragmentation to give water loss, make the negative ion spectra more suitable for qualitative and quantitative work. The thermospray mass spectra of a number of lipoxygenase products of arachidonic acid are reported below, with the advantages and drawbacks of the method compared with conventional mass spectrometric techniques.

HYDROXYEICOSATETRAENOIC ACIDS

There are various lipoxygenase enzymes present in different cells, which catalyse the formation of a number of isomeric hydroxy-eicosatetraenoic acids (HETEs). Many of these isomers possess weak chemotactic bioactivity for human neutrophils (Goetzl, Woods and Gorman, 1977), and some (for example 15-HETE) have been implicated in the control of the important 5-lipoxygenase pathway (Vanderhoek, Bryant and Bailey, 1980). Although g.c.m.s. methodology has been applied to the analysis of these species (Woollard and Mallet, 1984), in our hands the technique has not proved entirely suitable, notably because of problems of thermal instability and poor gas chromatography. The thermospray behaviour of the HETEs was examined (Richmond, Clarke, Watson, Chappell, Dollery and Taylor, 1986). Each HETE generates an intense positive ion mass spectrum corresponding to water loss from the protonated molecular ion (Figure 2b, $(M+H)^+ - H_2O$: m/z 303). The deprotonated molecular ion is the base peak in the negative ion mass spectrum (Figure 2a); the absolute mass spectrometric response of m/z 319 was proportional to the amount of sample loaded onto the h.p.l.c. The negative ion mode was used to monitor the h.p.l.c. eluant of HETEs generated by ionophore A23187 stimulated human inflammatory cells (Figure 3); 5-, 9- and 11-HETEs were observed. Thermospray l.c.m.s. was used to monitor HETE production from these cells in the presence of cyclosporin. Anecdotal evidence had linked cyclosporin treatment with a remission of the symptoms of psoriasis (Mueller and Hermann, 1979). As HETEs had previously been observed in psoriatic scale (Woollard and Mallet, 1984), it was considered that cyclosporin could act via inhibition of lipoxygenase. In fact, using the semi-quantitative thermospray m.s. assay, no significant change in the ionophore-stimulated release of HETEs was observed.

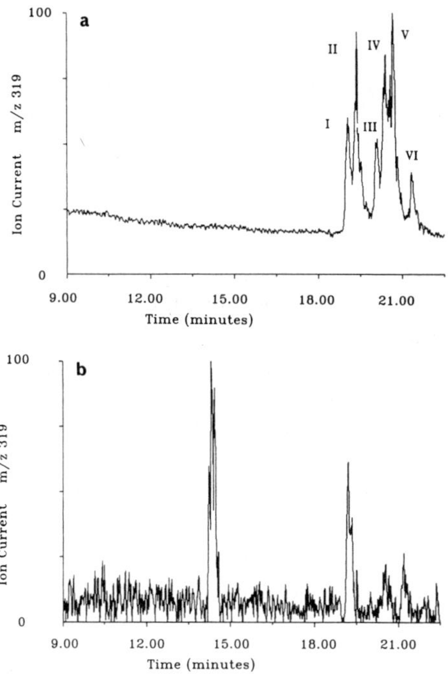

Figure 3(a) H.p.l.c. - negative ion thermospray mass spectrum of 6 authentic HETE isomers. I: 14/15 HETEs (not resolved); II: 11-HETE; III: 8-HETE; IV: 9-HETE; V: 12-HETE; VI: 5-HETE. (b) H.p.l.c. profile of HETEs generated from human inflammatory cells. 5-, 9- and 11-HETEs are present.

LEUKOTRIENES

The leukotrienes (LTs) are perhaps the most interesting lipoxygenase metabolites, arising via peroxidation at the 5 position of arachidonic acid (Samuelsson, 1983). There are two distinct sub-classes in this family arising from a common epoxide intermediate (leukotriene A_4):

1. The dihydroxyeicosatetraenoic acids including leukotriene B_4 (5-S, 12-R-dihydroxy 6,8,10,14-Z,E,E,Z-eicosatetraenoic acid), a potent chemotactic agent believed to be important in recruiting neutrophils to sites of infection.

2. The peptido-leukotrienes (LTC_4, D_4 and E_4), constituents of Slow-Reacting Substance of Anaphylaxis, a putative mediator of bronchospasm in asthma, are formed through the glutathione detoxication pathway. Leukotrienes C_4, D_4, E_4 are respectively 5-(S)-hydroxy 6-(R)-{S-glutathionyl, S-cysteinyl-glycinyl and S-cysteinyl} 7,9,11,14-E,E,Z,Z-eicosatetraenoic acids. Although leukotriene B_4 will chromatograph on g.c. as a suitable derivative (and indeed, there are g.c.m.s. assays available for this species (MacDermot et al. 1985)), there have been no reports that the peptidoleukotrienes are amenable to g.c., presumably because of their thermal instability at the column operating temperatures. Molecular weight data had previously been obtained on peptido-LTs by f.a.b. mass spectrometry (Morris et al. 1981); in biological samples, detection was limited to microgramme amounts by the requirement for prior h.p.l.c. purification. In contrast, the leukotrienes all generate intense negative ion thermospray mass spectra post h.p.l.c. at sub-microgramme levels. In each case, the deprotonated molecular ions $((M-H)^-)$ were the base peak and little fragmentation occurred - Leukotriene B_4 (m/z 335); Leukotriene C_4 (m/z 624); Leukotriene D_4 (m/z 495); Leukotriene E_4 (m/z 438).

The positive ion spectra $((M+H)^+)$ are also generated. In buffer, the leukotrienes may be detected at the 5-10 ng level on column. To investigate the thermospray behaviour of LTs in complex matrices, 1.2 µg each of LTC_4, D_4, E_4 and B_4 were added to 2.5 ml of whole blood. The blood was mixed with nordihydroguaritic acid/serine borate/cysteine in isotonic saline to inhibit formation and metabolism post collection (Heavey, Richmond, Turner, Kobza-Black, Taylor, Chappell, Barrow and Dollery, 1986). Further, to reduce handling losses, the plasma was extracted and chromatographed in one step using a column switching device (Taylor, Chappell, Clarke, Heavey, Turner, Watson, and Dollery, 1986). The negative ion thermospray spectra were obtained directly post h.p.l.c. Based on the known recoveries of immunoreactive LTs from blood, it was estimated that 300 ng of each leukotriene were presented to the mass spectrometer post h.p.l.c. Each of the four LTs could be detected (Figure 4a) and spectra containing the molecular ions could readily be obtained (Figure 4b). Thus molecular weight data are readily generated on relatively small amounts of these species. The data compare extremely favorably with the other soft ionisation techniques (such as f.a.b.m.s.) and point the way for structural studies on putative leukotriene metabolites such as glucuronides.

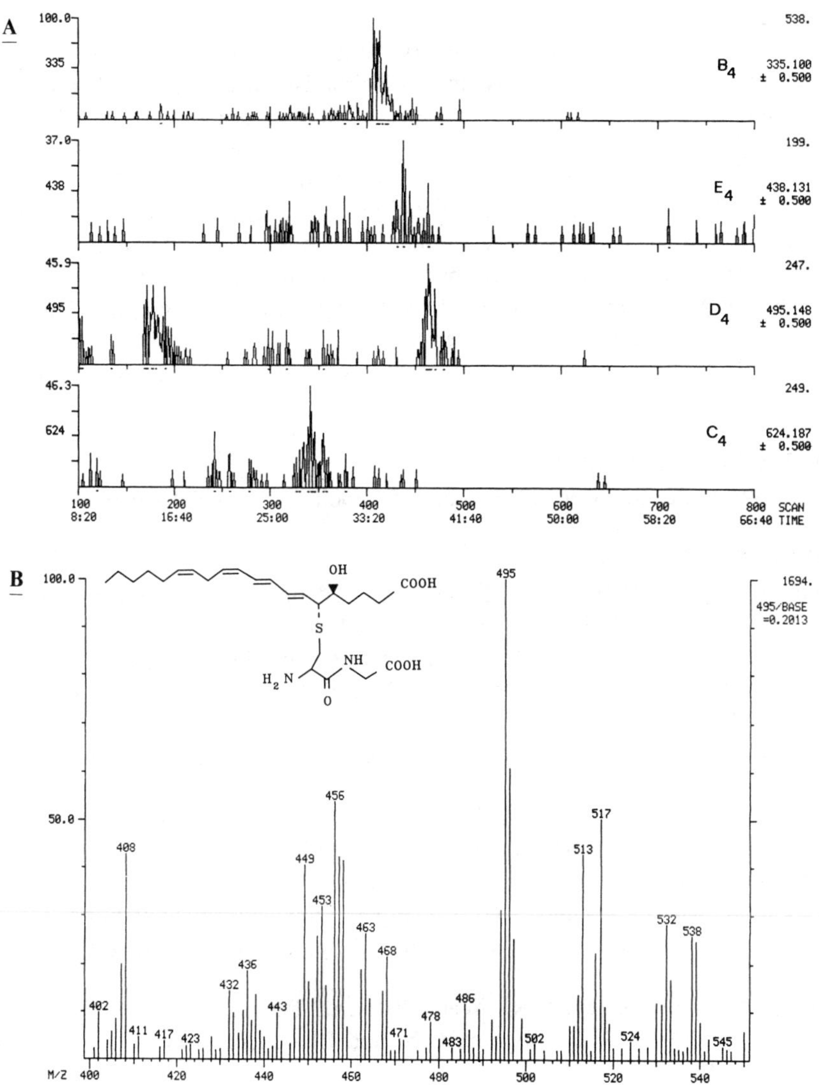

Figure 4(a) H.p.l.c.-negative ion thermospray mass spectrometric profile of exogenous leukotrienes C_4, D_4, E_4 and B_4 following extraction from whole blood. The four species were monitored at m/z 624, 495, 438 and 335 respectively. (b) Molecular ion region of the negative ion thermospray mass spectrum of leukotriene D_4 (extracted from whole blood) showing the deprotonated molecular ion at m/z 495.

CONCLUSION

Thermospray mass spectrometry, although commercially available for less than two years, is already showing promise as an analytical tool in the eicosanoid area. The clearest benefit of the technique is the ability to generate molecular weight data on polar, thermally labile substances in complex biological matrices without the need for derivatisation, and with minimal handling. The major limitation at present is that of sensitivity: although full spectra can be obtained at the sub-microgramme level, the limits of detection for assay purposes are insufficient (at 1-5 ng) for thermospray to compete effectively with h.p.l.c.-r.i.a. or g.c.m.s. However, as the thermospray interface is refined to increase sensitivity, the technique should become an important and widespread tool for both qualitative and quantitative analysis.

ACKNOWLEDGEMENTS

We thank the Medical Research Council for financial support.

REFERENCES

Barrow, S.E., Heavey, D., Ennis, M., Chappell, G.C., Blair, I.A. and Dollery, C.T. (1982) J. Chromatogr. 239, 71-80.

Blakey C.R. and Vestal, M.L. (1984) Anal. Chem. 55, 750-754.

Borgeat, P. and Samuelsson, B. (1979) J. Biol. Chem. 254, 2643-2646.

Catlow, D.A. (1985) J. Chromatogr. 323, 163-170.

Goetzl, E.J., Woods, J.M. and Gorman, R.R. (1977) J. Clin. Invest. 59, 179-183.

Hamberg, M. and Samuelsson, B. (1966) J. Biol. Chem. 241, 257-263.

Heavey, D.J., Richmond, R., Turner, N., Kobza-Black, A., Taylor, G.W., Chappell, C.G., Barrow, S.E. and Dollery, C.T. (1986) in "Leukotrienes; their biological significance", ed. P.J. Piper, Raven Press, New York.

Liberato, D.J., Feneslau, C.C., Vestal, M.L. and Yergey, A.L. (1983) Anal. Chem. 55, 1741-1744.

MacDermot, J., Kelsey, C.R., Waddell, K.A., Richmond, R., Knight, R.K., Cole, P.J., Dollery, C.T., Landon, C.N. and Blair, I.A. (1984) Prostaglandins 27, 163-179.

Morris, H.R., Dell, A., Judkins, M.B., McDowell, R.A., Panico, M. and Taylor, G.W. (1981) in "Peptides: synthesis-structure-function" (Proc. VIIth Amer. Peptide Symp.), ed. D.H.Rich and E.Gross, pp. 745-755, Pierce Chemical Co. USA.

Morris, H.R., Taylor, G.W., Piper, P.J. and Tippins, J.R. (1980) Nature 285, 104-106.

Mueller, W. and Hermann, B. (1979) N. Engl. J. Med. 301, 555-562.

Richmond, R., Clarke, S.R., Watson, D., Chappell, G.C., Dollery, C.T. and Taylor G.W. (1986) Biochim. Biophys. Acta 881, 159-166.

Samuelsson, B. (1983) Science 220, 568-575.

Taylor, G.W., Chappell, G.C., Clarke, S.R., Heavey, D.J., Turner, N.C., Watson, D. and Dollery, C.T. (1986) in "Leukotrienes: their biological significance" ed. P.J.Piper, Raven Press in the press.

Vanderhoek, J.Y., Bryant, R.W. and Bailey, J.M. (1980) J. Biol. Chem. 255, 10064-10065.

Vestal, M.L. (1984) Science 226, 275-281.

Watson, D., Taylor, G.W. and Murray, S. (1985) Biomed. Mass Spectrom. 12, 610-615.

Watson, D., Taylor, G.W. and Murray, S. (1986) Biomed. Environmen. Mass Spectrom. 13, 65-69.

Woollard, P.M. and Mallet, A.I. (1984) J.Chromatogr. 306, 1-21.

ELEVATED INTRACELLULAR CALCIUM STIMULATES PROSTACYCLIN PRODUCTION IN ENDOTHELIAL CELLS

Trevor J. Hallam

Department of Cellular Pharmacology, Smith Kline & French Research Ltd.,
The Frythe, Welwyn, Herts. AL6 9AR, U.K.

INTRODUCTION

Endothelial cells line the inner surface of all the blood vessels and form a permeability barrier against the loss of fluid and cells from the circulating blood into the surrounding tissues. The activity of other hormonally-sensitive cells which are in close proximity can also be regulated by substances released by the endothelial cell. These cells include the smooth muscle cells which are a major structural component of the artery and arteriole walls controlling the diameter and therefore the resistance to flow through these vessels, and the blood platelets which act to form a plug where vessels are damaged. The inhibitory actions are thought to be accounted for by two active lipid metabolites. Prostacyclin, PGI_2, is formed by endothelial cells on stimulation with various agonists (Weksler et al., 1978; Pearson et al. 1983). When PGI_2 interacts with specific receptor sites on the surface of the platelet it raises the cellular cyclic AMP levels which via a protein kinase inhibits platelet activation at a very early stage (Hallam et al. 1984) and the interaction of the platelet with the damaged vessel wall. Another factor, as yet unidentified, but termed endothelial-derived relaxing factor (EDRF) because it can cause the relaxation of pre-constricted smooth muscle, is also known to be released from endothelial cells on stimulation (DeMey & Vanhoutte, 1981; Furchgott, 1984). EDRF has a half-life of about 6 seconds as estimated by bio-assay and is also thought to be the product of arachidonic acid metabolism (Griffith et al. 1984), possibly via a cytochrome P450 activity.

The intracellular mechanism which controls the formation and release of PGI_2 and EDRF from the endothelial cell has so far not been determined. The rate-limiting step controlling the formation of prostaglandin endoperoxides has widely been believed to be the activation of the first enzyme in the cascade i.e. a phospholipase A_2 which selectively hydrolyses the phospholipid molecule at the 2-acyl position on the glyceryl backbone. The enzyme has been thought to be solely regulated by an increase in the Ca^{2+} ion concentration. However, it has not been until very recently that the techniques have been available to actually measure the existing Ca^{2+} ion concentration within the cytoplasm of the cell during stimulation with an agonist, allowing simultaneous analysis of phospholipids and their metabolites to be made. Most of what is known about the activation of phospholipase A_2 in intact cells has been inferred from experiments where calcium-selective ionophores have been used. Ionophores increase the permeability of the plasma membrane and allow the cytoplasmic free calcium ion concentration, $[Ca^{2+}]_i$, to rise from its normal resting level of around $0.1\mu M$ as the internal concentration of Ca^{2+} ions try to equilibrate with the concentration extracellularly, ca. 1mM. In most experiments $[Ca^{2+}]_i$ would rise to very high concentrations well above the maximally stimulated levels with physiological agonists. Indeed all that is known in the endothelial cell is that calcium ionophores can stimulate the production of PGI_2 from cultured cells (Weksler et al. 1978) and that calcium ionophores can stimulate the relaxation of pre-constricted aortic rings which have an intact endothelial cell layer in a standard organ-bath (Furchgott, 1984). One further observation which appears to support a role for more physiological concentrations of cytoplasmic free calcium in the control of arachidonic acid metabolite formation and/or release, albeit rather circumstantial, is that the action of agonists that can cause relaxation of these rings in an endothelial cell-dependent manner can be inhibited by removing the extracellular bathing Ca^{2+} ions (Furchgott, 1984).

With the design and development of fluorescent indicators that can report the concentration of Ca^{2+} and a method of selectively loading these dyes into the cytoplasm of intact functioning cells (Tsien et al.1982,1985), it has now become possible to determine whether a particular cell response can be triggered by an increase in $[Ca^{2+}]_i$. Graded doses of calcium ionophores can be used to determine the threshold $[Ca^{2+}]_i$ to elicit the response, and it can be determined whether the agonists that normally cause a response do so by elevating $[Ca^{2+}]_i$ to sufficient levels (e.g., Hallam et al., 1985). Although an increase in $[Ca^{2+}]_i$ can cause a response, for example the release of PGI_2, it is conceivable that under normal conditions with the physiological stimulant, an intracellular excitatory pathway other than elevated $[Ca^{2+}]_i$ may be responsible.

In this paper I have made use of the highly fluorescent Ca^{2+} indicator fura-2 which can be loaded selectively into the cytoplasm of intact endothelial cells grown to a confluent monolayer on glass coverslips. This study using human umbilical vein endothelial cells shows the first measurements of cytoplasmic free calcium concentration during stimulation of the cells with the physiological agonist, thrombin. The effect of thrombin both on the release of PGI_2 and on $[Ca^{2+}]_i$ are described in detail.

METHODS

Endothelial cells were stripped from the veins of freshly removed human umbilical chords by partial digestion of the supporting matrix with collagenase. The cells were then collected by centrifugation at 100g for 5 min. and resuspended in Dulbecco's modified Eagles'/HCO_3^- medium with 10% foetal calf serum and 10% new-born calf serum containing penicillin and streptomycin (Pearson et al., 1978). The cells were then plated into $25cm^2$ growth area tissue culture flasks and incubated at 37^oC in a 5% CO_2-gassed humid incubator. After 6 days the cells formed a confluent layer and at this stage the cells were once more detached from the flasks with trypsin/EDTA and re-plated onto glass coverslips. After 24 hours in the same growth media the confluent layer of cells was now loaded with fura-2 by incubating the coverslips with the membrane-permeant penta-acetoxymethyl ester form of fura-2 for 45 min at 37^oC. During this period the esterified dye is hydrolysed back to its original Ca^{2+}-sensitive membrane-impermeant form by cytosolic esterases. At this stage the coverslips of confluent cells, containing around 50-100μM fura-2, were removed and placed in a physiological saline consisting of 145mM NaCl, 5mM KCl, 1mM $MgSO_4$, 10mM Hepes, 10mM glucose, pH 7.4 at 37^oC.

The coverslips were placed across the diagonal of a quartz cuvette containing the above physiological saline but with either 1mM $CaCl_2$ or 1mM EGTA. The cuvette was then placed in a thermostatted holder at 37^oC in a Spex dual-wavelength excitation fluorescence spectrophotometer. Fluorescence of the fura-2 was then recorded during the experiment with two alternating excitation wavelengths of 340nm and 380nm; 500nm emission. While using an excitation of 340nm gives an increase in the emitted light at 500nm with increasing $[Ca^{2+}]_i$, 380nm excitation gives a decrease in the light output with increasing $[Ca^{2+}]_i$. A ratio of the two outputs therefore gives a greater sensitivity and $[Ca^{2+}]_i$ can be determined directly from these ratios. A full discussion of calibration procedures is given elsewhere (Tsien et al., 1985; Hallam and Pearson, 1986).

PGI_2 formation and release was estimated by determining the total amounts of its stable metabolite 6-keto-$PGF_{1\alpha}$ in the bathing solution at fixed time points after stimulation. Aliquots of the supernatant were taken and later analysed by radioimmunoassay (Ager et al., 1982).

The experiments shown are single experiments typical of several.

RESULTS AND DISCUSSION

Figure 1 shows the effect of thrombin on the fluorescence of human umbilical vein endothelial cells loaded with fura-2 in the presence of 1mM $CaCl_2$ in the bathing medium. The ratio of the fluorescence emitted at each 0.5s time-point with excitation at 340nm and 380nm is calculated (data not

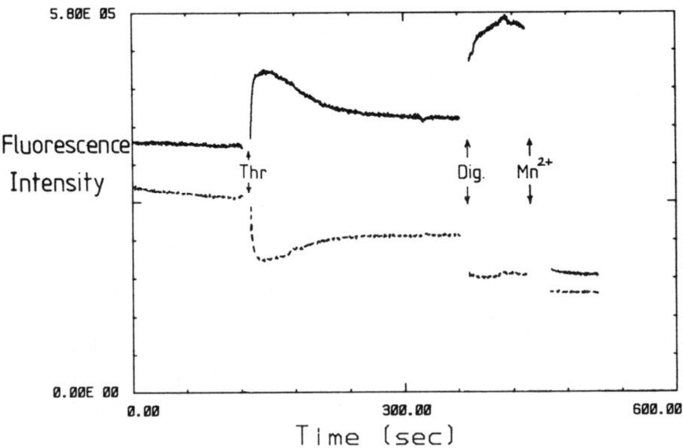

Figure 1. Raw fluorescence data from a coverslip containing a confluent layer of human umbilical chord endothelial cells loaded with the fluorescent indicator fura-2. The top trace shows the fluorescence intensity collected at 500nm with a 340nm excitation; the lower trace shows the intensity collected at 500nm with a 380nm excitation. Thrombin, 0.5U/ml, digitonin, 50μM and $MnCl_2$, 2mM, were added as indicated in the presence of 1mM $CaCl_2$. Digitonin and Mn^{2+} ions were added to quench the intracellular dye and reveal the cell autofluorescence.

shown) and each ratio value then converted into a known free calcium concentration. This value represents the cytoplasmic free calcium concentration as an average of all the approx. 5×10^4 cells in the incident excitation light-path. Figure 2A shows the result of such a manipulation of the data collected in Fig. 1. Thrombin, 0.5U/ml, stimulated a very rapid increase in $[Ca^{2+}]_i$ beginning within 2 seconds and peaking within 10-15 seconds. $[Ca^{2+}]_i$ increased about 30-fold from the resting level of around 100nM, typical of most other cells, to a peak of between 2-3μM. $[Ca^{2+}]_i$ then declined again until a new steady-state level was achieved of around 1μM. This level was maintained for many minutes and was dependent on the continued presence of the agonist (data not shown).

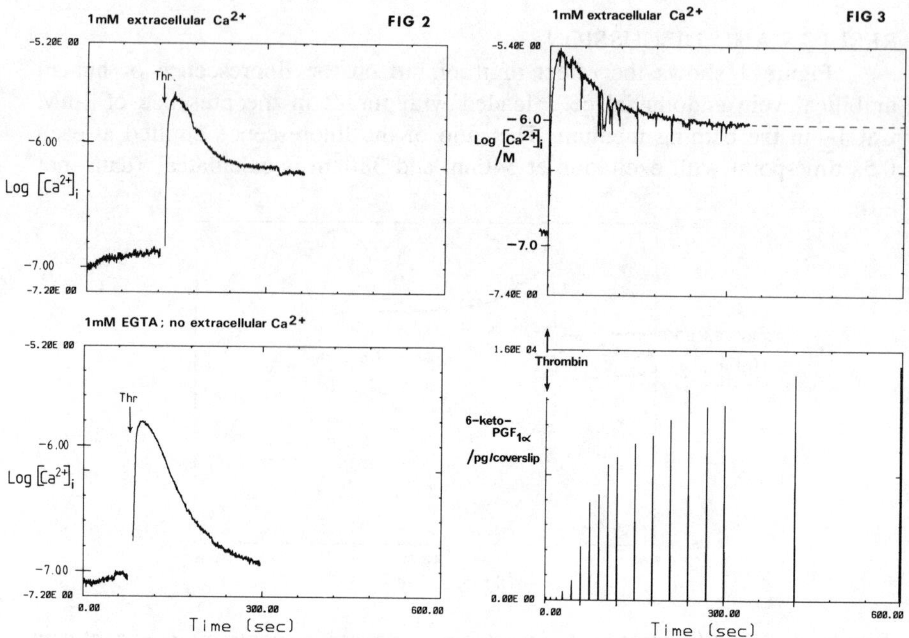

Figure 2. The effect of thrombin, 0.5U/ml, on $[Ca^{2+}]_i$: (A) in the presence of 1mM extracellular free calcium, (B) in the absence of extracellular calcium with 1mM EGTA.

Figure 3. Stimulated $[Ca^{2+}]$: (A) and 6-keto-PGF$_{1\alpha}$ accumulation (B) with 0.5U/ml thrombin in the presence of extracellular free calcium.

In the absence of extracellular bathing Ca^{2+} ions, i.e. with no added $CaCl_2$ and in the presence of a chelator, 1mM EGTA, the initial thrombin-stimulated elevation in $[Ca^{2+}]_i$ was very similar to that seen in the presence of extracellular Ca^{2+} ; see Fig.2B. This suggests that the initial increase in $[Ca^{2+}]_i$ is caused by the discharge of Ca^{2+} ions from an internal store in the cell into the cytoplasm. However, under these conditions with no added extracellular calcium present, $[Ca^{2+}]_i$ declined back to the basal level. The simplest interpretation of this data is that the difference between the two $[Ca^{2+}]_i$ transients is due to an influx of Ca^{2+} across the plasma membrane.

To estimate the amount of prostacyclin formed and released from the endothelial cells during this stimulation, aliquots of the bathing medium were removed at various intervals and the accumulation of the down-stream product of PGI_2, 6-keto-$PGF_{1\alpha}$, determined. The results of such experiments are shown in Figure 3. Fig. 3A shows $[Ca^{2+}]_i$ following stimulation with 0.5U/ml thrombin added at the zero time-point; Fig. 3B shows the simultaneously accumulating 6-keto-$PGF_{1\alpha}$ in the same cuvette. After an initial delay of approximately 10 seconds or so there is a marked increase in the formation and release of PGI_2. Although the rate of production is maximal between 45 and 90 seconds after stimulation, 6-keto-$PGF_{1\alpha}$ carries on accumulating throughout the 10 minutes of stimulation observed. Figs.4A and 4B show the results from a similar experiment except

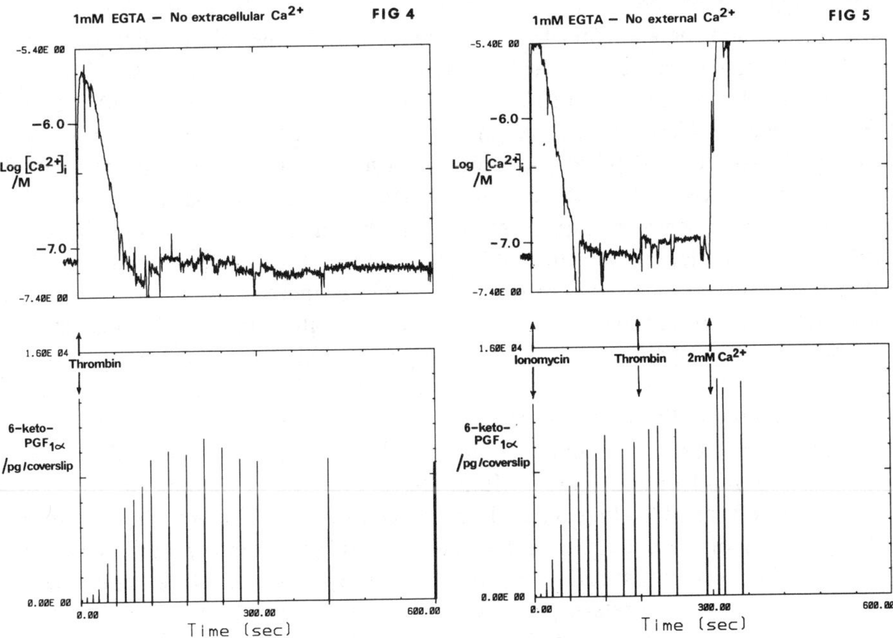

Figure 4. Stimulated $[Ca^{2+}]_i$ (A) and 6-keto-$PGF_{1\alpha}$ accumulation (B) with 0.5U/ml thrombin in the absence of extracellular calcium.

Figure 5. The effect of ionomycin, 4µM, followed by thrombin, 0.5U/ml, followed by 2mM $CaCl_2$ on $[Ca^{2+}]_i$ (A) and 6-keto-$PGF_{1\alpha}$ accumulation (B) in the absence of extracellular calcium with 1mM EGTA.

here the stimulation is in the presence of 1mM EGTA. Again a short delay of some 10-20 seconds is observed before any dramatic increase in 6-keto-$PGF_{1\alpha}$ is observed. PGI_2 formation then increases maximally over the next 45-90 s. After 2 minutes stimulation with thrombin the levels of 6-keto-$PGF_{1\alpha}$ plateau indicating no further production or release of PGI_2.

It is clear from these data that the initial increase in PGI_2 formation is very similar whether in the presence or absence of extracellular Ca^{2+}. This observation is entirely consistent with the similarity in the observed $[Ca^{2+}]$. transient in Ca^{2+}-free or Ca^{2+}-containing medium. Whereas in the presence of extracellular Ca^{2+} the stimulated $[Ca^{2+}]_i$ remains elevated, albeit at a reduced steady-state level of around 1μM, production and release of PGI_2 is also maintained at a slower rate. In Ca^{2+}-free medium the $[Ca^{2+}]_i$ returns rapidly to basal over 1-2 minutes and the production of PGI_2 is halted. The data suggest that the threshold concentration of $[Ca^{2+}]_i$ required to cause the release and metabolism of arachidonic acid is below micromolar and that the stimulated increase in $[Ca^{2+}]_i$ resulting from the release of Ca^{2+} from the intracellular store is sufficient to exceed this concentration and cause PGI_2 release. However, although the data appear consistent with the proposal that elevated $[Ca^{2+}]_i$ causes the observed PGI_2 accumulation in the extracellular medium, both responses being better maintained in Ca^{2+}-medium, the observed responses can only be interpreted as being parallel effects on stimulation with thrombin. The data in Figs. 3 & 4 do not prove that the observed elevations in $[Ca^{2+}]_i$ cause the increase in PGI_2 formation and release. To approach the problem of causality the calcium-selective ionophore, ionomycin, was used. Ionomycin translocates Ca^{2+} ions across the phospholipid membranes of intact functioning cells down the concentration gradient and completely by-passes the cell surface receptors. Hence in the absence of extracellular Ca^{2+} ions ionomycin should release Ca^{2+} from intracellular stores within the cell into the cytoplasm. Fig. 5A shows the effect of adding an optimal concentration of ionomycin, 4μM, in the presence of no extracellular free calcium. $[Ca^{2+}]_i$ rapidly increases to a peak within several seconds presumably because under these conditions ionomycin causes the release of calcium from the intracellular store. $[Ca^{2+}]_i$ then declines rapidly back to basal levels. In the continued presence of the ionophore the intracellular store would not be expected to be able to re-accumulate the calcium and it would now be actively pumped out of the cell by plasma membrane Ca^{2+}-ATPases. In any event the resulting elevation in $[Ca^{2+}]_i$ is similar to that observed on stimulation with thrombin under similar conditions. Fig. 5B shows the simultaneous accumulation of 6-keto-$PGF_{1\alpha}$ in the bathing medium. Again the time-course and extent of 6-keto-$PGF_{1\alpha}$ are very similar to those stimulated by thrombin under similar conditions. This result demonstrates that the total amount of Ca^{2+} ions stored in the intracellular pool is sufficient when rapidly discharged to elevate $[Ca^{2+}]_i$ and cause an increase in the PGI_2 formation and release. A subsequent addition of thrombin after ionomycin has very little or no further effect on $[Ca^{2+}]_i$ suggesting that the intracellular pool of Ca^{2+} was completely discharged by the ionophore. Convincingly no further increase in 6-keto-$PGF_{1\alpha}$ is observed which strongly supports the notion that the only intracellular mechanism for activating PGI_2 production following occupation of the receptor by agonist is the elevation in $[Ca^{2+}]_i$. To control the experiment and make sure that the substrate had not become limiting, i.e., that all the phospholipid arachidonic acid had not been released before the addition of thrombin, extracellular calcium ions were added back to the bathing medium. As the cells were still in the presence of ionophore and thrombin, $[Ca^{2+}]_i$ again increased to above $1\mu M$. A further increase in the extracellular levels of 6-keto-$PGF_{1\alpha}$ could now be measured.

CONCLUSIONS

In conclusion, the results presented here with the endothelial cell are entirely consistent with the concept that an agonist causes the stimulated production of prostaglandin endoperoxides by elevating the cytoplasmic free calcium concentration. The data do not demonstrate where Ca^{2+} acts to activate the pathway but since the addition of free arachidonate to endothelial cell monolayers leads to immediate conversion to PGI_2 (data not shown) it is likely that the controlled step is the initial hydrolysis of the phospholipid to release the free arachidonic acid. The only other cell where the role of elevated $[Ca^{2+}]_i$ in causing the release of arachidonic acid has been studied in detail, is in the blood platelet by Pollock and co-workers (1986; and see chapter in this book). In the platelet it appears that elevated $[Ca^{2+}]_i$ is not the only pathway for stimulating the release of arachidonic acid and subsequent formation of prostaglandin endoperoxides and thromboxane A_2. In the endothelial cell it remains to be determined whether raised $[Ca^{2+}]_i$ causes simultaneous production of PGI_2 and EDRF or whether one pathway contains a further regulatory step controlled by $[Ca^{2+}]_i$ or another intracellular messenger. It may make good physiological sense to inhibit further platelet recruitment into the site of vessel injury once the plug has formed with PGI_2, and simultaneously to ensure flow by dilating the vessel locally with EDRF.

ACKNOWLEDGEMENT

I thank Lindsey Needham for the radio-immunoassays and much helpful discussion and Jeremy Pearson and John Gordon for the use of facilities at the MRC Clinical Research Centre, Harrow.

REFERENCES
Ager, A., Gordon, J.L., Moncada, S., Pearson, J.D., Salmon, J.A. & Trevethick, M.A. (1982) J. Cell.Physiol. 110, 9-16
DeMey, J.G. & Vanhoutte, P.M. (1981) J. Physiol. 316, 347-355
Furchgott, R.F. (1984) Ann. Rev. Pharmacol. Toxicol. 24, 175-197
Griffith, T.M., Edwards, D.H., Lewis, M.J., Newby, A.C. & Henderson, A.H. (1984) Nature 308, 645-647
Hallam, T.J., Sanchez, A. & Rink, T.J. (1984) In: "Prostaglandins and Membrane Ion Transport", P. Braquet, ed.; Raven Press, N.Y. pp. 157-163
Hallam, T.J. & Rink, T.J. (1985) J. Physiol. 368, 131-146
Hallam, T.J. & Pearson, J.D. (1986) FEBS Letts. in press
Pearson, J.D., Carleton, J.S., Hutchings, A. & Gordon, J.L. (1978) Biochem. J. 170, 265-271
Pearson, J.D., Slakey, L.L. & Gordon, J.L. (1983) Biochem. J. 170, 265-271
Pollock, W.K., Irvine, R.F. & Rink, T.J. (1986) Eur. J. Pharmacol. submitted
Tsien, R.Y., Pozzan, T. & Rink, T.J. (1982) J. Cell Biol. 94, 325-334
Tsien, R.Y., Rink, T.J. & Poenie, M. (1985) Cell Calcium 6, 145-157
Weksler, B.B., Ley, C.W. & Jaffe, E.A. (1978) J. Clin. Invest. 62, 923-930

THROMBOXANE A_2 GENERATION BY COLLAGEN STIMULATED HUMAN PLATELETS: CAUSE OR CONSEQUENCE OF CHANGES IN CYTOSOLIC FREE CALCIUM?

W. Kenneth Pollock

Department of Biochemistry, AFRC Institute of Animal
Physiology, Babraham, Cambridge

INTRODUCTION

Blood platelets may be activated in vivo or in vitro by physiological agents such as collagen and thrombin. In response to such agonists platelets synthesise and secrete biologically active lipids that are derived from arachidonic acid, including prostaglandin endoperoxides and thromboxane A_2 (TXA_2) (Hamberg and Samuelsson, 1974; Hamberg et al., 1975). Once secreted these lipids serve to reinforce the response to initial stimuli by activating other platelets via interaction with thromboxane receptors on the platelet plasma membrane (Armstrong et al., 1983).

The rate determining step in TXA_2 biosynthesis is the liberation of free arachidonic acid from membrane phospholipids into which it is esterified, predominantly at the sn-2 position (see Irvine, 1982 for review). In platelets such deacylation may involve several processes of which phospholipase PLA_2 activity on membrane phospholipids may be the most important. PLA_2 type enzymes have been reported to be Ca^{2+}-calmodulin dependent, being activated by an elevation in cytosolic free calcium $[Ca^{2+}]_i$ (Wong and Cheung, 1979; Van den Bosch, 1980). We would therefore predict that the liberation of arachidonic acid and its subsequent conversion to TXA_2 could result from agonist-evoked increases in $[Ca^{2+}]_i$. Paradoxically however the mechanisms underlying TXA_2-induced platelet activation also involve an elevation in $[Ca^{2+}]_i$ (Pollock et al., 1984).

Collagen is a powerful platelet agonist that depends upon TXA_2 generation to exert its full biological activity on platelets. We therefore investigated whether arachidonic acid liberation and TXA_2 biosynthesis in response to this agonist is a cause or a consequence of changes in $[Ca^{2+}]_i$.

METHODS

Platelet $[Ca^{2+}]_i$ was monitored fluorimetrically in intact platelets loaded, as previously described, with a fluorescent Ca^{2+} indicator dye, either quin2 (Hallam et al., 1984) or fura-2 (Pollock et al., 1986), and suspended in HEPES buffered saline. Platelet shape change and aggregation was monitored by optical density changes of the cell suspension. Liberation of arachidonic acid by collagen was measured isotopically as the appearance of free $[^3H]$arachidonate from prelabelled platelets. TXA_2 generation was measured by radio-immunoassay of the stable metabolite TXB_2.

RESULTS AND DISCUSSION

In Figure 1 typical fluorescence and optical density records from

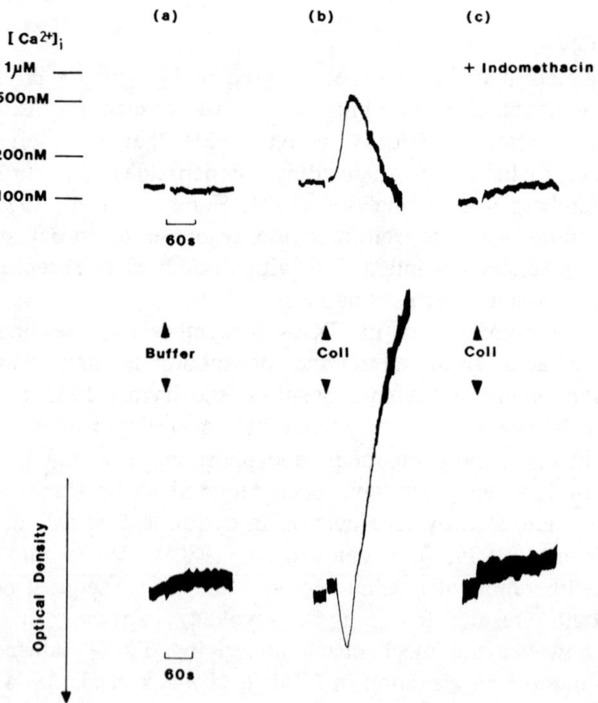

Figure 1. Fluorescence and optical density records from fura-2-loaded platelets stimulated by collagen in the absence and presence of indomethacin.

suspensions of fura-2-loaded platelets prelabelled with [^{3}H]arachidonate are shown. From the calibration of the fluorescence signal the mean resting $[Ca^{2+}]_i$ was 115 $\pm$ 18nM (S.E., $\underline{n}$=12). Following the addition of 10μg collagen/ml (fig. 1b) there is a 10-20s lag phase prior to an elevation in $[Ca^{2+}]$ up to a maximum value of 527 $\pm$ 66nM (S.E., $\underline{n}$=3). This elevation in $[Ca^{2+}]_i$ was accompanied by platelet shape change and aggregation as seen from the optical density changes. The formation of platelet aggregates quenches the fluorescence causing the observed fall in the signal in figure 1b.

Indomethacin is an inhibitor of the enzyme cyclooxygenase and so prevents the conversion of arachidonic acid to TXA_2. In the presence of 10μM indomethacin (fig. 1c) collagen raised $[Ca^{2+}]_i$ barely at all to 123 $\pm$ 19nM (S.E., $\underline{n}$=3) and caused no measurable platelet aggregation. The [^{3}H]arachidonate levels corresponding to these $[Ca^{2+}]_i$ changes are summarised in Table 1. In both the absence and presence of indomethacin, collagen caused substantial (2.5 - 3 fold) arachidonate liberation indicating that the initial deacylation response to collagen can occur at near to resting $[Ca^{2+}]_i$ levels.

Table 1. Free [^{3}H]arachidonate and $[Ca^{2+}]_i$ levels from collagen stimulated platelets.

Additions	($\underline{n}$)	$[Ca^{2+}]_i$	[^{3}H]arachidonate/c.p.m.
Controls	(12)	115 ± 18	2,991 ± 660
Collagen	(3)	527 ± 66	11,930 ± 959
Collagen + Indomethacin	(3)	112 ± 19	11,328 ± 3,089

Figure 2 shows fluorescence and optical density changes from quin2 loaded platelets. Samples were taken for TXB_2 analysis at TX-1 and TX-2. In Figure 2a 3μg collagen/ml raised $[Ca^{2+}]_i$ from 80 $\pm$ 3nM to a peak value of 740 $\pm$ 145nM (S.E., $\underline{n}$=4) at which point the fluorescence falls as the platelets aggregate. The corresponding TXB_2 levels increased from 0.125ng to 19ng/10^8 cells (Table 2).   The compound EP092 is a selective thromboxane receptor antagonist (Armstrong et al., 1985) and was used to block any effect of TXA_2 without inhibiting TXA_2 formation. In the presence of 3μM EP092 the maximum $[Ca^{2+}]_i$ attained in response to collagen was 121 $\pm$ 9nM (fig. 2b). However these cells still managed to generate 10ng of TXB_2 as seen in Table 2. These results clearly demonstrate that collagen induced elevations in $[Ca^{2+}]_i$ are mediated by TXA_2 acting at cell surface receptors.

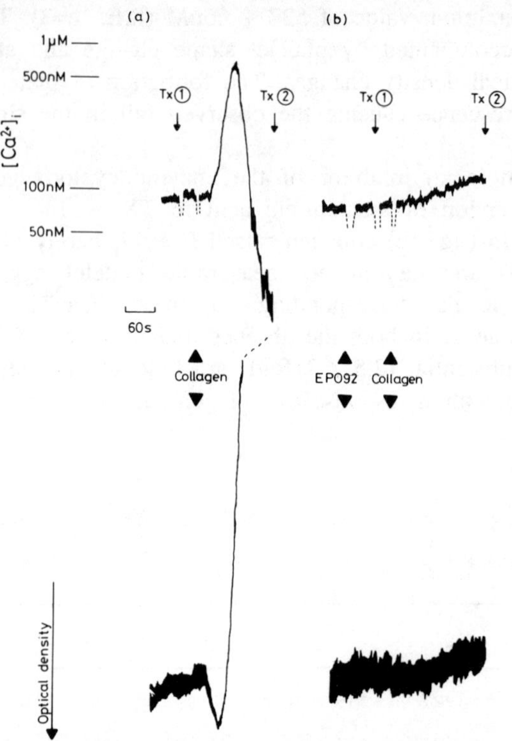

Figure 2. Fluoresence and optical density records from quin2-loaded platelets stimulated by collagen in the absence and presence of the thromboxane receptor antagonist EP092.

In assessing the role of Ca^{2+} in the regulation of PLA_2 activity many previous studies have utilised broken cells or isolated membranes, resuspended in buffers of defined Ca^{2+} concentration. The use of quin2 or fura-2 allows a direct measure of $[Ca^{2+}]_i$ in intact cells during stimulation to be correlated with arachidonate liberation and thromboxane formation. In fura-2 loaded platelets the Ca^{2+} ionophore ionomycin can liberate substantial amounts of arachidonate only when $[Ca^{2+}]_i$ is in the micromolar range (Pollock et al., 1986) ; this may reflect Ca^{2+}-calmodulin dependent PLA_2 activity. By contrast the results shown here with collagen clearly

Table 2. TXB_2 and $[Ca^{2+}]_i$ levels from collagen stimulated platelets.

(n)	$[Ca^{2+}]_i$: nM		TXB_2: ng/10^8 cells	
	resting	maximum	resting	maximum
Collagen	4	80 ± 3	740 ± 145	0.125 ± 0.02	19.0 ± 3.8
+ EPO92	4		121 ± 9		10.0 ± 2.5

show that in response to this agonist the mechanisms involved in arachidonate liberation and subsequent TXA_2 formation are essentially independent of any changes in $[Ca^{2+}]_i$. Other additional mechanisms must therefore exist which either utilise resting $[Ca^{2+}]_i$ or are completely Ca^{2+} independent. One possible mechanism could be membrane perturbation allowing access for the enzyme to the phospholipid substrate, as proposed by Dawson (1982). Another possibility for $[Ca^{2+}]_i$ independent deacylation could involve activation of protein kinase C (PKC). This protein kinase phosphorylates a number of 40K (M_r) proteins in the absence of an elevation in $[Ca^{2+}]_i$. It has recently been shown in platelets (Touqui et al., 1986) that a 40K endogenous PLA_2 inhibitory protein, lipocortin, is a substrate for PKC that is inactivated by phosphorylation. Whether this mechanism is employed by collagen is unknown. However we have shown that pharmacological activation of PKC by phorbol myristate acetate fails to evoke arachidonate liberation from platelets at concentrations that promote platelet aggregation (Pollock et al., 1986).

Hence it would appear that collagen induced platelet activation is at least a 2-stage process. In the first instance collagen adhesion to platelets promotes arachidonic acid liberation in the abscence of an elevation in $[Ca^{2+}]_i$ by an as yet unknown mechanism. Then TXA_2 once secreted interacts with cell surface receptors to raise $[Ca^{2+}]_i$ which may contribute to the mechanisms involved in promoting further platelet aggregation.

ACKNOWLEDGEMENTS

This study was supported by funds from the M.R.C. to Drs. Robin Irvine and Tim Rink, and was partly carried out in the Physiological Laboratory, University of Cambridge, U.K. Thanks to Dr. R.L. Jones, University of Edinburgh, for the EP092.

REFERENCES

Armstrong R.A., Jones R.L. and Wilson N.H. (1985) Br. J. Pharmacol.79, 953.

Armstrong R.A., Jones R.L., Peesapati,V., Will S. and Wilson N.H. (1985) Br. J. Pharmacol. 84, 595.

Dawson R.M.C. (1982) J. Am. Oil Chem. Soc. 59, 401.

Hallam T.J., Sanchez A. and Rink T.J. (1984) Biochem. J. 218, 819.

Hamberg M. and Samuelsson B. (1974) Proc. Natl. Acad. Sci. (USA) 71, 3400.

Hamberg M., Svensson J. and Samuelsson B. (1975) Proc. Natl. Acad. Sci. (USA) 72, 2994.

Irvine R.F. (1982) Biochem. J. 204, 3.

Pollock W.K., Armstrong R.A., Brydon L.J., Jones R.L. and MacIntyre D.E. (1984) Biochem. J. 219, 833.

Pollock W.K., Rink T.J. and Irvine R.F. (1986) Biochem. J. 235, 869.

Rink T. J. and Pozzan T. (1985) Cell Calcium 6, 133.

Touqui L., Rothut B., Shaw A.M., Fradin A., Vargaftig B.B. and Rousso-Marie F. (1986) Nature 321, 177.

Van den Bosch H. (1980) Biochim. Biophys. Acta 604, 191.

Wong P. and Cheung W.Y. (1979) Biochem. Biophys. Res. Commun. 90, 473.

THE CHEMISTRY OF ANACARDIC ACIDS

J.H.P. Tyman and N. Visani

Department of Chemistry, Brunel University, Uxbridge, Middlesex, UB8 3PH

INTRODUCTION

Anacardic acids are 6-substituted salicylic acids (1) and can be regarded as phenolic lipids in which the polar head group is a phenolic acid. In recent years increasing numbers of this group have been found as shown in Table 1. This review briefly describes sources, analysis, utilisation and synthetic studies of these lipids.

TABLE 1

Side chain R in (1)	Source	Reference
(17:1), (17:2)	Pentaspadon motleyi	Lamberton, 1959
(15:0),(15:1),	Cashew, Anacardium	Tyman,1979
(15:2),(15:3)	occidentale	
(15:1)	Ginkgo biloba	Loev, 1958
(13:0),(13:1),	Pistacia vera	Yalpani, 1983
(15:0),(15:1)		
(11:0)	Anacardium	Sharma, 1966
	giganteum	
C_5,C_6-diynyl	Chrysanthemum	Bohlmann, 1962
(as O-methyl ethers)	frutescens	
(2:0)	Type of beetle	Benn, 1983
Cl	Penicillium	Harborne, 1964
	species	

The major source is the cashew, the shell of which contains up to 50% of C_{15} phenolic material (natural cashew nutshell liquid, CNSL) comprising more than 70% of C_{15} anacardic acids accompanied by 20% of cardol (2) and smaller proportions of 2-methylcardol and cardanol (3). In all these related compounds the side chains exist primarily with 8'($\underline{Z}$), 11'($\underline{Z}$) and 14-position double bonds in the (15:1), (15:2) and (15:3) constituents.

(1)   (2)   (3)   (4)   (5)

By 1990 it has been predicted (Ohler, 1979) that 1×10^6 tonnes of cashews are likely to be processed annually, mostly by the hot oilbath method (Tyman, 1980) which decarboxylates the anacardic acid to give cardanol (3), the main component of the commercially available cashew nut-shell liquid (technical CNSL), sometimes referred to as raw CNSL. Some natural CNSL is potentially available by solvent extraction after cold cutting of the shell (Grimminger, 1984).

C_{11} and C_{15}-anacardic acids can be conveniently isolated from phenolic extracts by precipitation as metal salts, notably the lead salt, isolation and acidic regeneration (Lam, 1981). By HPLC with the reversed phase method natural CNSL has been analysed quantitatively (Tyman, 1984) and with the preparative method the (15:0), (15:1), (15:2) and (15:3) constituents have been separated (Lloyd, 1980). Argentation thin-layer chromatography serves as a useful monitoring technique (Tyman, 1967; Tyman, 1986). Biosynthetic studies with <u>Ginkgo</u> <u>biloba</u> (Schlenk, 1974) have indicated separate pathways for the polyketide-derived ring and for the side chain. The transformation of the polyketide to the phenolic acid has been discussed (Scott, 1971) and the mode of formation of 6-methylsalicylic acid examined by the use of ^{2}H NMR (Abell, 1982). The C_7-anacardic acid is considered to be the biosynthetic precursor of flavoglaucin.

UTILISATION

Anacardic acids have been used in synthetic, biological and technical work. The ethyl ester $\underline{O}$-methyl ether of (1; R=OMe) has been converted to its carbanion (Hauser, 1979) used in other condensation reactions (Yamomoto, 1971; Gerhard, 1975) and in natural product transformations (Inoue, 1954). Antibacterial activity for C_{15}-anacardic acids (Biswas, 1958) and for the (15:1) constituent in <u>Ginkgo</u> <u>biloba</u> (Adawarkar, 1981) has been reported. The C_{15}-anacardic acids have been described as possessing molluscicidal activity (Sullivan, 1982). In an attempt to study the mechanism of this action the inhibition of prostacyclin (PGI$_2$) action in human blood platelets by competition of the unsaturated C_{15}-anacardic acids for the PGI$_2$

receptor site has been investigated (Lloyd, 1980; see note added in proof). It is not known, however, whether PGI_2 receptor sites are present and if anacardic acids combine with them when these compounds are used as molluscicides. The C_{11} and the C_{15}-anacardic acids in the form of their aldoximes have useful solvent extraction properties for copper (II) (Tyman, 1983). By polymerisation of natural CNSL in concentrated alkaline solution with formaldehyde in the presence of wood chips, materials of constructional use have been fabricated (Hawkes, 1980). 6-n-Pentadecyl-salicylic acid O-methyl ether and the 6-n-butyl compound have been attached to 6-aminopenicillanic acid emulating the well-known 2,6-dimethoxybenzoyl derivative (Beecham, 1974). By catalytic reduction or chemically with diimide, the C_{15}-anacardic acids can be converted into the saturated member (Iddenten, 1983). Allergenic action of the C_{15}-anacardic acids has been investigated (Johnson, 1972).

The isomeric acids (4) and (5) do not apparently occur naturally although they could arise by the novel combination of two acetoacetate units and by dehydroxylation of an orsellinic acid or a precursor respectively. The former is obtained by the Kolbe reaction on cardanol (Hawkes, 1980).

SYNTHESIS

(A) Compounds with a saturated side chain.

Although the structure of the (15:0)-anacardic acid was found some years ago (Backer, 1941), it was synthesised only comparatively recently by the thermolysis of the basic copper salt of 2-pentadecylbenzoic acid (Durrani, 1979) and rather more effectively by the interaction of o-methoxy or m-methoxy fluorobenzene with pentadecyllithium in a general regiospecific synthesis (Tyman, 1979) shown in Scheme 1.

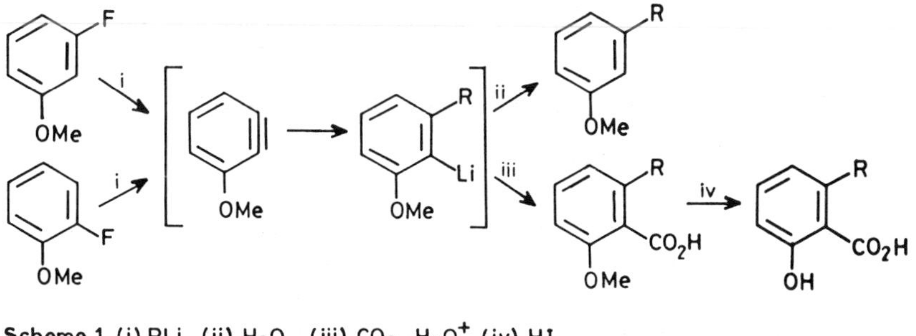

Scheme 1. (i) RLi, (ii) H_2O, (iii) CO_2, H_3O^+ (iv) HI

Alternative syntheses have been developed based on the formation of ethyl 6-methylsalicylate O-methyl ether by the Michael addition of sodio ethylacetoacetate with but-2-enal followed by cyclisation (Hauser, 1980) and alkylation of the carbanion formed with lithium diethylamide in tetrahydrofuran containing hexamethylphosphoric triamide (Tyman, 1985) (Scheme 2).

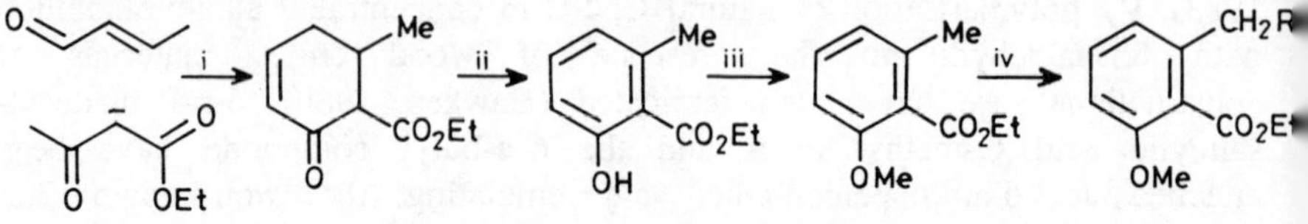

Scheme 2. (i) EtOH, Δ, H_3O^+ (ii) Br_2, Δ (iii) PTC or Me_2SO_4/K_2CO_3 (iv) LDA, THP/HMPA, RX

Other routes to 6-methylsalicylic acid have utilised the Diels-Alder addition (Young, 1963) of 1-methoxybuta-1,3-diene and methyl but-2-enoate (Durrani, 1974) followed by aromatisation. Cyclo-aromatisation of 4-methoxybut-3-en-2-one has been employed in a regiospecific approach (Chan, 1980) in a synthesis of methyl 6-methylsalicylate and the 4-methyl isomer (Scheme 3).

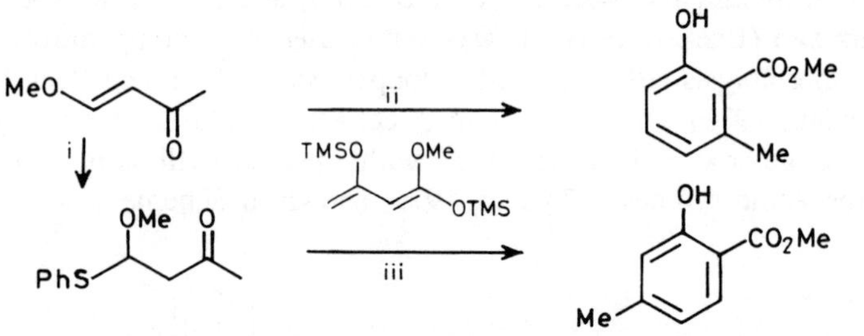

Scheme 3. (i) PhSH, pyr. (ii) $TiCl_4$, $Ti(OPr^i)_4$, $CH_2Cl_2(-40°)$

(iii) $TiCl_4$, $CH_2Cl_2(-78°)$

A biogenetically modelled synthesis of 6-methylsalicylic acid has been achieved (Harris, 1974) based on the interaction of the dilithio salt of t-butylacetoacetate with sodio formylacetone to give the tetraketide isolated as the cyclohexanone. Acidic dehydration yields the acid or the ester with aqueous alkali as shown (Scheme 4). No details have been published.

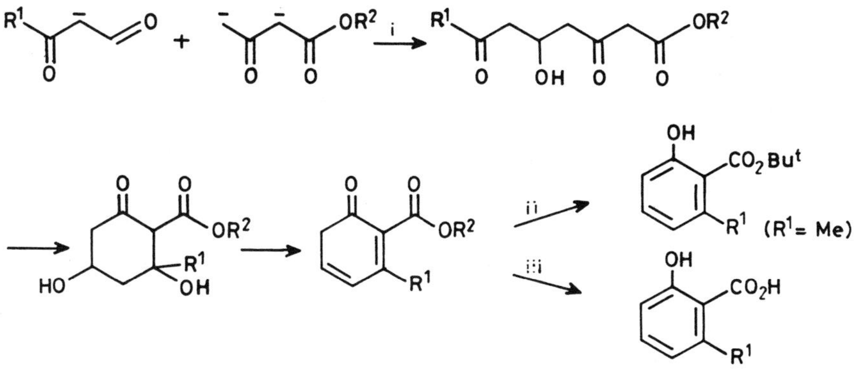

Scheme 4. (i) THF, 15°, NaOAc (ii) aq. KOH (iii) HCl, dioxan

(B) Compounds with an unsaturated side-chain.

In connection with biological studies, allergenic work and spectroscopic investigations, the synthesis of the (15:1), (15:2) and (15:3)-constituents in their various stereoisomeric forms is of interest. Methyl or ethyl 6-(7-hydroxy-heptyl)-salicylate is a key intermediate for this purpose and has been obtained by the aryne method (Durrani, 1982) and the alkylation route (Visani, 1986). Similar general methodology has been employed from then on as with (15:1) and (15:2)-cardanol (Caplin, 1982), and (15:1) and (15:2)-cardol dimethyl ether (Baylis, 1981).

(15:1)-Anacardic acid has been synthesised as the 8($\underline{Z}$) isomer by the interaction of methyl 6-(7-bromoheptyl)salicylate with 1-octinyllithium followed by catalytic hydrogenation with a palladium/barium sulphate catalyst regulated by quinoline (Tyman, 1976).

The 8($\underline{Z}$),11($\underline{Z}$) diene has been synthesised (Visani, 1986) from methyl 6-(7-hydroxy-heptyl)salicylate by reaction successively with a C_3 and a C_5 synthon, rather than by conversion to an Ar C_9 terminal acetylene followed by alkylation with a C_6 synthon, on account of an earlier observed isomerisation of the starting material in this approach (Tyman, 1975). The preferred route is shown in Scheme 5.

Routes to the 8($\underline{E}$),11($\underline{E}$) isomer involved chemical reduction of the diynyl intermediate. The synthesis of the 8($\underline{Z}$),11($\underline{E}$) and the 8($\underline{E}$),11($\underline{Z}$) isomers has involved selective procedures with the $\underline{E}$ and $\underline{Z}$-reduction products of the Ar C_{10} compound, methyl 6-(10-hydroxydec-9-ynyl)salicylate. The synthesis of the triene (1; R=$C_{15}H_{25}$) is being examined (Visani, 1986).

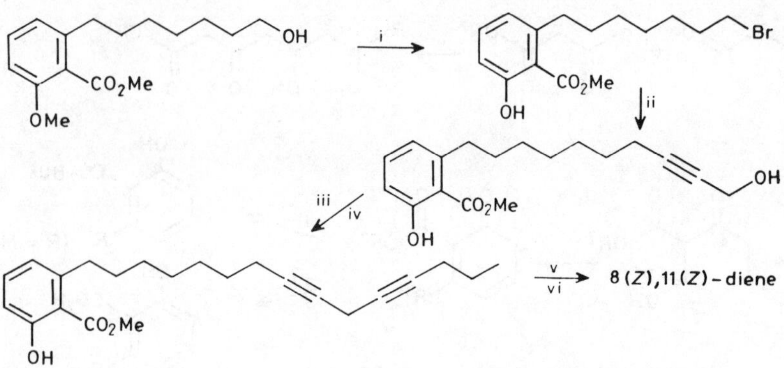

Scheme 5. (i) BBr$_3$, CH$_2$Cl$_2$ ($-80° \rightarrow 0°$) (ii) LiC$\equiv$CCH$_2$OCH(CH$_3$)OEt (from LiNH$_2$), THF, HMPA, H$_3$O$^+$ (iii) PBr$_3$ (iv) BrMgC$\equiv$CC$_3$H$_7$ THF, CuCl (v) Pd/BaSO$_4$, H$_2$, quin., MeOH (vi) MeOH, KOH

Note added in proof:

G.A. Krishna (1986, personal communication) has investigated the effect of anacardic acid on the cyclooxygenase in human platelets.

In general, the triene analogue of anacardic acid appears to be somewhat more potent than the diene or monoene analogues. Anacardic acid inhibits the effect of prostacyclin on both cyclic AMP formation as well as the inhibitory effect of PGI$_2$ on platelet aggregation induced by arachidonic acid. The concentration of the triene analogue of anacardic acid required to inhibit by 50% the effect of 200 nM PGI$_2$ on cyclic AMP is in the range of 10-50 μM. Binding studies with [^{3}H]-PGI$_2$ indicate that anacardic acid is able to displace the bound PGI$_2$ from platelet membranes. The IC$_{50}$ for anacardic acid (triene) to displace PGI$_2$ is also in the range of 10-100 μM. Moreover, the IC$_{50}$ for the inhibition of thromboxane synthesis by anacardic acid in platelet membranes is also in the range 10-100 μM. Anacardic acid inhibits both PGE$_2$ and PGF$_{2\alpha}$ synthesis in platelets, but to a lesser extent than thromboxane synthesis. The inhibitory effect of anacardic acid on thromboxane synthesis appears to predominate.

The EC$_{50}$ for PGI$_2$ to increase cyclic AMP levels in platelets is in the range of 200 nM. The concentration of anacardic acid required to inhibit 50% of the PGI$_2$ effect is in the range 10-50 μM. It thus appears that more than 1000 times the concentration of anacardic acid is required to compete for PGI$_2$ at the receptor. Even though the triene analogue of anacardic acid is more potent than the monoene, diene and saturated analogues, these compounds also have some inhibitory effects on the action of PGI$_2$ on platelets.

REFERENCES

Abell, C., Garson, M.J., Leeper, F.J. and Staunton, J. (1982) J. Chem. Soc., Chem. Comm., 1011-1013.

Adawarkar, P.D. and Elsohly, M.A. (1981) Fitoterapia, 52: 129-135.

Backer, H.J. and Haack, N.H. (1941) Rec. Trav. Chim., 60: 661-677.

Baylis, C.J., Odle, S.W.D. and Tyman, J.H.P. (1981) J. Chem. Soc., Perkin 1, 132-141.

Beecham Res. Ltd. and Tyman, J.H.P. (1974) (unpublished data).

Benn, M.L. (private communication, 1983).

Biswas, A. and Ray, A. (1958) Nature, 182: 1299-1300.

Bohlmann, F. and Kleine, K.W. (1962) Chem. Ber., 95: 602-610.

Caplin, J. and Tyman, J.H.P. (1982) J. Chem. Res. (M), 319-327.

Chan, T-H. and Brownbridge, P. (1980) J. Am. Chem. Soc., 102: 3534-3538.

Durrani, A.A. (1974) PhD thesis, Brunel University.

Durrani, A.A. and Tyman, J.H.P. (1979) J. Chem. Soc., Perkin 1, 2069-2078.

Durrani, A.A., Goh, C.S. and Tyman, J.H.P. (1982) Lipids, 17: 561-569.

Gerhard, A., Muntwyler, R. and Keller-Schierlein (1975) Helv. Chim. Acta, 58: 1323-38.

Grimminger, M. (1984) Technical Brochure of Bühler-Miag (Eng.).

Harborne, J. B., The Biosynthesis of Phenolic Compounds (1964) Academic Press, New York.

Harris, T.M., Murray, T.P., Harris, C.M. and Gumulka, M. (1974) J. Chem. Soc., Chem. Comm., 362-363.

Hauser, F.M. and Pogany, S.A. (1980) Synthesis, 814-815.

Hauser, F.M. and Rhee, R.P. (1979) J. Am. Chem. Soc., 99: 4533-34.

Hawkes, A. J., Durrani, A.A. and Tyman, J.H.P. (1980) Eur. Pat. 0015761.

Iddenten, S. J.A., Barker, J., Mehet, S. and Tyman, J.H.P. (1983) (unpublished data).

Inoue, H. (1954) Pharm. Bull. (Japan), 2: 359-367.

Johnson, R.A., Baer, H., Kirkpatrick, C.H., Dawson, C.R. and Khurana, R.G., (1972) J. Allergy Chim. Immunol., 49: 27-35.

Lam, Soot Kiong and Tyman, J.H.P. (1981) J. Chem. Soc., 1941-1952.

Lamberton, J.A. (1959) Austral. J. Chem., 12: 234-239.

Lloyd, H.A., Denny, C. and Krishna, G., (1980) J. Liq. Chromatogr., 3: 1497-1504.

Loev, B. and Dawson, C.R. (1958) J. Amer. Chem. Soc., 80: 643-645.

Ohler, J. G., Cashew (1979) Royal Tropical Institute, Amsterdam, p.21.

Schlenk, H., Gellerman, J.L. and Anderson, W.H. (1974) Lipids, 9: 722-725.

Scott, A.I., Guilford, H. and Skingle, D. (1971), Tetrahedron, 27: 3039-3049.

Sharma, N.K. and Sharma, V.N. (1966) Indian J. Chem., 4: 99, 504.

Sullivan, J.T., Richards, C. S., Lloyd, H.A. and Krishna, G. (1982) Planta Med., 44: 175-177.

Tyman, J.H.P. and Morris, L. J. (1967) J. Chromatogr., 26: 287-288.

Tyman, J.H.P. (1975) Syn. Comm., 5: 21-26.

Tyman, J.H.P. (1976) J. Org. Chem., 41: 894.

Tyman, J.H.P. and Durrani, A.A. (1979) J. Chem. Soc., Perkin 1, 2079-2087.

Tyman, J.H.P. (1979) Chem. Soc. Rev., 8: 499-537.

Tyman, J.H.P. (1980) Chem. and Ind. (London), 59-62.

Tyman, J.H.P. (1983) UK Pat. 2104516 (9.3.83).

Tyman, J.H.P. (1984) J. Chromatogr., 303: 137-150.

Tyman, J.H.P. and Visani, N. (1985) 3rd FECS Int. Symposium on the Chemistry and Biotechnology of Natural Products, Sofia.

Tyman, J.H.P., The Industrial Application of QTLC (1986) M. Dekker, New York, Ch. 5, ed. L. Treiber.

Yalpani, M. and Tyman, J.H.P. (1983) Phytochemistry, 22: 2263-2266.

Yamomoto, K., Irie, H. and Uyeo (1971) Yagugaku Zasshi, 91: 257-268.

Young, S. T., Turner, J.R. and Tarbell, D. S. (1963) J. Org. Chem., 28: 928-932.

Visani, N. and Tyman, J.H.P. (1986) (unpublished work).

<u>Section</u> <u>3</u>

GLYCOLIPIDS

SYNTHETIC STUDIES ON THE MAJOR SEROLOGICALLY ACTIVE GLYCOLIPID FROM <u>MYCOBACTERIUM</u> <u>LEPRAE</u> (THE LEPROSY BACILLUS)

Jill Gigg, Roy Gigg, Sheila Payne and Robert Conant

Laboratory of Lipid and General Chemistry, National Institute for Medical Research, Mill Hill, London NW7 1AA, U.K.

"What strange ideas people have about leprosy, doctor. They learn about it from the Bible. Like sex." Graham Greene, 'A Burnt-Out Case', 1961.

INTRODUCTION

Leprosy (or 'Hanseniasis') is a disease of great antiquity with a rich cultural heritage and mythology. Ancient historical references, including the Bible, have given the disease a special place in many cultures and the disease has evoked images of horror and fascination whilst the afflicted have been subject to isolation, ostracism and even murder (Weymouth, 1938; Brody, 1974).

"For, over the years, 'leprosy' has acquired so many emotional connotations that its victims are not only the victims of <u>Mycobacterium leprae</u> but also of man's hate, fear, disgust and pity. The very stealth of its approach suggested a state of corruption which was not attributable to any one specific sin. Consequently its victims had been 'struck down by God', which, being a vague phrase, was impossible to prove or disprove and, therefore, carried a deadly stigma with it. With syphilis you knew that a man had done wrong; with leprosy you did not and you could allow your imagination to run riot over every crime it was possible to commit, and when the possible was exhausted you could start again with the impossible" (Feeney, 1964).

Armauer Hansen recognised the responsible microorganism, <u>Mycobacterium</u> <u>leprae</u>, by light microscopy of skin biopsies of leprosy patients in Norway in 1873 (Hansen, 1874) and this was one of the first identifications of a microbial pathogen of man and the first acid-fast bacterium to be described. Despite this early discovery the microorganism has never been cultured in vitro (Baker, 1983) and it was only in 1974 that it was finally established as a mycobacterium (Etemadi and Convit, 1974; Stanford et al., 1975).

In 1943 dapsone [bis(4-aminophenyl)sulphone] was shown to be an effective, safe and inexpensive chemotherapeutic agent for controlling the multiplication of the leprosy bacillus but spontaneously arising resistant mutants occur and have led to the development of dapsone-resistant leprosy which is now a severe problem in some places. Although multidrug therapy is still effective the other available drugs (rifampicin and clofazimine) are much more expensive than dapsone and hence there is a great need for a vaccination programme.

Moreover, the social stigma of leprosy inhibits persons from seeking medical assistance before the disease is advanced. M. leprae multiplies very slowly, doubling its number every twelve days compared with as little as twenty minutes for Escherichia coli for example, and the time from infection to the appearance of clinical leprosy is estimated to be three to ten years. During this time the people with 'subclinical' infection may spread the disease and there is thus a great need for a simple test for infection. It is generally estimated (Binford et al., 1982; Bloom and Godal, 1983) that there are twice as many unreported cases as recorded cases and the total world estimate based on this assumption is 15 million cases.

Because of the lack of an animal host there had been a limited supply of the microorganisms, from human tissue, for scientific study. In 1960 it was found (Shepard, 1960) that the microorganism would multiply to a limited extent in the mouse foot pad but a major breakthrough came in 1971 when it was established (Kirchheimer and Storrs, 1971; Storrs, 1971) that leprosy bacilli from human tissue would multiply extensively in the experimentally infected nine-banded armadillo (Dasypus novemcinctus Linn., Fig. 1) thus providing a potential source of large amounts of the microorganism for studies of its biochemistry and possibly for the production of vaccines.

Although it was traditionally believed that natural M. leprae infection was specific to man alone, it has now been observed in a chimpanzee (Pan troglodytes) (Leininger et al., 1980), a mangabey monkey (Cercocebus torquatus atys) (Meyers et al., 1985; Wolf et al., 1985) from West Africa and in the wild armadillos of Louisiana (Smith et al., 1978) and of the Texas Gulf coast of the U.S.A. where approximately five per cent (Smith et al., 1983) of the wild armadillos caught were found to be infected.

This has occasioned some concern since; "In the last ten years, the armadillo has become the mascot of the Texas counterculture, resulting in

Dasypus novemcinctus Linn.
(Nine-banded armadillo)

Figure 1

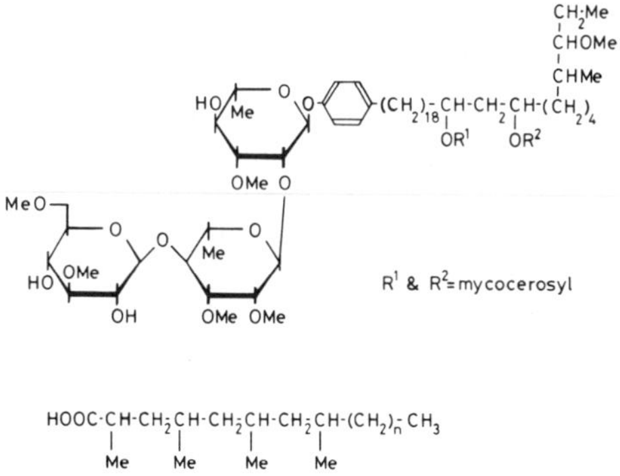

$$R^1 \ \& \ R^2 = \text{mycocerosyl}$$

$$HOOC\text{-}CH\text{-}CH_2CH\text{-}CH_2CH\text{-}CH_2CH\text{-}(CH_2)_n CH_3$$

MYCOCEROSIC ACIDS n = 17, 19 or 21

Phenolic Glycolipid - I

Figure 2

increased armadillo-human contact by means of armadillo races and beauty contests. Armadillo memorabilia such as mounted armadillos, armadillo boots, vests, watchbands etc., sell readily and are produced in many places. Moreover, armadillo meat is a delicacy to some and a staple to many Texans." (Smith et al., 1983).

The World Health Organisation Immunology of Leprosy Programme (WHO/IMMLEP) was initiated in 1974 and exploited the availability of M. leprae-infected armadillo tissue which provided a greater opportunity for research and Philip Draper (Anon, 1980) developed procedures for isolating the M. leprae from infected armadillo tissues which were stored for the WHO at the Laboratory for Leprosy and Mycobacterial Research in the NIMR, Mill Hill under the control of Dr R. J. W. Rees. With the premise that induction of immunological reactivity (by cell mediated mechanisms) to M. leprae antigens might lead to protection against clinical leprosy, vaccination programmes using whole M. leprae were initiated (Bloom, 1983a; Noordeen and Sansarricq, 1984; Rees, 1983). Various protein antigens carrying distinct antigenic determinants, expressed by M. leprae but not by other mycobacteria, have been identified by the monoclonal antibody technique and these M. leprae-specific determinants may help to further research on prophylactic or therapeutic immunisation against leprosy (Ivanyi, 1984; Young, D. B. et al., 1984, 1985c; Sengupta and Sinha, 1984; Engers et al., 1985; Seckl, 1985; Atlaw et al., 1985; Kolk et al., 1985; Engers et al., 1986). If any of these are useful as possible vaccines then recombinant DNA technology to produce the M. leprae antigens in Escherichia coli, which has already been demonstrated (Young, R.A. et al., 1985; Jacobs et al., 1986), will be useful and remove the dependence on armadillos for the production of materials. T-Cell clones, able to recognise M. leprae-specific antigens, have also been demonstrated (Mustafa et al., 1986; Ottenhoff et al., 1986).

Investigations of the lipids from armadillo-derived M. leprae by Patrick Brennan and his co-workers in 1980 (Brennan and Barrow, 1980) demonstrated that a certain fraction showed lines of precipitation with anti-sera from lepromatous patients which did not react with antisera from patients with other mycobacterial infections. The pure lipid was shown to be serologically active when incorporated into liposomes (Payne et al., 1982) and was characterised (Hunter and Brennan, 1981; Hunter et al., 1982; Tarelli et al., 1984) as the structure shown in Fig. 2 (although the absolute configurations of the sugars were not determined), in which the oligosaccharide portion has the sequence 3,6-di-O-methyl-beta-gluco-pyranose-(1→4)-2,3-di-O-methyl-alpha-rhamnopyranose-(1→2)-3-O-methyl-alpha-rhamnopyranose.

This glycolipid ('phenolic glycolipid I') is present in relatively high amounts (ca. 2% of the dry weight) in M. leprae isolated from infected armadillo liver and is also present in high amounts in tissue surrounding the bacilli (2.2mg/g) (Hunter and Brennan, 1981). Although the lipid portion of the molecule, a diacyl phenolic phthiocerol, with mycocerosic acids attached as acyl groups is typical of mycobacterial lipids (Demarteau-Ginsburg and Lederer, 1963; Gastambide-Odier and Sarda, 1970), the unique sugar composition (3,6-di-O-methylglucose has not been isolated from any other natural source and 2,3-di-O-methylrhamnose and 3-O-methylrhamnose have not been encountered in the same oligosaccharide) suggested a role for the glycolipid in the specific serological diagnosis of leprosy. The hydrophobicity of the molecule presented problems in the aqueous serological assays although the enzyme-linked immunosorbent assay (ELISA) (Hunter et al., 1982; Cho et al., 1983; Brett et al., 1983; Young, D. B. and Buchanan, 1983; Buchanan et al., 1983; Brett et al., 1984), immunodiffusion (Payne et al., 1982) and immunoblotting (Young, D. B. et al., 1985a, 1985b) have been successful.

Much smaller quantities of two similar glycolipids, one containing 6-O-methylglucose in place of 3,6-di-O-methylglucose ('phenolic glycolipid III') (Hunter and Brennan, 1983) and the other containing 3-O-methylrhamnose in place of 2,3-di-O-methylrhamnose ('phenolic glycolipid II') (Fujiwara et al., 1984) have also been isolated from M. leprae lipids probably being metabolic byproducts of 'phenolic glycolipid I'.

The structural studies on the glycolipid (Fig. 2), described above, were carried out on material isolated from armadillo-derived M. leprae but subsequent studies have shown that identical material is present in formalin-fixed human lepromatous liver (Izumi et al., 1985; Vemuri et al., 1985; Hunter et al., 1985), large stocks of which exist in national leprosy hospitals. Earlier work (Young, D. B., 1981) had also shown the presence of a novel lipid in skin biopsies from leprosy patients and the presence of the glycolipid (Fig. 2) in the sera of lepromatous leprosy patients has also been established (Young, D. B. et al., 1985d; Cho et al., 1986b).

With the structure of the glycolipid established, apart from the absolute configurations of the sugars, the scene was set for synthetic studies in order to determine the immunodominant portion of the molecule.

SYNTHETIC STUDIES ON PHENOLIC GLYCOLIPID I at Mill Hill

For the synthesis (Gigg, R. et al., 1983) of the terminal disaccharide (10, Fig. 3) of the glycolipid, 3,6-di-O-methyl-D-glucose, which has been prepared by a variety of routes (Gigg, R. et al., 1983; Chatterjee et al., 1985), was converted into the glucosyl chloride (2) by acetylation and subsequent reaction of the acetate (1) with hydrogen chloride in glacial acetic acid. L-Rhamnose was converted by Fischer glycosidation into allyl-L-rhamnopyranoside (3), which gave the isopropylidene derivative (4) on treatment with 2,2-dimethoxypropane and an acid catalyst. Condensation of the alcohol (4) and the glucosyl chloride (2) in the presence of mercury(II) cyanide in acetonitrile gave predominantly the crystalline beta-linked disaccharide derivative (5) together with a small amount of the syrupy alpha-linked disaccharide derivative. Hydrolysis of the acetate groups from (5) gave the diol (6), the ^{1}H-NMR spectrum of which confirmed the beta-linkage of the 3,6-di-$\underline{O}$-methyl-D-glucopyranoside residue.

Benzylation of the diol (6) gave (7) and the isopropylidene group was removed by acidic hydrolysis to give the diol (8), which was methylated to give the tetra-$\underline{O}$-methyl ether (9). Hydrogenolysis of (9) removed the benzyl groups and reduced the allyl group to give propyl 4-$\underline{O}$-(3,6-di-$\underline{O}$-methyl-beta-D-glucopyranosyl)-2,3-di-$\underline{O}$-methyl-alpha-L-rhamnopyranoside (10). In the ELISA test this compound inhibited the interaction of the glycolipid from <u>M. leprae</u> with the antibody produced in rabbits, against the glycolipid, to a much greater extent than did methyl 3,6-di-$\underline{O}$-methyl-beta-D-glucopyranoside or 3,6-di-$\underline{O}$-methyl-D-glucose (Brett et al., 1984).

Since the structural studies (Hunter et al., 1982; Tarelli et al., 1984) of the glycolipid had not established the absolute configurations of the sugars, we also decided to prepare (Gigg, J. et al., 1985a) the corresponding disaccharide (16, Fig. 4) containing D-rhamnose, in place of L-rhamnose, to see if this was superior to the disaccharide (10) in the inhibition of the interactions of the glycolipid with the antibody in the ELISA test. For this purpose (Fig. 4) it was necessary to prepare the corresponding protected derivative (14) of D-rhamnose. D-Rhamnose is not readily available and therefore it was prepared from D-mannose (Gigg, J. et al., 1985a) via the allyl 2,3:4,6-di-$\underline{O}$-isopropylidene-alpha-D-mannopyranoside (11). Preferential hydrolysis of the 4,6-$\underline{O}$-isopropylidene group from (11) gave the diol (12) which was readily converted into the 6-$\underline{O}$-tosyl derivative (13) and this on reduction with lithium aluminium hydride gave the required allyl 2,3-$\underline{O}$-isopropylidene-alpha-D-rhamnopyranoside (14). Condensation of (14) with the 3,6-di-$\underline{O}$-methylglucosyl chloride (2) gave the crystalline beta-linked disaccharide derivative (15), which was converted into the required disaccharide (16) as described above for the conversion of compound (5) into the disaccharide (10). In the ELISA test, the D-rhamnose-containing disaccharide (16) was not superior to the L-rhamnose-containing disaccharide (10) for the inhibition of interaction of the glycolipid with the antibody, giving us an indication that the immunodominant sugar was the 3,6-di-$\underline{O}$-methyl-D-glucose.

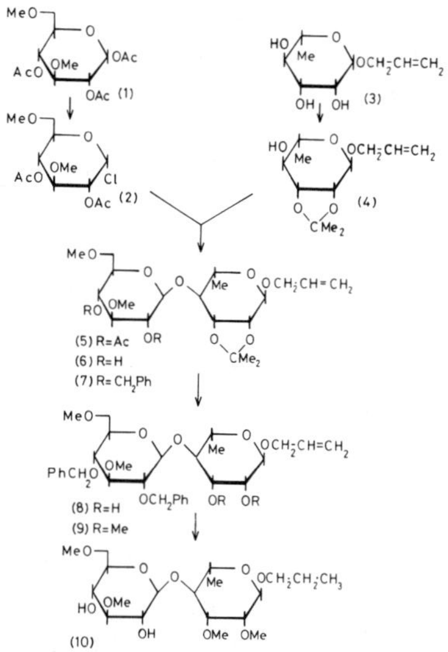

Figure 3

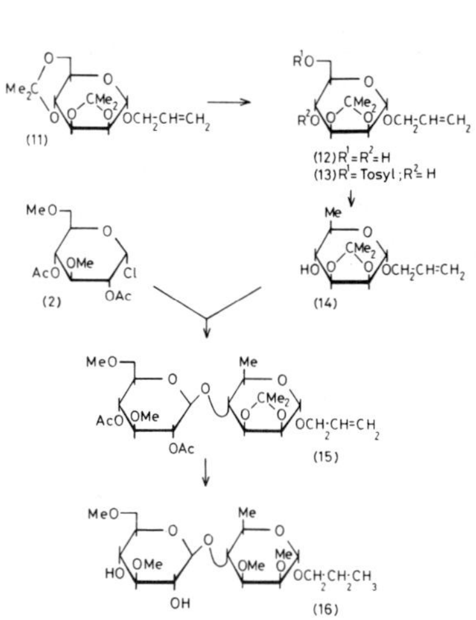

Figure 4

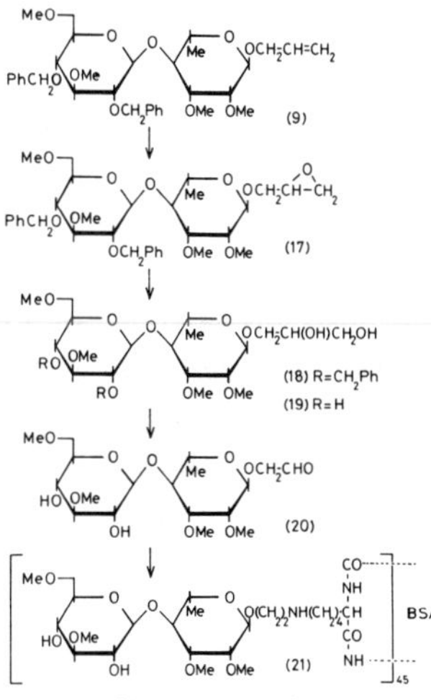

BSA = bovine serum albumin
Figure 5

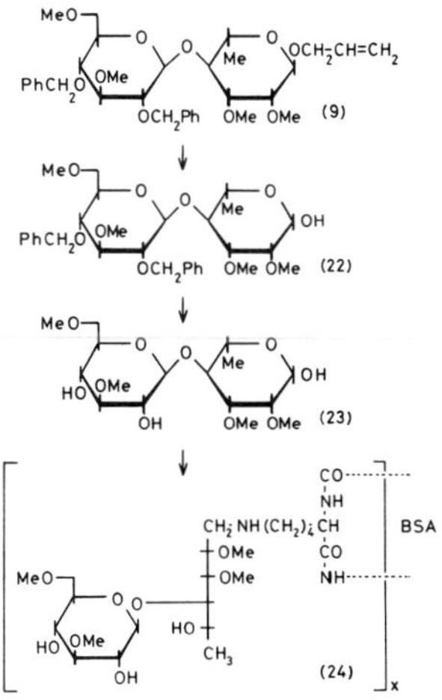

BSA = bovine serum albumin
Figure 6

In the ELISA test for the serodiagnosis of antibodies to the glycolipid in the sera of leprosy patients, the glycolipid is adsorbed on the plastic microtitre ELISA test plates. Some dificulty has been experienced in this procedure because of the physical properties of the glycolipid, but it has been improved to some extent by removal of the mycocerosyl groups esterified to the phthiocerol residue (Young, D. B. and Buchanan, 1983). Proteins adhere much more readily to the plate and therefore a conjugate of our haptenic disaccharide with bovine serum albumin was prepared.

For this purpose (Fig. 5), the allyl glycoside of the disaccharide (9) was converted into the epoxide (17), by the action of m-chloroperbenzoic acid, and this on basic hydrolysis gave the glycerol glycoside (18). Removal of the benzyl groups from the disaccharide derivative (18) by hydrogenolysis gave the disaccharide (19) and treatment of (19) with sodium metaperiodate gave the aldehyde (20) suitable for coupling to protein by reductive amination (Gray, 1978).

Bovine serum albumin contains 59 lysine residues, out of a total of 581 amino-acids (Peters, 1975), with epsilon-amino groups available for condensation. Reaction of the aldehyde (20) with bovine serum albumin in the presence of sodium cyanoborohydride in phosphate buffer (pH 7.5) gave a coupled product (21) (Gigg, J. et al., 1986) in which ca. 45 of the lysine residues had been substituted, as determined by amino-acid analyses on the freeze-dried product.

This product performed well in the ELISA test for the serodiagnosis of antibodies to the natural glycolipid in the sera of leprosy patients and was much more convenient to use than the glycolipid or its deacylated derivative. A large batch (900mg) of the conjugate (21) (sufficient for 1.8 million test doses) has been prepared for serodiagnosis of leprosy by WHO/IMMLEP-supported investigators.

A different conjugate (24, Fig. 6) was also prepared from the allyl glycoside (9). Removal of the allyl group (Gigg, J. and Gigg, R. 1966; Gigg, R. and Warren, 1968) from (9) gave the free disaccharide (22) and the benzyl groups were removed from this by hydrogenolysis to give the free terminal disaccharide (23) of the glycolipid. The free aldehyde group of the reducing 2,3-di-O-methyl-L-rhamnose was used to couple (23) to the amino-groups of lysine in bovine serum albumin, using the reductive amination technique, to give the conjugate (24). The conjugate (24) is effectively only a monosaccharide conjugate since the stereochemistry of the terminal 2,3-di-O-methylrhamnose has been destroyed in the coupling reaction. The conjugate (24) was found to be less efficient than conjugate (21, Fig. 5) in the ELISA tests indicating that the linkage region between the two sugars may be part of the haptenic portion of the molecule.

Two further routes (Gigg, J. et al., 1986) (Fig. 7) to the aldehyde (20) were also investigated so that the unprotected allyl glycoside (27) of the disaccharide could be prepared. Reaction of the diol (6) with p-methoxybenzyl chloride and sodium hydride in N,N-dimethylformamide gave the bis-p-methoxybenzyl ether (25) and this was converted into the corresponding tetramethyl derivative (26) as described previously. Since the p-methoxybenzyl groups can be removed oxidatively, by dichloro-dicyanoquinone (Oikawa et al., 1982; Yonemitsu, 1985), without affecting the double bond of the allyl group, compound (26) was readily converted into the allyl disaccharide (27). Compound (27) could be converted into the aldehyde (20) via the epoxide (29) as described previously or by ozonolysis.

The epoxide (29) should also be suitable for coupling directly with the lysine residues of bovine serum albumin and for the preparation of affinity columns. Kochetkov and his coworkers (Chernyak et al., 1984) have also shown that allyl glycosides can be copolymerised with acrylamide to give useful immunologically active polymers and the allyl disaccharide (27) will be used for this purpose also. Since the p-methoxybenzyl ether can be removed without affecting the allyl group, a further route (Fig. 7) to the allyl disaccharide (27) was also developed (Gigg, J. et al., 1986). The alcohol (4) was converted into the p-methoxybenzyl ether (31) and removal of the isopropylidene group gave the diol (32) which was converted into the dimethyl ether (33). Removal of the p-methoxybenzyl group with dichlorodicyanoquinone gave the alcohol (30) which was condensed with the 3,6-di-O-methylglucosyl chloride (2) to give directly the acetate (28) of the allyl disaccharide (27).

For the synthesis (Gigg, J. et al., 1985b) of a trisaccharide derivative (46, Figs. 8 and 9) suitable for coupling, if necessary, to a protein, the alcohol (4) was converted into the benzyl ether (34) and subsequent hydrolysis of the isopropylidene group gave the diol (35). The diol (35) was converted (David and Hanessian, 1985) into the dibutylstannylene derivative (36) and this on reaction with methyl iodide in N,N-dimethylformamide led to monomethylation on the equatorial 3-hydroxyl group to give the methyl ether (37). Condensation of the alcohol (37) with 2,3,4-tri-O-acetyl-L-rhamnopyranosyl chloride gave the disaccharide derivative (38) and this on basic hydrolysis gave (39) which was converted into the isopropylidene derivative (40) on reaction with 2,2-dimethoxypropane and an acid catalyst.

Further elaboration of the disaccharide derivative (40) into the trisaccharide (46, Fig. 9) was achieved by condensation of the 3,6-di-O-methylglucosyl chloride (2) with the disaccharide derivative (40) to give the crystalline derivative (41) of the beta-linked glucose derivative. Basic hydrolysis of (41) gave the diol (42) and this was benzylated to give (43).

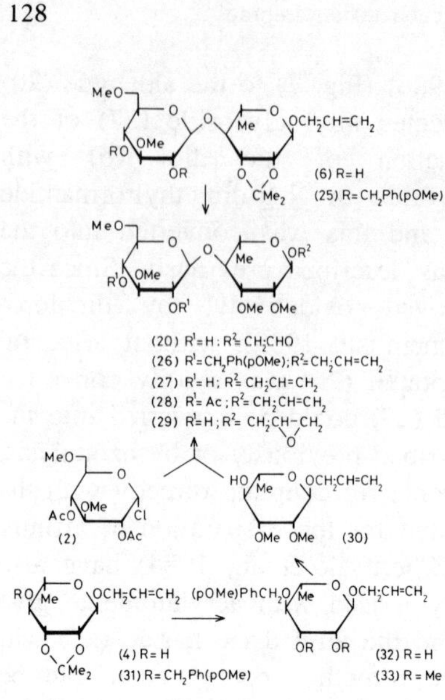

Figure 7

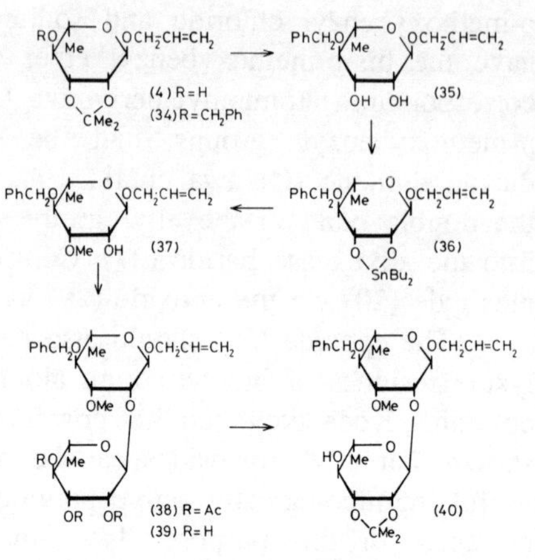

Figure 8

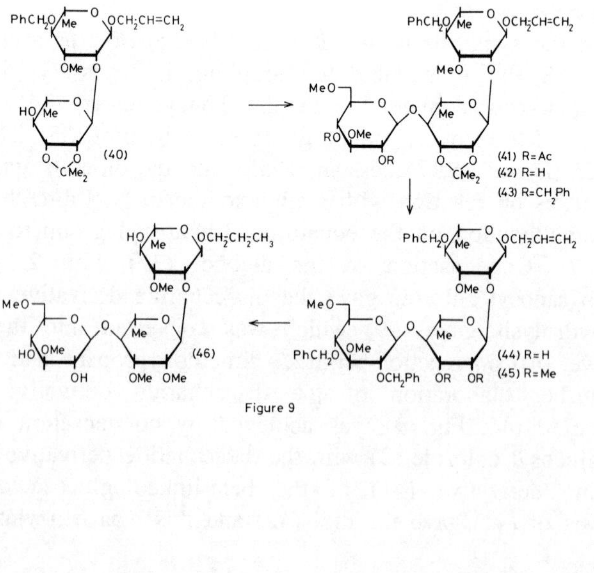

Figure 9

Hydrolysis of the isopropylidene group from (43) gave the diol (44) which was converted into the methyl ether (45). Removal of the three benzyl groups by hydrogenolysis (and concomitant reduction of the allyl group) gave the required propyl trisaccharide (46) containing the entire sequence of sugars of the natural glycolipid (assuming that the rhamnose residues have the L-configuration).

However, the trisaccharide derivative (46) was no more efficient than the disaccharide (10, Fig. 3) in inhibition of the interaction of the natural glycolipid with its antibody in the ELISA test.

In case the phenyl group of the intact glycolipid should be important in the immunological reactions, the corresponding phenyl trisaccharide (48, Fig. 10) was prepared (Gigg, J. et al., 1986) by an exactly analogous procedure to that described in Figs. 8 and 9 for the propyl trisaccharide but using, instead of the allyl glycoside (4) as the starting material, the phenyl glycoside (47, Fig. 10). The ^{1}H-NMR spectrum of the synthetic phenyl trisaccharide (48) showed identical peaks to the corresponding peaks of the natural glycolipid (Fig. 10) but the phenyl trisaccharide (48) was again no more active than the disaccharide (10, Fig. 3) in the competitive bioassays using the ELISA test.

The results of the immunological tests on these synthetic oligosaccharides suggested that the immunodominant sugar of the sequence was the 3,6-di-O-methyl-beta-D-glucopyranose residue and therefore some simple derivatives of 3,6-di-O-methyl-beta-D-glucopyranose (54 and 60, Fig. 11) were prepared (Gigg, J. et al., 1986) with spacer arms suitable for coupling to proteins.

The commercially available alcohols (49) and (55) were used as starting materials. Compound (49) was converted into the isopropylidene derivative (50) and this was condensed with the 3,6-di-O-methylglucosyl chloride (2) to give the acetate (51), which was hydrolysed by base to the diol (52). Hydrolysis of the isopropylidene group from (52) gave (53) which was cleaved with sodium metaperiodate to give the aldehyde (54).

Similarly the alcohol (55) was converted into the beta-glucosides (56) and (57) and the latter was converted into the epoxide (58) which gave the alcohol (59) on basic hydrolysis. Sodium metaperiodate converted the alcohol (59) into the aldehyde (60).

Conjugates of the aldehydes (54) and (60) with bovine serum albumin were prepared by the reductive amination technique and the conjugates were found to be somewhat inferior to the disaccharide conjugate (21, Fig. 5) in the tests for antibody to the glycolipid in the sera of leprosy patients and therefore the conjugate (21) was preferred for serodiagnosis (Brett et al., 1986).

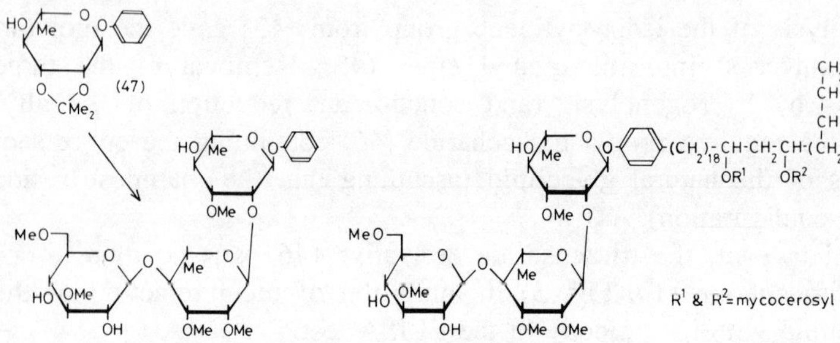

SYNTHETIC TRISACCHARIDE (48)   NATURAL GLYCOLIPID

¹H-N.m.r δ				
5·48	H-1 aromatic rhamnose	5·45	5·45
5·10	H-1 rhamnose	5·05	5·10
4·41 (J=7Hz)	H-1 glucose	4·40 (7·5)	4·42 (7·6)
4·23	H-2 aromatic rhamnose	4·25	
3·74	H-2 rhamnose	3·75	Hunter et al.(1982)

Tarelli et al. (1984)

Figure 10

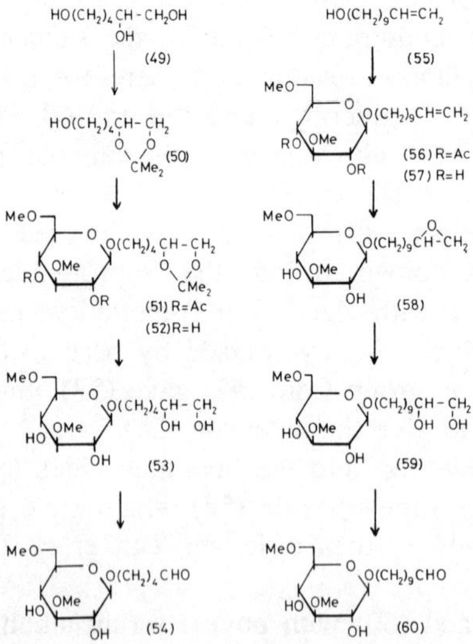

Figure 11

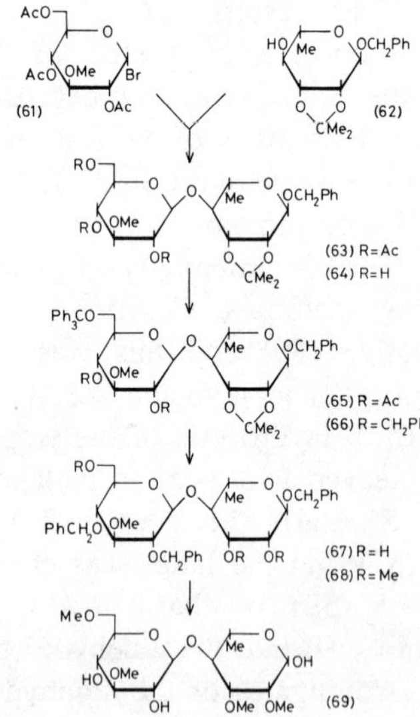

Figure 12

SYNTHETIC STUDIES ON PHENOLIC GLYCOLIPID I by Brennan and his Associates

Brennan and his associates have also carried out extensive synthetic work on the oligosaccharide portion of the glycolipid. For the synthesis (Fujiwara et al., 1984; Cho et al., 1984; Fujiwara et al., 1986) of the disaccharide (69, Fig. 12), 2,4,6-tri-O-acetyl-3-O-methyl-alpha-D-gluco-pyranosyl bromide (61) was condensed with benzyl 2,3-O-isopropylidene-alpha-L-rhamnopyranoside (62) in the presence of mercury(II) cyanide to give the crystalline disaccharide derivative (63). This was hydrolysed with base to give the triol (64) which was tritylated and subsequently acetylated to give the acetate (65). This on benzylation, under basic conditions, gave the benzyl ether (66), which on acidic hydrolysis gave the triol (67) and this was converted to (68) on methylation. Hydrogenolytic removal of the three benzyl groups from (68) gave the free disaccharide (69). Analogous disaccharides, containing beta-D-glucose or 6-O-methyl-beta-D-glucose in place of the 3,6-di-O-methyl-beta-D-glucose, were also prepared (Fujiwara et al., 1986). The disaccharide (69) was coupled (Fujiwara et al., 1984; Cho et al., 1984) to bovine serum albumin, using reductive amination, as described above for the same disaccharide (23) prepared as described in Fig. 6.

For the synthesis (Fujiwara et al., 1984) of a derivative of the trisaccharide (77, Fig. 13), Brennan and his associates converted the alcohol (62) into the benzyl ether (70) which was hydrolysed to the diol (71). Phase transfer allylation of (71) gave predominantly the 2-O-allyl ether (72) which was methylated to (73) and the allyl group removed to give the alcohol (74). This was condensed with a bromide (76), derived from the free disaccharide (69) via the corresponding acetate (75). The resulting mixture of isomers (77) (due to the mixture of anomers formed at the rhamnose-rhamnose linkage) was separated by thin-layer chromatography and subsequent deacetylation and debenzylation gave the free trisaccharide (78) corresponding to the oligosaccharide sequence of the natural glycolipid. This trisaccharide (78) was also shown not to be superior to the free disaccharide (69) in the competition of binding of the glycolipid to its antibody in the ELISA test.

Various disaccharides containing a p-hydroxyphenylpropionate aglycone (Fig. 14), suitable for coupling to protein, have also been synthesised (Fujiwara et al., 1985). Methyl 3-(p-hydroxyphenyl)-propionate was condensed with tri-O-acetyl-L-rhamnopyranosyl bromide and the product deacetylated to give the glycoside (79) which was converted into the isopropylidene derivative (80). This was condensed with 2,4,6-tri-O-acetyl-3-O-methyl-D-glucosyl bromide to give the disaccharide derivative (81) which was deacetylated to give (82). The trityl derivative (83) of (82) was

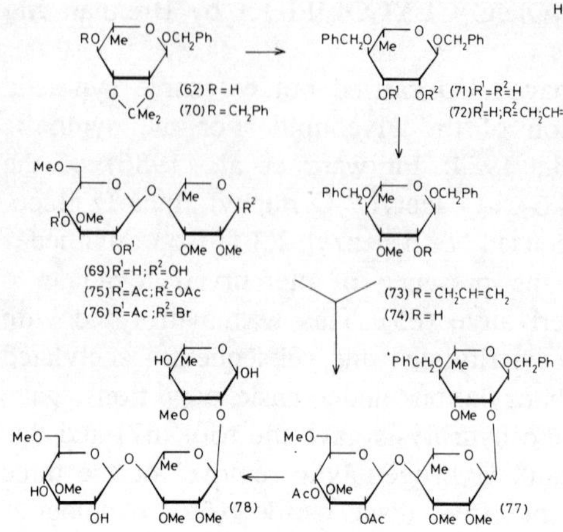

Figure 13

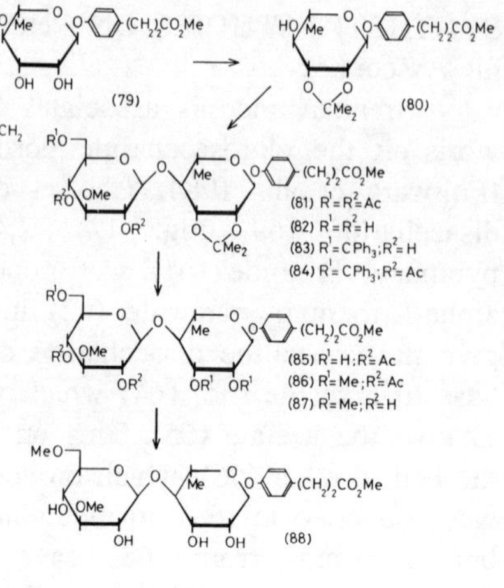

Figure 14

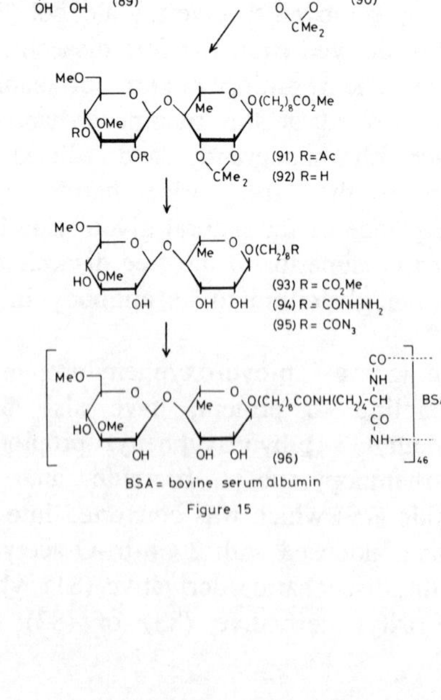

BSA = bovine serum albumin

Figure 15

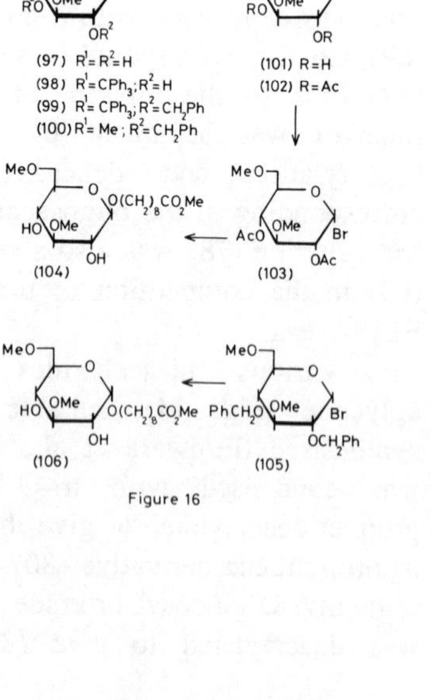

Figure 16

acetylated to give (84) and this on acidic hydrolysis gave predominantly the triol (85) which was methylated with diazomethane in the presence of boron trifluoride-etherate to give (86). Hydrolysis of (86) with sodium methoxide in methanol gave the required disaccharide (87) containing the terminal two sugars of the native glycolipid.

The corresponding disaccharide (88) (lacking O-methyl groups on the rhamnose molecule) was also prepared in this work. Both compounds (87) and (88) were active in the ELISA test as competitive inhibitors of the reaction between the glycolipid and its antibody. Interestingly, other derivatives of (87) containing acetyl groups on either the 2- or 4-hydroxy groups were also active as was the corresponding derivative of (87) containing an alpha-linked 3,6-di-O-methyl-D-glucopyranose residue.

A further disaccharide derivative suitable for conjugation to protein has also been described (Chatterjee et al., 1985) by Brennan and his associates (Fig. 15). Tri-O-acetyl-L-rhamnopyranosyl bromide was condensed with 8-methoxycarbonyloctan-1-ol and the product hydrolysed with sodium methoxide to give the L-rhamnose derivative (89), which was converted into the isopropylidene derivative (90). This was condensed with 2,4-di-O-acetyl-3,6-di-O-methyl-alpha-D-glucopyranosyl bromide in the presence of mercury(II) cyanide and mercury(II) bromide to give the disaccharide derivative (91) which was hydrolysed with sodium methoxide to give the diol (92). This on acidic hydrolysis gave 8-methoxycarbonyloctyl 4-O-(3,6-di-O-methyl-beta-D-glucopyranosyl)-alpha-L-rhamnopyranoside (93).

Following the general technique developed by Lemieux and his coworkers (Lemieux et al., 1975a) for producing spacer-arm glycosides for protein conjugation, compound (93) was converted by reaction with hydrazine hydrate into the hydrazide (94) and this was treated with tert-butyl nitrite in N,N-dimethylformamide to give the acyl azide (95). The azide was used directly and coupled to the lysine residues of bovine serum albumin, substituting ca. 46 of these, to give the conjugate (96).

Conjugates of 3,6-di-O-methyl-beta-D-glucopyranose with proteins were also prepared (Chatterjee et al., 1985) (Fig. 16). 2,4,6-Tri-O-acetyl-3-O-methyl-alpha-D-glucopyranosyl bromide was condensed with benzyl alcohol to give the beta-benzyl glycoside, which was deacetylated to give the triol (97). This was converted into the trityl derivative (98) which was benzylated to give (99); subsequent removal of the trityl group and methylation of the free 6-hydroxy group gave the 3,6-di-O-methyl derivative (100). The benzyl groups were removed by hydrogenolysis to give 3,6-di-O-methylglucose (101) and this was acetylated and the acetate (102) was converted into the glucosyl bromide (103). Condensation of (103) with

8-methoxycarbonyloctan-1-ol and subsequent hydrolysis of the product with sodium methoxide gave the required beta-glucoside (104).

The corresponding alpha-linked glucoside (106, Fig. 16) was also prepared (Chatterjee et al., 1985) by condensing 2,4-di-O-benzyl-3,6-di-O-methyl-alpha-D-glucopyranosyl bromide (105) and 8-methoxycarbonyloctan-1-ol in the presence of tetrabutylammonium bromide following the general method of Lemieux and his coworkers (Lemieux et al., 1975b).

Both of the monosaccharide derivatives (104) and (106) were coupled (Chatterjee et al., 1985) with bovine serum albumin and the alpha-linked conjugate (106) was only half as effective as the beta-linked conjugate (104) in the ELISA tests, but the disaccharide conjugate (96, Fig. 15) was better than the monosaccharide conjugate (104) for the serological detection of leprosy.

CONCLUSIONS

Various synthetic strategies have been employed to provide water-soluble conjugates of parts of the oligosaccharide portion of 'phenolic glycolipid I' with protein, which are superior to the water-insoluble natural glycolipid for the serodiagnosis of leprosy by the conventional ELISA method or by diffusion-in-gel ELISA (DIG-ELISA) (Cho et al., 1986a). These conjugates may allow the serodiagnosis of M. leprae infection (Anon, 1986) in patients before clinical signs of the disease are visible and thus allow a chemotherapeutic regime to be initiated to prevent infection by 'subclinical' carriers.

The synthetic oligosaccharides should also help to define the specificities of the monoclonal antibodies that have been raised against the glycolipid (Ivanyi, 1984; Young, D. B. et al., 1984; Engers et al., 1985; Seckl, 1985; Atlaw et al., 1985; Kolk et al., 1985; Engers et al., 1986) and help in studies of lymphocyte suppression (Bloom, 1983b; Colston, 1984) induced by the glycolipid (Mehra et al., 1984).

ACKNOWLEDGEMENTS

The immunoassays of the synthetic materials were performed at the Laboratory of Leprosy and Mycobacterial Research at the NIMR in the laboratory of Dr. Sara Brett. This investigation received financial support from the Immunology of Leprosy (IMMLEP) component of the UNDP/World Bank/WHO Special Programme for Research and Training in Tropical Diseases.

REFERENCES

Anon (1980) UNDP/World Bank/WHO Special Programme for Research and Training in Tropical Diseases. Report of the Fifth Meeting of the Scientific Working Group on the Immunology of Leprosy (IMMLEP). TDR/IMMLEP-SWG(5)/80.3, Annex 4, p. 23. WHO, Geneva.

Anon (1986) Lancet I: 533-535.

Atlaw, T., Kozbor, D. and Roder, J. C. (1985) Infect. Immun. 49: 104-110.

Baker, C. J. (1983) Int. J. Leprosy 51: 397-403.

Binford, C.H., Meyers, W. M. and Walsh, G. P. (1982) J. Amer. Med. Assoc. 247: 2283-2292.

Bloom, B. R. (1983a) Int. J. Leprosy 51: 505-509.

Bloom, B. R. (1983b) Nature (London) 303: 284-285.

Bloom, B. R. and Godal, T. (1983) Rev. Infect. Diseases 5: 765-780.

Brennan, P. J. and Barrow, W. W. (1980) Int. J. Leprosy 48: 382-387.

Brett, S. J., Draper, P., Payne, S. N. and Rees, R. J. W. (1983) Clin. Exp. Immunol. 52: 271-279.

Brett, S. J., Payne, S. N., Draper, P. and Gigg, R. (1984) Clin. Exp. Immunol. 56: 89-96.

Brett, S.J., Payne, S.N., Gigg, J., Burgess, P. and Gigg, R. (1986) Clin. Exp. Immunol. 64: 476-483.

Brody, S. N. (1974) 'The Disease of the Soul. Leprosy in Medieval Literature', Cornell University Press.

Buchanan, T. M., Young, D. B., Miller, R. A. and Khanolkar, S. R. (1983) Int. J. Leprosy 51: 524-530.

Chatterjee, D., Douglas, J. T., Cho, S.-N., Rea, T. H., Gelber, R. H., Aspinall, G. O. and Brennan, P. J. (1985) Glycoconjugate J. 2: 187-208.

Chernyak, A. Ya., Levinsky, A. B., Dmitriev, B. A. and Kochetkov, N. K. (1984) Carbohydr. Res. 128: 269-282.

Cho, S.-N., Yanagihara, D. L., Hunter, S. W., Gelber, R. H. and Brennan, P. J. (1983) Infect. Immun. 41: 1077-1083.

Cho, S.-N., Fujiwara, T., Hunter, S. W., Rea, T. H., Gelber, R. H. and Brennan, P. J. (1984) J. Infect. Diseases 150: 311-322.

Cho, S.-N., Chatterjee, D. and Brennan, P. J.(1986a) Am. J. Trop. Med. Hyg. 35: 167-172.

Cho, S.-N., Hunter, S.W., Gelber, R.H., Rea, T.H. and Brennan, P.J. (1986b) J. Infect. Diseases 153: 560-569.

Colston, M. J. (1984) Immunol. Today 5: 199-200.

David, S. and Hanessian, S. (1985) Tetrahedron 41: 643-663.

Demarteau-Ginsburg, H. and Lederer, E. (1963) Biochim. Biophys. Acta 70: 442-451.

Engers, H. D., Bloom, B. R. and Godal, T. (1985) Immunol. Today 6: 345-348.

Engers, H. D., Houba, V., Bennedsen, J., Buchanan, T. M., Chaparas, S. D., Kadival, G., Closs, O., David, J. R., van Embden, J. D. A., Godal, T., Mustafa, S. A., Ivanyi, J., Young, D. B., Kaufmann, S. H. E., Khomenko, A. G., Kolk, A. H. J., Kubin, M., Louis, J. A., Minden, P., Shinnick, T. M., Trnka, L. and Young, R. A. (1986) Infect. Immun. 51: 718-720.

Etemadi, A. H. and Convit, J. (1974) Infect. Immun. 10: 236-239.

Feeney, P. (1964) 'The Fight Against Leprosy', Elek Books, London.

Fujiwara, T., Hunter, S. W., Cho, S.-N., Aspinall, G. O. and Brennan P. J. (1984) Infect. Immun. 43: 245-252.

Fujiwara, T., Izumi, S. and Brennan P. J. (1985) Agric. Biol. Chem. 49: 2301-2308.

Fujiwara, T., Hunter S. W. and Brennan P. J. (1986) Carbohydr. Res. 148: 287-298.

Gastambide-Odier, M. and Sarda, P. (1970) Pneumonologie 142: 241-255.

Gigg, J. and Gigg, R. (1966) J. Chem. Soc. (C): 82-86.

Gigg, J., Gigg, R., Payne, S. and Conant, R. (1985a) Carbohydr. Res. 141: 91-97.

Gigg, J., Gigg, R., Payne, S. and Conant, R. (1985b) Chem. Phys. Lipids 38: 299-307.

Gigg, J., Gigg, R., Payne, S. and Conant, R. (1986) J. Chem. Soc. Perkin Trans. I, in the press.

Gigg, R. and Warren, C. D. (1968) J. Chem. Soc. (C): 1903-1911.

Gigg, R., Payne, S. and Conant, R. (1983) J. Carbohydr. Chem. 2: 207-223.

Gray, G. R. (1978) Methods Enzymol. 50: 155-160.

Hansen, G. A. (1874) Norsk. Mag. Laegevidensk 4: 1; see Hansen, G. A. (1955) Int. J. Leprosy 23: 307-309 for a reprint of the original paper.

Hunter, S. W. and Brennan P. J. (1981) J. Bacteriol. 147: 728-735.

Hunter, S. W., Fujiwara, T. and Brennan, P. J. (1982) J. Biol. Chem. 257: 15072-15078.

Hunter, S. W. and Brennan, P. J. (1983) J. Biol. Chem. 258: 7556-7562.

Hunter, S. W., Stewart, B. S. and Brennan, P. J. (1985) Int. J. Leprosy 53: 484-486.

Ivanyi, J. (1984) Leprosy Rev. 55: 1-9.

Izumi, S., Sugiyama,K., Fujiwara, T., Hunter, S. W. and Brennan, P. J. (1985) J. Clin. Microbiol. 22: 680-682.

Jacobs, W. R., Docherty, M. A., Curtiss, R. and Clark-Curtiss, J. E. (1986) Proc. Nat. Acad. Sci. 83: 1926-1930.

Kirchheimer, W. F. and Storrs, E. E. (1971) Int. J. Leprosy 39: 693-702.

Kolk, A. H. J., Ho, M. L., Klatser, P. R., Eggelte, T. A. and Portaels, F. (1985) Ann. Inst. Pasteur (Microbiol.) 136B: 217-224.

Leininger, J. R., Donham, K. J. and Meyers, W. M. (1980) Int. J. Leprosy 48: 414-421.

Lemieux, R. U., Bundle, D. R. and Baker, D. A. (1975a) J. Amer. Chem. Soc. 97: 4076-4083.

Lemieux, R. U., Hendriks, K. B., Stick, R. V. and James, K. (1975b) J. Amer. Chem. Soc. 97: 4056-4062.

Mehra, V., Brennan, P. J., Rada, E., Convit, J. and Bloom, B. R. (1984) Nature (London) 308: 194-196.

Meyers, W. M., Walsh, G. P., Brown, H. L., Binford, C. H., Imes, G. D., Hadfield, T. L., Schlagel, C. J., Fukunishi, Y., Gerone, P. J., Wolf, R. H., Gormus, B. J., Martin, L. N., Harboe, M. and Imaeda, T. (1985) Int. J. Leprosy 53: 1-14.

Mustafa, A. S., Gill, H. K., Nerland, A., Britton, W. J., Mehra, V., Bloom, B. R., Young, R. A. and Godal, T. (1986) Nature (London) 319: 63-66.

Noordeen, S. K. and Sansarricq, H. (1984) Bull. W.H.O. 62: 1-6.

Oikawa, Y., Yoshioka, T. and Yonemitsu, O. (1982) Tetrahedron Lett. 23: 885-888.

Ottenhoff, T. H. M., Klatser, P. R., Ivanyi, J., Elferink, D. G., de Witt, M. Y. L. and de Vries, R. R. P. (1986) Nature (London) 319: 66-68.

Payne, S. N., Draper, P. and Rees, R. J. W. (1982) Int. J. Leprosy 50: 220-221.

Peters, T. (1975) in 'The Plasma Proteins', 2nd Ed. Vol. 1, pp. 133-181. Ed. Putnam, F. W., Academic Press.

Rees, R. J. W. (1983) Int. J. Leprosy 51: 515-518.

Seckl, M. J. (1985) Int. J. Leprosy 53: 618-640.

Sengupta, U. and Sinha, S. (1984) Int. J. Leprosy 56: 727-741.

Shepard, C. C. (1960) J. Exp. Med. 112: 445-454.

Smith, J. H., File, S. K., Nagy, B. A., Folse, D. S., Buckner, J. A., Webb, L. J. and Beverding, A. M. (1978) J. Reticuloendothelial Soc. 24: 705-719.

Smith, J. H., Folse, D. S., Long, E. G., Christie, J. D., Crouse, D. T., Tewes, M. E., Gatson, A. M., Ehrhardt, R. L., File, S. K. and Kelly M. T. (1983) J. Reticuloendothelial Soc. 34: 75-88.

Stanford, J. L., Rook, G. A. W., Convit, J., Godal, T., Kronvall, G., Rees, R. J. W. and Walsh, G. P. (1975) Brit. J. Exp. Path. 56: 579-585.

Storrs, E. E. (1971) Int. J. Leprosy 39: 703-714.

Tarelli, E., Draper, P. and Payne, S. N. (1984) Carbohydr. Res. 131: 346-352.

Vemuri, N., Khandke, L., Mahadevan, P. R., Hunter, S. W. and Brennan, P. J. (1985) Int. J. Leprosy 53: 487-489.

Weymouth, A. (1938) 'Through the Leper-squint', Selwyn and Blount, London.

Wolf, R. H., Gormus, B. J., Martin, L. M., Baskin, G. B., Walsh, G. P., Meyers, W. M. and Binford, C. H. (1985) Science 227: 529-531.

Yonemitsu, O. (1985) J. Synth. Org. Chem. (Japan) 43: 691-702.

Young, D. B. (1981) Int. J. Leprosy 49: 198-204.

Young, D. B. and Buchanan, T. M. (1983) Science 221: 1057-1059.

Young, D. B., Khanolkar, S. R., Barg, L. L. and Buchanan, T. M. (1984) Infect. Immun. 43: 183-188.

Young, D. B., Fohn, M. J. and Buchanan, T. M. (1985a) J. Immunol. Methods 79: 205-211.

Young, D. B., Fohn, M. J., Khanolkar, S. R. and Buchanan, T. M. (1985b) Leprosy Rev. 56: 193-198.

Young, D. B., Fohn, M. J., Khanolkar, S. R. and Buchanan, T. M. (1985c) Clin. Exp. Immunol. 60: 546-552.

Young, D. B., Harnisch, J. B., Knight, J. and Buchanan, T. M. (1985d) J. Inf. Diseases 152: 1078-1081.

Young, R. A., Mehra, V., Sweetser, D., Buchanan, T., Clark-Curtiss, J., Davis, R. W. and Bloom B. R. (1985) Nature (London) 316: 450-452.

ANALYSIS OF MYCOBACTERIAL MYCOLIC ACIDS

D.E. Minnikin[1], G. Dobson[1,2], J.H. Parlett[1,2], A.K. Datta[1],
S. Megan Minnikin[1] and M. Goodfellow[2]

Departments of [1]Organic Chemistry and [2]Microbiology, The University,
Newcastle upon Tyne, UK

INTRODUCTION

Mycolic acids are high molecular weight 2-alkyl branched, 3-hydroxy, long-chain fatty acids forming a covalent basal layer in the outer membranes of mycobacteria (Minnikin, 1982). These acids occur as complex mixtures differing in the type of additional oxygen function and skeletal units such as methyl branches, $\underline{E}$ and $\underline{Z}$ double bonds and cyclopropane rings. The main structural types (Dobson et al., 1985) are the so-called α-mycolates (I) and α′-mycolates (II) with no oxygen functions in addition to the 3-hydroxy acid unit, methoxymycolates (III), ketomycolates (IV), epoxymycolates (V) and wax-ester mycolates (VI) which release ω-carboxymycolates and 2-alkanols on hydrolysis. The analysis of these relatively intractable compounds requires efficient extraction procedures and the preparation of good all-round derivatives suitable for chromatographic profiles and spectroscopic analysis.

MYCOLATE EXTRACTION

Mycolic acids were originally released by heating under reflux with methanolic potassium hydroxide, followed by acidification and esterification with diazomethane, but a degree of C-2 epimerisation was often a complication (see Dobson et al., 1985). A small-scale acid methanolysis procedure was developed for systematic analysis (Minnikin et al., 1975, 1980) of many strains, particularly to distinguish mycobacteria from nocardiae. Acidic methods, however, degrade epoxymycolates (V) and can cause partial rupture of cyclopropane rings (Dobson et al., 1985). An attractive approach was to release mycolic acids with methanolic tetramethylammonium hydroxide and convert the resulting salts to methyl esters by adding iodomethane in dimethylformamide, the reaction being driven to completion by the precipitation of tetramethylammonium iodide

(Minnikin et al., 1982, 1984). This method has the advantage that basic conditions are maintained throughout, but only methyl esters are available since substantial transesterification takes place during the first stage and wet samples are not completely derivatised. Another route to methyl esters involves hydrolysis with potassium hydroxide in methyl cellosolve (2-methoxyethanol), followed by esterification with diazomethane (Daffé et al., 1983). This latter procedure causes no epimerisation of mycolates but diazomethane is a potent carcinogen and methyl cellosolve is a teratogen.

Derivatisation methods which utilise phase-transfer catalysed esterification offer the possibility of preparing a range of different derivatives. Mycolic acids have been treated with an alkaline solution of tetrabutylammonium hydrogen sulphate, followed by reaction in dichloromethane with iodomethane or 4-nitrobenzyl bromide to give methyl or 4-nitrobenzyl esters, respectively (Minnikin et al., 1985a, b). Unpublished results have shown that the procedure can be streamlined by using 15% aqueous tetrabutylammonium hydroxide at 100°C for the hydrolysis; addition of a dichloromethane solution of the appropriate halide then gives the corresponding ester. Using this very simple methodology, mycolic acids are currently being converted into methyl, pentafluorobenzyl and 9-methylanthracenyl esters for purposes to be discussed later.

CHROMATOGRAPHY OF MYCOLATES

Column chromatography was extensively used in early studies (see Asselineau, 1966) to fractionate mycolate esters but the true complexity of the natural mixtures only became clear with the introduction of thin-layer chromatography (TLC) by Lanéelle (1963). Single dimensional TLC of methyl mycolates provides resolution of the main mycolate types (Minnikin et al., 1980; Daffé et al., 1983) but more than one system may be necessary to separate critical pairs. Two-dimensional TLC of mycolate methyl esters gives distinct patterns (Minnikin et al., 1980, 1984, 1985a; Dobson et al., 1985) but, if both α′- (II) and methoxymycolates (III) are present, clear confirmation is given by TLC of t-butyldimethylsilyl ether derivatives (Dobson et al., 1985). Satisfactory two-dimensional TLC systems are being developed for the resolution of pentafluorobenzyl and 9-methylanthracenyl esters of mycolic acids.

High performance liquid chromatography (HPLC) of mycolates was introduced by Qureshi et al., (1978) and Steck et al., (1978), using p-bromophenacyl esters to achieve satisfactory separations in normal and reversed-phase modes. The preparation of these derivatives was not easy for the rapid small-scale processing of multiple samples and many interfering by-products were produced. A selection of ultra-violet absorbing derivatives have been investigated and pentafluorobenzoates of methyl esters (Draper et al., 1982) and t-butyldimethylsilyl ethers of 4-nitrobenzyl esters (Minnikin et al., 1985b) have been used for the analyses of the mycolic acids of <u>Mycobacterium</u> <u>leprae</u>. In unpublished studies (D.E. & S.M. Minnikin), t-butyldiphenylsilyl ethers of methyl esters were chromatographically suitable but the ultra-violet absorption was weak and mass spectra were uninformative. The strategy currently being followed is to use phase-transfer catalysis to prepare suitable ultra-violet absorbing esters; pentafluorobenzyl (Gyllenhaal & Ehrsson, 1975) for general use and 9-methylanthracenyl (Korte, 1982) for high sensitivity, including fluorescence detection. Pentafluorobenzyl esters are prepared easily without by-products, taking care to avoid the powerful lachrymatory properties of pentafluorobenzyl bromide. These derivatives chromatograph well on TLC and HPLC and have the advantage that they are also excellent derivatives for the gas chromatography of non-hydroxylated fatty acids with sensitive electron capture detection. The preparation of esters from 9-chloromethylanthracene, however, produces a number of coloured by-products, but these can be removed from the mycolate derivatives by washing a petroleum ether solution with acetonitrile. The chromatographic behaviour of these latter derivatives is adequate to exploit their exceptional detection sensitivity.

Pyrolysis gas chromatography of methyl mycolates results in the release of long-chain esters, corresponding to the chain in the 2-position and the first two carbons, and high molecular weight aldehydes (see Minnikin, 1982). This is a convenient method for determining the length of the chain in 2-position which varies in mycolates from different species (Daffé et al., 1983). Intact mycolic esters can be analysed by combined gas chromatography-mass spectrometry of trimethylsilyl ether derivatives (Yano, 1985) to provide a wealth of structural information.

CONCLUSIONS

The extraction and chromatographic methods outlined above provide diagnostic profiles, suitable for comparing the mycolates from different species, and purified mycolic acid derivatives for detailed spectroscopic and chemical analysis. Mass spectrometry of methyl or pentafluorobenzyl esters provide molecular weights and informative fragmentation patterns; further conversion to trimethylsilyl (Yano, 1985) or t-butyldimethylsilyl (Dobson et al., 1985; Minnikin et al., 1982) ethers gives characteristic M-15 or M-57 peaks, respectively. High-field nuclear magnetic resonance spectroscopy is providing access to the detailed skeletal composition of mycolates (Minnikin et al., 1982).

The systematic detailed analysis of mycolic acids from the whole range of mycobacterial species is consolidating their structural role in an unusual outer membrane system in which they interact with a range of complex free lipids (Minnikin, 1982).

FORMULAE

I $\quad CH_3 \cdot (CH_2)_l \; X \cdot (CH_2)_m \cdot Y \cdot (CH_2)_n \cdot \overset{\displaystyle OH}{CH} \cdot \overset{\displaystyle COOH}{CH} \cdot (CH_2)_x \cdot CH_3$

$X = \underline{cis}\ -CH=CH-,\ -\overset{\textstyle CH_2}{CH\!-\!\!-\!CH}-\ ;\ Y = X\ or\ \underline{trans}\ -CH=CH.\overset{\displaystyle CH_3}{CH}-\ with\ (CH_2)_{m-1}$

$\underline{l} = 15,17,19;\ \underline{m} = 14,16;\ \underline{n} = 11,13,15,17;\ \underline{x} = 19,21,23$

II $\quad CH_3 \cdot (CH_2)_l \cdot CH = CH \cdot (CH_2)_m \cdot \overset{\displaystyle OH}{CH} \cdot \overset{\displaystyle COOH}{CH} \cdot (CH_2)_x \cdot CH_3$

$\underline{l} = \underline{m} = 17;\ \underline{x} = 21$

III $\quad CH_3 \cdot (CH_2)_l \cdot \overset{\displaystyle CH_3}{CH} \cdot \overset{\displaystyle OCH_3}{CH} \cdot (C_y H_{2y-2}) \cdot \overset{\displaystyle OH}{CH} \cdot \overset{\displaystyle COOH}{CH} \cdot (CH_2)_x \cdot CH_3$

IV $\quad CH_3 \cdot (CH_2)_l \cdot \overset{\displaystyle CH_3}{CH} \cdot \overset{\displaystyle O}{C} \cdot (C_y H_{2y-2}) \cdot \overset{\displaystyle OH}{CH} \cdot \overset{\displaystyle COOH}{CH} \cdot (CH_2)_x \cdot CH_3$

V $\quad CH_3 \cdot (CH_2)_l \cdot \overset{\displaystyle CH_3}{CH} \cdot \overset{\displaystyle O}{CH\!-\!\!-\!CH} \cdot (C_y H_{2y-2}) \cdot \overset{\displaystyle OH}{CH} \cdot \overset{\displaystyle COOH}{CH} \cdot (CH_2)_x \cdot CH_3$

VI $\quad CH_3 \cdot (CH_2)_l \cdot \overset{\displaystyle CH_3}{CH} \cdot O \cdot \overset{\displaystyle O}{C} \cdot (C_y H_{2y-2}) \cdot \overset{\displaystyle OH}{CH} \cdot \overset{\displaystyle COOH}{CH} \cdot (CH_2)_x \cdot CH_3$

$\underline{l} = 15,17;\ \underline{y} = 32\text{-}39;\ \underline{x} = 19,21,23$

ACKNOWLEDGEMENTS

Support is acknowledged from the Medical Research Council (G974/522/S; G8216538), Science and Engineering Research Council (GRA 88651) and the British Leprosy Relief Association.

REFERENCES

Asselineau, J. (1966) The Bacterial Lipids, Hermann, Paris.

Daffé, M., Lanéelle, M.A., Asselineau, C., Levy-Frebault, V. & David, H. (1983) Ann. Microbiol. (Inst. Pasteur) 134B: 241-256.

Dobson, G., Minnikin, D.E., Minnikin, S.M., Parlett, J.H., Goodfellow, M., Ridell, M. & Magnusson, M. (1985) in Chemical Methods in Bacterial Systematics (Goodfellow, M. & Minnikin, D.E., eds.) pp. 237-265 Academic Press, London.

Draper, P., Dobson, G., Minnikin, D.E. & Minnikin, S.M. (1982) Ann. Microbiol. (Inst. Pasteur) 133B: 39-47.

Gyllenhaal, O. & Ehrsson, H. (1975) J. Chromatogr. 107: 327-333.

Korte, W.D. (1982) J. Chromatogr. 243: 153-157.

Lanéelle, G. (1963) C.R. Acad. Sci. 257: 781-783.

Minnikin, D.E. (1982) in The Biology of the Mycobacteria (Ratledge, C. & Stanford, J.L., eds.) pp. 95-184 Academic Press, London.

Minnikin. D.E., Alshamaony, L. & Goodfellow, M. (1975) J. Gen. Microbiol. 88: 200-204.

Minnikin. D.E., Hutchinson, I.G., Caldicott, A.B. & Goodfellow, M. (1980) J. Chromatogr. 188: 221-233.

Minnikin, D.E., Minnikin, S.M. & Goodfellow, M. (1982) Biochim. Biophys. Acta 712: 616-620.

Minnikin, D.E., Minnikin, S.M., Parlett, J.H., Goodfellow, M. & Magnusson, M. (1984) Arch. Microbiol. 139: 225-231.

Minnikin, D.E., Minnikin, S.M., Parlett, J.H. & Goodfellow, M. (1985a) Zbl. Bakt. Hyg. A 259: 446-460.

Minnikin, D.E., Dobson G., Goodfellow, M., Draper, P. & Magnusson, M. (1985b) J. Gen. Microbiol. 131: 2013-2021.

Qureshi, N., Takayama, K., Jordi, C. & Schnoes, H.K. (1978) J. Biol, Chem. 253: 5411-5417.

Steck, P.A., Schwartz, B.A., Rosendahl, M.S. & Gray, G.R. (1978) J. Biol. Chem. 253: 5625-5629.

Yano, I. (1985) in Rapid Methods and Automation in Microbiology and Immunology (Habermehl, K.O., ed.) pp. 239-247 Springer, Berlin.

THIN-LAYER CHROMATOGRAPHY IN COMBINATION WITH IMMUNOSTAINING FOR ANALYSIS OF MYCOBACTERIAL LIPID ANTIGENS

D.E. Minnikin[1], Malin Ridell[2], Inger Mattsby-Baltzer[3] & J.H. Parlett[1]

[1]Department of Organic Chemistry, The University, Newcastle upon Tyne, UK, [2]Department of Medical Microbiology and [3]Department of Clinical Bacteriology, University of Gothenburg, Gothenburg, Sweden

INTRODUCTION

The cell envelopes of mycobacteria are rich in complex free lipids, ranging in character from inert waxes to antigenic glycolipids (Minnikin, 1982). The best-studied examples of the latter class are the glycopeptidolipid (I) agglutinins (Brennan, 1984) and glycosyl phenolphthiocerol dimycocerosates (II) such as that from <u>Mycobacterium</u> <u>leprae</u> (Hunter et al., 1982). The recognition of the antigenicity of such lipids has not been very easy in the past and general methods for this purpose are long overdue. Radio-immunology has been the main method used to detect directly the antigenicity of mammalian lipids separated by thin-layer chromatography (TLC) (Brockhaus et al., 1981; Magnani et al., 1982; Mori et al., 1982; Symington et al., 1984; Fredman et al., 1986). The lipid part of bacterial

```
I     Fatty acyl - Phe - allo Thr - Ala -alaninol - sugar ┐
                                  |                        |
                                  |                        |
                                  O                        ├─ (Acetyl)
                                  |                        |         n
                                  |                        |
                      Basal sugar unit - Specific sugars ┘
```

$$\text{II} \quad R-O-\!\!\bigcirc\!\!-(CH_2)_{17} \cdot \underset{OR'}{CH} \cdot CH_2 \cdot \underset{OR'}{CH} \cdot (CH_2)_4 \cdot \underset{CH_3}{CH} \cdot \underset{OCH_3}{CH} \cdot CH_2 \cdot CH_3$$

$$\overset{\beta}{} \qquad \overset{\alpha}{} \qquad \overset{\alpha}{}$$

$$R = 3.6\text{-}Me_2\text{-}Glu(1\rightarrow 4)2.3\text{-}Me_2\text{-}Rha(1\rightarrow 2)3\text{-}Me\text{-}Rha\ 1 \longrightarrow$$

$$R' = CH_3 \cdot (CH_2)_{17\text{-}21} \cdot \underset{CH_3}{CH} \cdot CH_2 \cdot \underset{CH_3}{CH} \cdot CH_2 \cdot \underset{CH_3}{CH} \cdot CO$$

lipopolysaccharides (Lipid A) has been studied by immuno-TLC with an enzyme-linked immunosorbent assay (ELISA) technique using a spectrophotometer to read the results (Mattsby-Baltzer & Alving, 1984). The glycosyl phenolphthiocerol dimycocerosate (II) from M. leprae has been detected immunochemically using TLC on silica gel plates (Young et al., 1984) and polysulphone membranes (Young et al., 1985); the former was subject to high background levels and the latter medium was unsuitable for the chromatography of polar lipid antigens. A recent communication (Ridell et al., 1986) introduced a general immunostaining technique for the direct detection of mycobacterial lipid antigens on TLC plates, using the glycopeptidolipids of M. scrofulaceum as the example. The present paper reviews this latter technique and its relation to other TLC methods for the identification of lipid antigens.

TLC IMMUNOSTAINING

Antisera against whole cells of mycobacteria, such as M. scrofulaceum and M. avium (Ridell et al., 1986), were produced in rabbits according to the agglutination method of Anz et al., (1969). Antiserum against intracellular material was prepared with the supernatant from X-pressed mycobacterial cells (Ridell, 1975).

Free lipids were extracted and separated by TLC (Minnikin et al., 1984; Dobson et al., 1985) on pieces (10 x 10 cm) cut from Whatman K6 TLC plates (10 x 20 cm) which were resistant to washing with aqueous solutions (Mattsby-Baltzer & Alving, 1984). After chromatography, dry TLC plates were kept for 1 h in 0.01M phosphate buffer solution (PBS), pH 7.2, containing 0.25% bovine serum albumin and then washed three times (5-10 min) in PBS containing 0.05% Tween 20. The plates were then placed for 5 h at room temperature in antiserum diluted (1/100 v/v) in PBS Tween. After washing as above, the plates were kept at room temperature overnight in anti-rabbit IgG horseradish peroxidase conjugate (Biorad) diluted (1/1000 v/v) in PBS Tween. The following day the plates were washed three times as above. The final stage involved treatment with 0.015% hydrogen peroxide and the Biorad HRP Colour Development Reagent, the essential ingredient being 4-chloro-1-naphthol. When blue spots became clearly visible (15-30 min), the plates were washed in distilled water and allowed to dry.

As an example of the method, the pattern of the lipids present in the polar lipid fraction of M. scrofulaceum (Ridell et al., 1986) are shown in Fig. 1. The components labelled NPGP are relatively non-polar, non-antigenic glycopeptidolipids and the spots marked PGP are surface glycolipid antigens, corresponding to deacetylated lipids identified in previous studies (Brennan, 1984) for this sero-variant. The expected phospholipids diphosphatidylglycerol (DPG), phosphatidylinositol (PI) and the phosphatidylinositol dimannosides (PIDM) (Dobson et al., 1985) were also present (Fig. 1a). Serological analysis of the TLC plates by the immunostaining method, with an antiserum against whole cells of the same strain of M. scrofulaceum, is shown in Fig. 1b. Positive reactions were obtained for the polar glycopeptidolipids (PGP) but not for the non-polar varieties (NPGP). This is in agreement with previous studies on this sero-variant (Brennan, 1984). Analysis of a TLC plate, prepared as in Fig. 1 but with an antiserum against whole cells of a sero-variant of M. avium gave no reaction, thus supporting the species specificity of the polar glycopeptidolipids (Brennan, 1984).

The family of phosphatidylinositol dimannosides (PIDM) are conside-red to be antigens (Bannerjee & Subrahmanyam, 1978) but only a very faint response, or none at all, was obtained with the antiserum against the whole cells (Fig. 1). These glycophospholipids are probably associated with the plasma membrane, rather than the outer membrane (Minnikin, 1982), and antisera against intracellular material may be more appropriate for assessing their antigenicity. Indeed, the less polar (A) of the PIDMs and two more polar pentamannosides (PIPMs) (Ridell et al., 1986) gave strong reactions with an antiserum against intracellular material from a different strain of M. scrofulaceum but no other lipids reacted with this serum. Since the

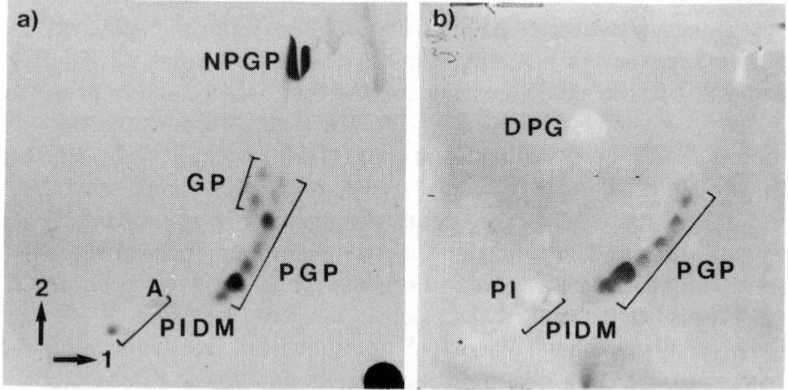

Figure 1. Two-dimensional TLC of polar lipids of M. scrofulaceum. 1, Chloroform-methanol-water (65:25:4, by vol.); 2, chloroform-acetic acid-methanol-water (80:15:12:4, by vol.). Detection: (a) 1-naphthol-sulphuric acid for glycolipids; (b) immunostaining with antiserum against whole cells of the same serovar. See text for abbreviations. Reproduced, with permission, from Ridell et al., (1986).

structures of PIDMs and PIPMs are considered to be essentially the same in all mycobacteria (Minnikin, 1982), the above result is expected.

CONCLUSIONS

The present method makes possible rapid objective surveys of lipid antigens without prior purification of each antigen from complex mixtures. An additional advantage of the method is that it avoids the problem of hydrophobicity which hindered studies of the antigenic properties of the glycosyl phenolphthiocerol (II) from M. leprae (Hunter et al., 1982). Minor antigenic lipid components can easily be recognised (Fig. 1); such a task would previously have necessitated the isolation of each individual lipid from a large amount of biomass.

The immunostaining technique requires repeated treatment with aqueous solutions and Whatman K6 plates were selected because of their excellent water resistance. The lipid patterns on these plates were generally the same as those recorded previously on Merck aluminium-backed sheets (Dobson et al., 1985). Plates which are less water-resistant can be treated with materials such as poly(isobutyl methacrylate) (Brockhaus et al., 1981; Hansson et al., 1985; Holmgren et al., 1985; Fredman et al., 1986) but in the first of these papers problems with detachment of the silica gel were encountered. The actual amount of the plastic coat is also important since too much may block access to the lipids and too little may give an unacceptable background level (Brockhaus et al., 1981; Hansson et al., 1985). In addition, little is known about the interactions between the plastic coat and different lipid types; the use of water-resistant plates avoids all the problems of plastic treatment.

The present method is a modified version of that used by Mattsby-Baltzer and Alving (1984) to analyse the antigenicity of Lipid A fractions from Gram-negative bacteria. The same Whatman K6 TLC plates were employed but the enzyme was alkaline phosphatase instead of peroxidase and the substrate was added using an agarose gel. Spectrophotometric detection has also been replaced by simple visual inspection (Fig. 1) made possible by the insolubility of the dye formed from the 4-chloro-1-naphthol reagent.

ACKNOWLEDGEMENTS

Support is acknowledged from the Swedish National Association against Heart and Chest Diseases (M.R.) and the Medical Research Council (G8216538) (D.E.M., J.H.P).

REFERENCES

Anz, W., Meissner, G. & Roder, W. (1969) Zentralbl. Bakt. Orig.1 211: 530-550.

Bannerjee, B. & Subrahmanyam, D. (1978) Immunochemistry 15, 359-363.

Brennan, P.J. (1984) in The Mycobacteria: a Sourcebook (Kubica, G.P. & Wayne, L.G.,eds.) pp. 467-489. Marcel Dekker, New York.

Brockhaus, M., Magnani, J.L., Blaszczyk, M., Steplewski, Z., Koprowski, H., Karlsson, K.A., Larson, G. & Ginsburg, V. (1981) J. Biol. Chem. 256: 13223-13225.

Dobson, G., Minnikin, D.E., Minnikin, S.M., Parlett, J.H., Goodfellow, M., Ridell, M. & Magnusson, M. (1985) in Chemical Methods in Bacterial Systematics (Goodfellow, M. & Minnikin, D.E.,eds.) pp. 237-265. Academic Press, London.

Fredman, P., Brezicka, T., Holmgren, J., Lindholm, L., Nilsson, O. & Svennerholm, L. (1986) Biochim. Biophys. Acta 875: 316-323.

Hansson, G.C., Karlsson, K.A., Larson, G., Stromberg, N. & Thurin, J. (1985) Anal. Biochem. 146: 158-163.

Holmgren, J., Lindblad, M., Fredman, P., Svennerholm, L. & Myrvold, H. (1985) Gastroenterology 89: 27-35.

Hunter, S.W., Fujiwara, T. & Brennan, P.J. (1982) J. Biol. Chem. 257: 15072-15078.

Magnani, J.L., Nilsson, B., Brockhaus, M., Zopf, D., Steplewski, A., Koprowski, H. & Ginsburg, V. (1982) J. Biol. Chem. 257: 14365-14369.

Mattsby-Baltzer, I. & Alving, C.R. (1984) Eur. J. Biochem. 138: 333-337.

Minnikin, D.E. (1982) in The Biology of the Mycobacteria (Ratledge, C. & Stanford, J.L.,eds.) pp. 95-184. Academic Press, London.

Minnikin, D.E., O'Donnell, A.G., Goodfellow, M., Alderson, G., Athalye, M., Schaal, A. & Parlett, J.H. (1984) J. Microbiol. Methods 2: 233-241.

Mori, E., Mori, T., Sanai, Y., & Nagai, Y., (1982) Biochem. Biophys. Res. Comm. 108: 926-932.

Ridell, M. (1975) Int. J. Syst. Bacteriol. 25: 124-132.

Ridell, M., Minnikin, D.E., Parlett, J.H. & Mattsby-Baltzer, I. (1986) Lett. Appl. Microbiol. 2: 89-92.

Symington, F.W. Bernstein, I.D. & Hakomori, S. (1984) J. Biol. Chem. 259: 6008-6012.

Young, D.B., Khanolkar, S.J., Barg, L.L. & Buchanan, T.M. (1984) Infect. Immun. 43: 183-188.

Young, D.B., Fohn, M.J. & Buchanan, T.M. (1985) J. Immunol. Methods 79: 205-211.

STRUCTURAL INVESTIGATIONS ON THE GLYCANS OF THE VARIANT SURFACE GLYCOPROTEIN OF <u>TRYPANOSOMA BRUCEI</u>

Heinz Egge, Johannes Gunawan, Jasna Peter-Katalinic, Institut für Physiologische Chemie, University of Bonn,

Brigitte Schmitz, Imogen A. Duncan and Roger A. Klein, Molteno Institute, University of Cambridge

INTRODUCTION

The parasitic protozoan <u>T. b.</u> <u>brucei</u> evades the immune system of the host by sequential expression of structurally and immunologically distinct clone-specific variant surface glycoproteins as reviewed (1,2). These VSGs, which form a dense cell surface coat on the slender form of the parasite present in the host's bloodstream, are anchored in the plasma membrane by a covalently linked phosphatidylinositol (PI). This PI, together with a glucosamine and ethanolamine containing glycan which forms the immunologically cross-reacting determinant (CRD), is in turn attached to the C-terminus of the peptide chain. The VSG from strain MITat 1.6 contains an additional glycan that is bound N-glycosidically to an asparagine residue about 60 amino-acids off the C-terminus.

Upon lysis of the trypanosome, the membrane form (mfVSG) is converted to the soluble form sVSG. It was demonstrated recently that this conversion is catalysed by an endogenous phospholipase C which releases dimyristoyldiglyceride (3,4). During the biosynthesis of the VSG at least four co- or post-translational modifications occur:

a) removal of the amino terminal signal peptide (5-7),

b) addition of non-immunogenic N-glycans linked to asparagine residues (8-11),

c) removal of a C-terminal hydrophobic peptide chain (5,10,12) and

d) addition of an immunologically cross-reacting determinant (CRD) that consists of carbohydrate, ethanolamine, glucosamine and PI as reviewed (13).

The glucosamine of the CRD is linked directly to the PI as evidenced by the release of PI after nitrous deamination of the glucosamine with concomitant formation of 2,5-anhydro-mannitol (14,15).

In this paper we present data on the N-glycans liberated by both <u>endo</u>-<u>N</u>-acetylglucosaminidases H and F and the glycopeptides separated after pronase digestion of sVSG derived from the Molteno Institute Trypanozoon antigenic type MITat 1.6. In the pronase digest an additional glycopeptide fraction could be separated that was partially characterized by fast atom bombardment mass spectrometry.

MATERIALS AND METHODS

The isolation of the sVSG followed the method described by Cross (16). The determination of carbohydrate constituents was performed by GLC of the alditol acetates essentially as described (17). sVSG (2.5 mg) was hydrolysed with 0.5 ml of 0.7 N H_2SO_4 in 85% acetic acid for 16 hours at 80°C. After neutralization with Dowex 1 x 8 in acetate form, the released sugars were reduced with $NaBH_4$ and peracetylated using pyridine, acetic acid anhydride 1:1 (v/v). GLC analysis was performed on a Carlo Erba gas chromatograph with a 1.5 m x 2 mm column packed with 3 % SP 2340 on Supelcoport. Temperature programme: 2°C/min from 150-250°C. GLC-MS analysis was done on an LKB 9000 mass spectrometer using a 25 m fused silica capillary BP 25 (SGE, Victoria, Australia) linked in an open coupling to the ion source. The temperature of the column was programmed from 150-250°C at 3°C/min, the restriction capillary of the open coupling was held at 250°C.

Treatment of this sVSG with <u>endo</u>-<u>N</u>-acetylglucosaminidases H and F followed the procedures outlined by Tarentino et al. (18) and Steube et al. (19) respectively. The liberated oligosaccharides were reduced and radiolabelled with $NaB^3H_4/NaBH_4$. Reduced oligosaccharides were separated on 2.5 cm x 200 cm columns of Bio-gel P-4 as described by Kobata (20). Similar conditions were used for the separation on Fractogel TSK HW-50S (E.Merck, Darmstadt, FRG). For the digestion with pronase the procedure described by Monsigny et al. was used (21). Released oligosaccharides and glycopeptides were permethylated using lithium sulfinyl carbanion essentially as described by Paz Parente et al. (22).

Permethylated samples were freed from reagents by chromatography on Sephadex LH-20 using chloroform/methanol 1:1 (v/v) as eluent. Mass spectrometric analysis of native and permethylated fractions was performed on a ZAB HF mass spectrometer (VG Analytical, Manchester, U.K.) equipped with an M-scan FAB gun (23).

RESULTS AND DISCUSSION

The general structure of mfVSG is shown in scheme 1 together with the possible sites of enzymic attack of the glycolytic and proteolytic enzymes used in this study. The GLC analysis of the alditol acetates

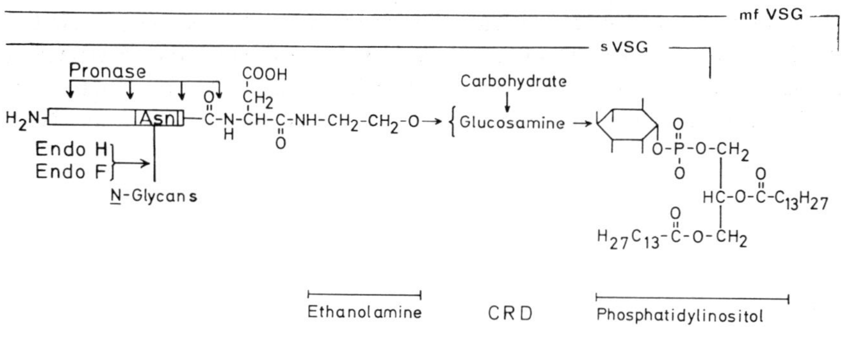

SCHEME 1

showed, that mannose and galactose were present in sVSG in a ratio of 2:1 besides minor amounts of glucosamine.

The experimental protocols for the endo-β-N-acetylglucosaminidase H

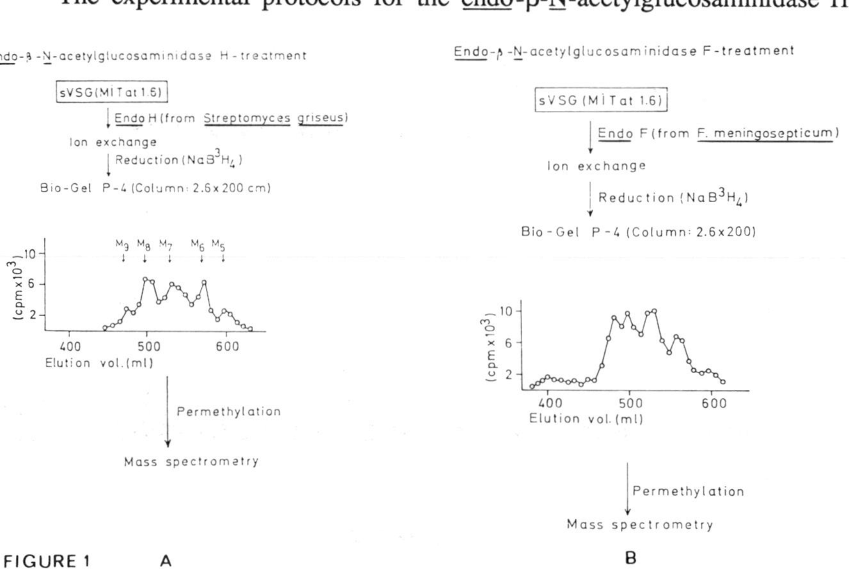

FIGURE 1   A   B

and F treatments are shown in Fig. 1a and 1b. In essence, the oligosaccharides released by both enzymes were identical according to the results of FAB-MS of the permethylated derivatives. Both peracetylated and permethylated derivatives revealed the presence of N-glycans of the high mannose type ranging from Man_{5-9}-GlcNAc-ol. FAB-MS of the permethylated derivatives provided evidence that the Man_{5-8}-GlcNAc-ol fractions were mixtures of isomers. As an example the FAB-MS of permethylated Man_7-GlcNAc-ol is shown in Fig.2. In the high mass range of the spectrum molecular ion related ions $(M+H)^+= 1736$ and $(M+Na)^+=$

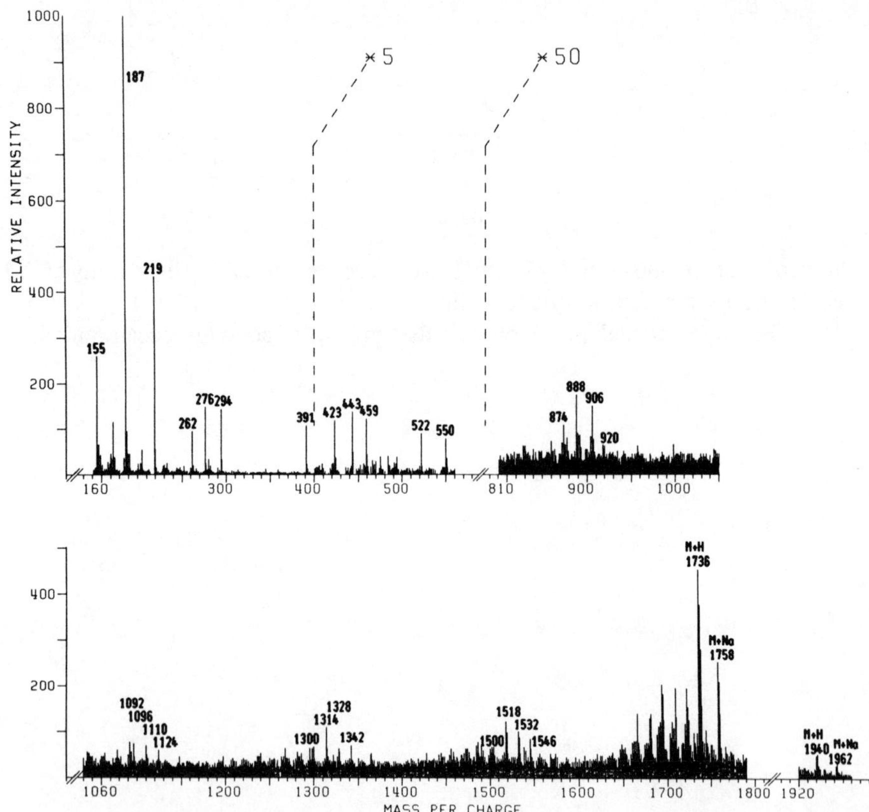

FIGURE 2

1758 are present. Fragmentation occurs by sequential elimination of hexose residues giving rise to ion clusters centered around m/z 1518 and 1314, 1110 and 906, each separated by 204 a.m.u. corresponding to one permethylated hexose unit.

In the low mass range, m/z 219 and 187 represent terminal hexose residues and m/z 276 and 294 the reduced N-acetyl-glucosaminitol. Two terminal hexoses give rise to m/z 423 and 391. The spectrum is in good agreement with the two structures shown in Scheme 2a and b. It is identical to the FAB spectrum obtained from the Man$_7$GlcNAc-ol isolated from the urine of a patient suffering from mannosidosis (24).

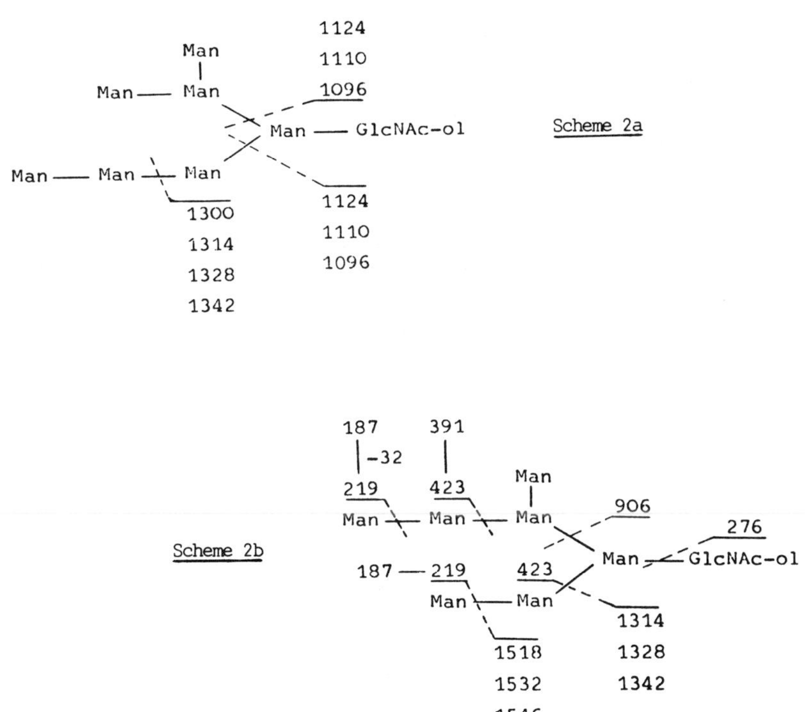

Among the fractions isolated after endo-F treatment, a very minor amount of a 'complex type' N-glycan could be identified by FAB-MS. It constituted less than 1 % of the total N-glycans. From the MS data of the permethylated compound the following structure is proposed:

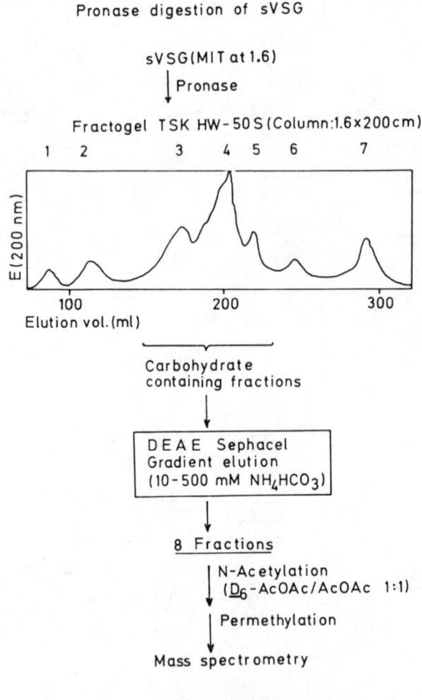

Pronase digestion was originally intended to be performed on the deglycosylated sVSG obtained after treatment with <u>endo</u>-β-<u>N</u>-acetylglucosaminidase F. However, the yields of deglycosylated sVSG were

SCHEME 3

extremely low, possibly due to denaturation of the proteins during the enzymic treatment. Therefore another experimental protocol was chosen as shown in Scheme 3. The pronase digest of sVSG was first fractionated on Fractogel TSK HW-50S in order to remove high molecular and low molecular components. The glycopeptides eluting between 150 and 250 ml water were further separated on DEAE Sephacel. At low ionic strength glycopeptides were eluted that corresponded to the high mannose glycans. The glycopeptides containing the CRD part were due to the presence of a terminal phosphate group expected to be eluted at higher ionic strength of the eluent. All carbohydrate-containing fractions of the DEAE column were N-acetylated with a 1:1 mixture of normal and perdeuterated acetic acid anhydride prior to permethylation.

Permethylated fraction 1 yielded FAB spectra that could be attributed to high mannose glycopeptides. The FAB mass spectrum of fraction 6, shown in Fig.3, revealed the deuterium isotope label in the ions at m/z 1982 and 1985, 2156 and 2159, and 2330 and 2333, indicating that originally free amino-groups have been N-acetylated.

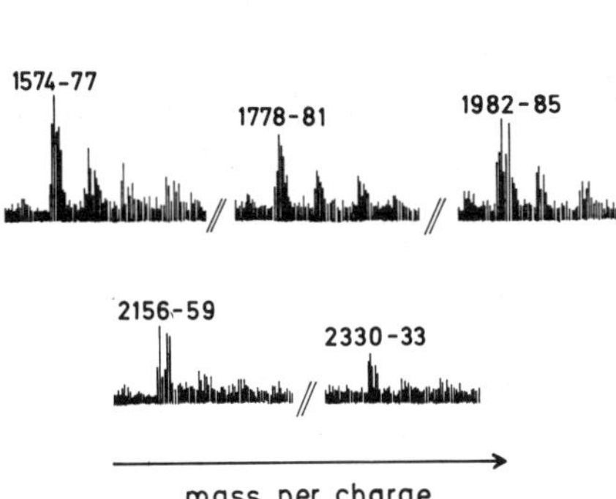

FIGURE 3

If molecular weights are calculated on the basis of the hypothetical structure shown in Scheme 1 with a C-terminal asparagine residue, values are obtained that differ by more than one hundred mass units from those found (Figure 3). The formula presented in Scheme 1 based on a model proposed by Ferguson et al. (13) is, however, incomplete in so far as it does not show how the carbohydrate residue is linked to the C-terminal amino acid. Assuming that the glucosamine and the galactose containing carbohydrate residue is linked glycosidically to the PI, the ethanolamine can either be linked to the carbohydrate by an ether linkage or by another as yet unidentified bifunctional unit. An ether linkage can be ruled out because the ethanolamine residue can be liberated by hydrolysis. Therefore it may be assumed that the difference between the observed and the calculated molecular weight is produced by a hitherto unidentified component.

ACKNOWLEDGEMENTS

This investigation received financial support from the UNDP/World Bank/WHO Special Programme for Research and Training in Tropical Diseases.

REFERENCES

1) Turner, M.J. (1985) Ann.Inst.Pasteur.Immunol. 136C, 41-49.

2) Turner, M.J. (1984) Phil.Trans.Roy.Soc. London. Ser.B. 307, 27-40.

3) Ferguson, M.A.J. and Cross, G.A.M. (1984) J.Biol.Chem. 259, 3011-3015.

4) Ferguson, M.A.J., Haldor, K. and Cross, G.A.M. (1985) J.Biol. Chem. 260, 4963-4968.

5) Boothroyd, J.C., Painter, C.A., Cross, G.A.M., Bernards, A. and Borst, P. (1981) Nucl.Acid Res. 8, 4735-3743.

6) McConnell, J., Gurnett, A.M., Cordingley, J.S., Walker, J.E. and Turner, M.J. (1981) Mol.Biochem.Parasitol. 4, 225-242.

7) McConnell, J., Cordingley, J.S. and Turner, M.J. (1982) Mol.Biochem.Parasitol. 5, 161-174.

8) Strickler, J.E. and Patton, C.L. (1980) Proc.Natl.Acad.Sci. U.S.A. 77, 1529-1533.

9) Rovis, L. and Dube, D.K. (1981) Mol.Biochem.Parasitol. 4, 77-93.

10) McConnell, J., Turner, M.J. and Rovis, L. (1983) Mol.Biochem. Parasitol. 8, 119-135.

11) Grab, D.J., Ito, S., Kara, U.A.K. and Rovis, L. (1984) J.Cell Biol. 99, 569-577.

12) Boothroyd, J.C., Cross, G.A.M, Hoeijmakers, J.H.J. and Borst, P. (1980) Nature, London 288, 624-626.

13) Low, M.G., Ferguson, M.A.J., Futerman, A.H. and Silman, J. (1986) TIBS 11, 212-215.

14) Ferguson, M.A.J., Low, M.G. and Cross, G.A.M. (1985) J.Biol.Chem. 260, 14547-14555.

15) Schmitz, B., Klein, R.A., Egge, H. and Peter-Katalinic, J. (1986) Biochem. Mol. Parasitology 20, 191-197.

16) Cross, G.A.M. (1984) J.Cell Biochem. 24, 79-90.

17) Hanfland, P., Egge, H., Dabrowski, U., Kuhn, S., Roelcke, D. and Dabrowski, J. (1981) Biochemistry 20, 5310-5319.

18) Tarentino, A.L., Plumer, T.H. jr. and Maley, F. (1974) J.Biochem. 249, 818-824.

19) Steube, K., Gross, V. and Heinrich, P.C. (1985) Biochemistry 24, 5587-5592.
20) Yamashita, K., Mizuochi, T. and Kobata, A. (1982) in: Methods Enzymol. (Ginsburg, V. ed.) 83, 105-126.
21) Monsigny, M., Adam-Chosson, A. and Montreuil, J. (1968) Bull.Soc.Chim. Biol. 50, 857-874.
22) Paz Parente, J., Cardon, P., Leroy, Y., Montreuil, J., Fournet, B. and Ricart, G. (1985) Carbohydr.Res. 141, 41-49.
23) Egge, H. and Peter-Katalinic, J. (1985) in: Mass spectrometry in the health and life sciences (Burlingame, A.L. and Castagnoli, N. eds.) Analytical Chemistry Symp.Ser. 24, 401-424.
24) Egge, H., Michalski, J.C. and Strecker, G. (1982) Arch. Biochem. Biophys. 213, 318-326.

A POSSIBLE ROLE FOR GLYCOLIPIDS IN PROKARYOTIC ORGANISMS

Norman Shaw

Department of Biochemistry, The University, Newcastle upon Tyne NE1 7RU

INTRODUCTION

Glycolipids of the glycosyldiacylglycerol type are common components of the bacterial membrane and are predominantly but not exclusively found in Gram-positive bacteria (Shaw, 1975). The glycosyldiacylglycerols of bacteria show considerably greater structural diversity than do similar lipids in plants where they were first discovered and this has proved to be of value in taxonomic studies (Shaw, 1974). Despite their prevalence, no conclusive evidence has yet been presented to support a specific function for these glycolipids. Earlier suggestions of a role in bacterial polysaccharide biosynthesis prompted by some similarity in structure of the carbohydrate head group of the glycolipids to repeating units in the polysaccharide were eliminated by the discovery of the polyisoprenoid intermediates. Although these glycolipids are also found as integral components of the membrane teichoic acids, sometimes called lipoteichoic acids, it seems unlikely that this is a major function as the proportion of the glycolipid present in these polymers represents only a small fraction of the total cellular concentration (Toon et al., 1972). The major proportion of the glycolipid is present in the membrane as unbound lipid, although in most bacteria this represents a minor proportion of the total lipid.

DISCUSSION

There are two groups of organisms where glycolipids are major components of the membrane; mycoplasma and L-forms, and thermophiles. A comparison of the lipids of various Gram-positive bacteria and their derived L-forms shows that in the L-form the proportion of glycolipid is increased to proportions similar to those found in many mycoplasma (Smith et al., 1973). Thus a characteristic feature of the membranes from many

"

organisms lacking a rigid cell wall is a high concentration of glycolipids. This was the first circumstantial evidence that glycolipids could play a role in maintaining the structural integrity of the membrane particularly in an adverse environment. Further evidence to support this hypothesis has come from continuing studies on the lipids of theromophilic bacteria. We have examined the lipids of <u>Thermus</u> <u>aquaticus</u> and other related extremely thermophilic organisms from a wide variety of hot springs and domestic hot water supplies. All the isolates contained one major glycolipid which accounted for half of the total lipid (Pask-Hughes & Shaw, 1982).

Although the structure of the carbohydrate head group varied in the different strains they had a number of interesting features in common. Firstly these glycolipids were all tetraglycosyldiacylglycerol derivatives and secondly they all contained a hexosamine to which was linked by an amide linkage a third fatty acid residue. This <u>N</u>-acylhexosamine residue was located in the penultimate position in the tetrasaccharide head group i.e. the glycolipids were diglycosyl-(<u>N</u>-acyl)-glycosaminylglucosyldiacylglycerols. Glycolipids of this size have only been observed previously as occasional minor components of some bacteria. All the <u>Thermus</u> spp. contain one major phospholipid which has not yet been identified. It is not one of the common bacterial phospholipids e.g. phosphatidylglycerol, phosphatidylethanolamine or bisphosphatidylglycerol but it shares a structural feature with the glycolipids, namely an <u>N</u>-acylhexosamine residue. <u>Thermus</u> is a Gram-negative organism and as such contains an inner (cytoplasmic) and outer membrane system which may be readily separated (Oshima & Osawa, 1983). We have prepared inner and outer membrane fractions in order to study the lipid distribution. These studies have shown that the unknown phospholipid is found predominantly in the inner membrane together with the glycolipid and that the outer membrane is virtually devoid of phospholipid and is composed essentially of glycolipid. The ability of <u>Thermus</u> spp. to maintain the structural integrity of their membranes at such high temperatures (70°C) cannot be ascribed solely to their fatty acids; C_{15} and C_{17} branched-chain acids are commonly found in mesophilic bacteria.

The stabilizing feature appears to be the high concentration of glycolipids. Similar concentrations of glycolipids have been found in other thermophilic bacteria (Langworthy et al., 1974), particularly <u>Sulfolobus</u> <u>acidocaldarius</u> and <u>Thermoplasma</u> <u>acidophilum</u>. The structural rigidity imparted to the outer membrane of <u>Thermus</u> spp. by the unique glycolipid composition was exemplified by an observation during the preparation of inner and outer membrane fractions. The latter involves lysis of the cell wall with lysozyme followed by rupture of the sphaeroplast to produce membrane fragments which may be separated by differential centrifugation.

Sphaeroplasts from most prokaryotic organisms have to be stabilised in an isotonic solution and subsequently disrupted by controlled dilution. However, sphaeroplasts from <u>Thermus</u> <u>aquaticus</u> were very resistant to disruption even when suspended in distilled water and it was found necessary to sonicate the suspension before centrifugation. On the basis of experiments with Electron Spin Resonance and Differential Scanning Calorimetry, Oshima & Osawa (1983) have concluded that the outer membranes from <u>T.</u> <u>thermophilus</u> are considerably less fluid than inner membrane preparations. Thus glycolipids play an important role in maintaining the structural integrity of the membrane in adverse conditions, e.g. at extremes of temperature in the case of some thermophiles or the absence of rigid cell walls in the case of mycoplasma and L-forms.

The molecular packing of a lipid bilayer composed essentially of glycolipids is a matter for conjecture. In the case of the glycolipids from <u>Thermus</u> spp. there is also the additional complication of three fatty acid residues, two located on the glycerol as usual but the third located some distance away on a sugar residue. In order that these three lipophilic residues may be located perpendicularly within the membrane the two internal sugar residues would lie parallel to the membrane surface with the outer disaccharide free to adopt any conformation above the surface within the constraints of the glycosidic linkages.

As discussed earlier relatively high concentrations of glycolipids are unusual in most mesophilic bacteria, so what of their role in these organisms? The lateral movement of lipids within the bilayer is well established, so small pockets of relatively high concentrations of glycolipids may easily be envisaged perhaps being produced at certain points in the growth cycle. Thus areas of changing fluidity may be generated without necessarily involving changes in fatty acid composition. This may be important in producing the correct environment for many membrane associated enzymes. Zerial, Gelman & Firschein (1978) have reported that a membrane-associated DNA polymerase extracted from Pneumococcus is specifically stimulated by the glycolipids from this organism: phospholipids are completely inactive. Thus the movement of glycolipids within the membrane provides the organism with a method for regulating the fluidity of specific areas which may be vital in order to carry out many enzyme-mediated processes.

ACKNOWLEDGEMENTS

The unpublished work described in this paper has been carried out in collaboration with R.Pask-Hughes and B.Bradley with the financial support of the SERC.

REFERENCES

Langworthy, T.A., Mayberry, W.R. & Smith, P.F. (1974) J.Bacteriol. 119, 106-116

Oshima, M. & Osawa, Y. (1983) J.Biochem. 93, 225-234

Pask-Hughes, R.A. & Shaw, N. (1982) J.Bacteriol. 149, 54-58

Shaw, N. (1974) Adv. in Appl.Microbiol. 17, 63-108

Shaw, N. (1975) Adv. in Microbial Physiol. 12, 141-167

Smith, P.F., Langworthy, T.A. & Mayberry, W.R. (1973) Ann.N.Y.Acad.Sci. 225, 22-27

Toon, P., Brown, P.E. & Baddiley, J. (1972) Biochem. J. 127, 399-409

Zerial, A., Gelman, I. & Firschein, W. (1978) J. Bacteriol., 135, 78-89

ENZYMOLOGY OF GLYCOSPHINGOLIPIDS : role of activator proteins in ganglioside metabolism

Konrad Sandhoff

Institut für Organische Chemie und Biochemie der Universität Bonn, F.R.Germany

SPHINGOLIPID METABOLISM

Glycosphingolipids are typical components of the outer leaflet of plasma membranes of animal cells. On the cell surface they form patterns specific for different cell types that change with cell differentiation and cell transformation (Hakomori, 1975; Feizi, 1985). Though their physiological function is still unclear, gangliosides, sialic acid containing glycosphingolipids, have been recognized as binding sites for toxins and viruses (Yamakawa and Nagai, 1978, Markwell et al., 1981) and as modulators of axonal sprouting and neuritogenesis (Di Gregorio et al., 1984; Ledeen, 1985). Gangliosides are especially enriched in the outer leaflet of neuronal plasma membranes where they might be indirectly involved in neurotransmission (Svennerholm, 1984).

Biosynthesis of sphingolipids is catalyzed by a group of membrane-bound transferases mainly associated with the membranes of the endoplasmic reticulum and the Golgi apparatus (Schachter and Roseman, 1980). Presumably by vesicle flow sphingolipids reach the plasma membrane. Here some of them such as sphingomyelin and oligosialylgangliosides can be modified by plasma membrane-bound enzymes which seem to be regulated by the physical properties of the membranes and by their organization (Scheel et al., 1982; Sandhoff, 1984a).

For final degradation sphingolipids are moved into the lysosomal compartment where they are degraded in an acidic environment by exo-hydrolases in a stepwise manner starting at the hydrophilic end of the amphiphilic molecules. The sequence of catabolic steps has been studied intensively in the last 20 years (for review see Stanbury et al., 1983; Sandhoff, 1977; Sandhoff and Christomanou, 1979; Conzelmann and Sandhoff, 1986) in order to understand the molecular basis of inherited lipidoses and their steadily increasing heterogeneity. As indicated in Fig. 1 almost each of the degrading steps can be deficient in a human lipid storage disease. Such a block causes the lysosomal accumulation of its poorly water-soluble lipid substrates.

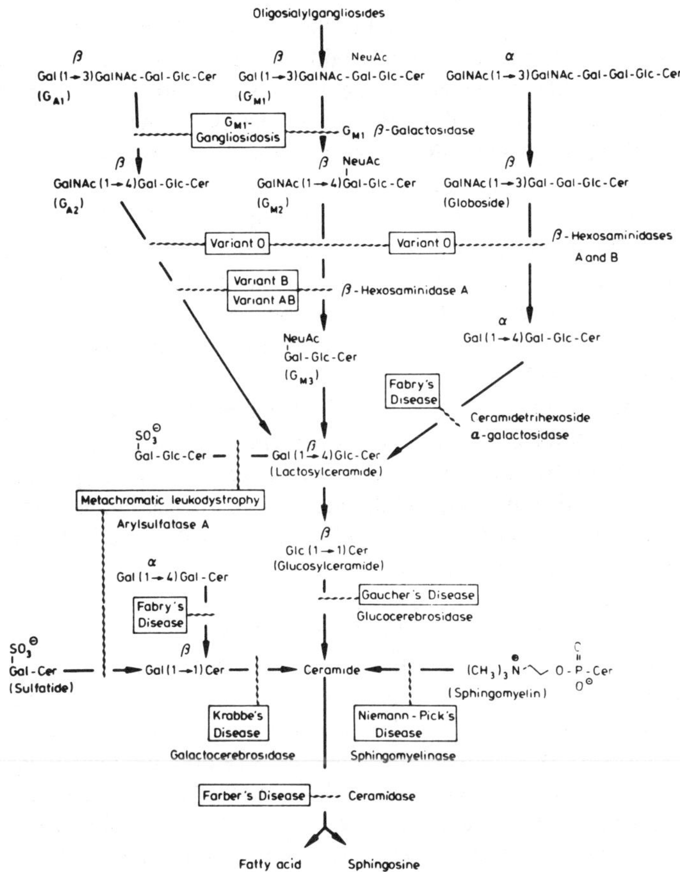

Fig. 1. Degradation scheme of sphingolipids denoting metabolic blocks of known diseases (Sandhoff and Christomanou, 1979). Cer = ceramide; Gal = D-galactose, GalNAc = 2-acetamido-2-deoxy-D-galactopyranoside, Glc = D-glucose, NeuAc = $\underline{N}$-acetylneuraminic acid. Variant B = Variant B of infantile G_{M2}-gangliosidosis, Tay-Sachs disease; Variant 0 = Variant 0 of infantile G_{M2}-gangliosidosis, Sandhoff disease, Sandhoff-Jatzkewitz disease; Variant 0 of juvenile G_{M2}-gangliosidosis, juvenile Sandhoff disease; Variant AB = Variant AB of infantile G_{M2}-gangliosidosis.

Despite this simple principle the diseases resulting from the deficiency in a single catabolic step are rather heterogeneous from the biochemical as well as from the clinical point of view (Stanbury et al., 1983). Several causes for these heterogeneities have been identified (von Figura and Hasilik, 1984; Neufeld et al., 1984):

1. Allelic mutations within the structural gene of a lysosomal hydrolase may result in gene products with different residual enzyme activities, with changed substrate specificities and reduced enzyme stabilities, or sometimes even in the absence of a mature lysosomal hydrolase.

2. Inherited defects in post-translational processing and targeting of lysosomal glycoproteins may result in misdirecting newly synthesized lysosomal enzymes to secretion as in mucolipidoses II and III.

3. Stability of lysosomal enzymes within the lysosomal compartment might be drastically reduced either due to a structural mutation of the enzyme itself or due to a defect of a non-enzymic protein cofactor which normally binds and stabilizes lysosomal enzymes against proteolysis (D'Azzo et al., 1982).

4. Finally, some water-soluble hydrolases require for activity towards glycolipid substrates the aid of cofactors, so-called activator proteins (Sandhoff, 1984 b). Two lipidoses, i.e., variant AB of G_{M2} gangliosidosis (Conzelmann and Sandhoff, 1978) and a juvenile variant of metachromatic leukodystrophy (Stevens et al., 1981), are caused by a deficiency of the respective activator proteins.

ENZYMOLOGY OF GLYCOSPHINGOLIPIDS

In aqueous solutions, sphingolipids do not exist as freely dissolved monomers but form larger aggregates. To study their interaction with catabolic or anabolic enzymes detergents have been used widely in in vitro studies in order to stimulate the enzymic reaction by dispersing the lipid aggregates (Gatt and Bahrenholz, 1973; Basu et al., 1984). However, in vivo such detergents are not available and it became clear that the interaction between membrane-bound glycosphingolipid substrates and metabolizing enzymes must be facilitated by other means. Studies on this subject were conducted mainly for enzymes involved in the catabolism of sphingolipids, where three modes of substrate-enzyme interaction have been identified in vitro in the absence of detergents:

1. Lipid substrate and catabolizing enzyme are components of the same membrane. In this case enzyme and substrate may interact by lateral diffusion within the plane of the membrane. The rate of glycolipid turnover is independent of the volume of the aqueous phase of the incubation mixture (Fig. 2) and the reaction follows two-dimensional Michaelis–Menten kinetics

(Fig. 3). This has been shown for the interaction of ganglioside G_{D1a} and sialidase of brain membranes (Scheel et al., 1982; 1985). Surprisingly, this reaction is stimulated by general anaesthetics which, however, do not increase the activity of the enzyme against water-soluble substrates (Sandhoff and Pallmann, 1978; Scheel et al., 1982).

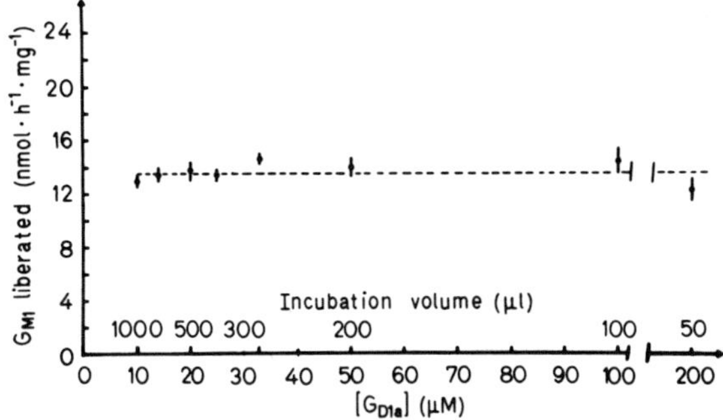

Fig. 2. Dependence of ganglioside G_{D1a} degradation by membrane-bound sialidase of calf brain on the incubation volume (Scheel et al., 1982). Incubation mixtures containing 10 nmol [^{3}H]ganglioside G_{D1a}, 2.5 μmol formate buffer, pH 3.6, 6.25 μmol NaCl and 100 μg protein in a volume of 50μl were diluted with formate buffer of the same pH and NaCl content to reach the indicated volumes. After incubation for 1 h at 37°C, [^{3}H]-ganglioside G_{M1} liberated was determined. The means of three determinations are given and the standard deviations indicated.

2. A water-soluble enzyme recognizes its glycolipid substrate on the surface of lipid bilayers. This situation has been demonstrated for the interaction of glucosylceramide inserted into unilamellar liposomes and highly purified water-soluble glucosylceramide β-glucosidase from human placenta (Sarmientos et al., 1986). In this case incorporation of acidic phospholipids into the substrate-containing liposomes caused a powerful stimulation (30-40fold) of glucosyl ceramide hydrolysis. On the other hand, the reaction was significantly dependent on structural variations in the lipophilic aglycon moiety of the substrate, e.g., derivatives containing L-threo-sphingosine were poorer substrates with higher apparent K_M-values than the corresponding naturally occurring D-erythro derivatives. These observations are compatible with the idea that the enzyme recognizes at least a portion of the aglycon, i.e., the hydrophobic moiety of its bilayer-bound lipid substrate. This view is supported by the observation that the

addition of ceramide to the lipid-bilayer inhibits the hydrolysis of liposomal glucosylceramide but not that of a water-soluble substrate, 4-methylumbelliferyl β-D-glucoside.

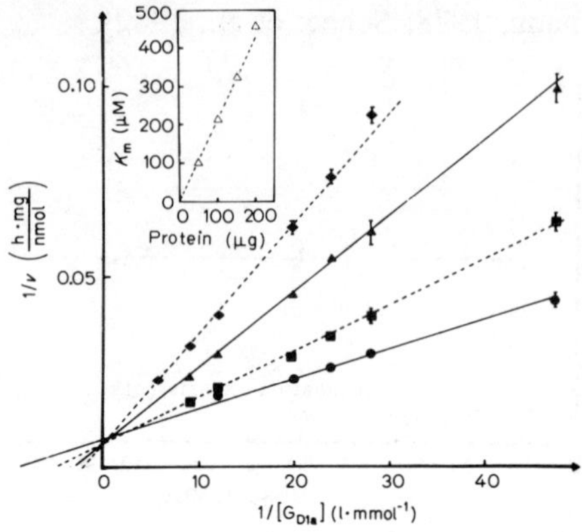

Fig. 3. Determination of the apparent kinetic constants, $\underline{K}_m$ and $\underline{V}_{max}$, for ganglioside G_{D1a} degradation in the presence of different amounts of membrane-bound sialidase of calf brain (Scheel et al., 1982). Incubation mixtures (100μl total volume) containing increasing [³H]ganglioside G_{D1a} concentrations as indicated in the figure, 5 μmol formate buffer, pH 3.6, and 12.5 μmol NaCl were incubated for 15 min at 37°C with (●) 50, (■) 100, (▲) 150 and (◆) 200 μg membrane protein, respectively. [³H]Ganglioside G_{M1} liberated was determined. The means of three determinations are given and the standard deviations indicated. Insert: $\underline{K}_m$ values obtained are plotted against the corresponding amount of membrane protein.

    3. Activator proteins facilitate the interaction between glycolipids and water-soluble lysosomal hydrolases. At physiological ionic strength some lysosomal hydrolases, such as hexosaminidase A and arylsulfatase A, do not recognize their glycolipid substrates when the latter are presented in micellar or liposomal structures unless appropriate detergents or activator proteins are added. The activator proteins identified so far (for review see Sandhoff, 1984 b; Conzelmann and Sandhoff, 1986; Li and Li, 1983) are lysosomal glycoproteins with isoelectric points between pH 4.3 and 4.8 and molecular weights in the range of 8,000 to 27,000. Two groups can be distinguished:
    a) cohydrolases which stimulate the activity of membrane-associated lysosomal hydrolases like glucosylceramidase, galactosylceramidase, and

sphingomyelinase (Ho and O'Brien, 1971; Wenger et al., 1982; Christomanou, 1980) and

b) lipid binding proteins which stimulate the degradation of glycosphingolipids such as sulfatides and gangliosides G_{M1} and G_{M2} by water-soluble lysosomal hydrolases but not that of water-soluble substrates.

THE GANGLIOSIDE G_{M2} ACTIVATOR PROTEIN

The G_{M2} activator is a ganglioside G_{M2}-binding protein which forms in <u>vitro</u> lipid/protein complexes with the stochiometry of 1:1 under suitable conditions. These complexes have been identified by isoelectric focusing, gel filtration and centrifugation studies (Conzelmann et al., 1982). Equilibrium

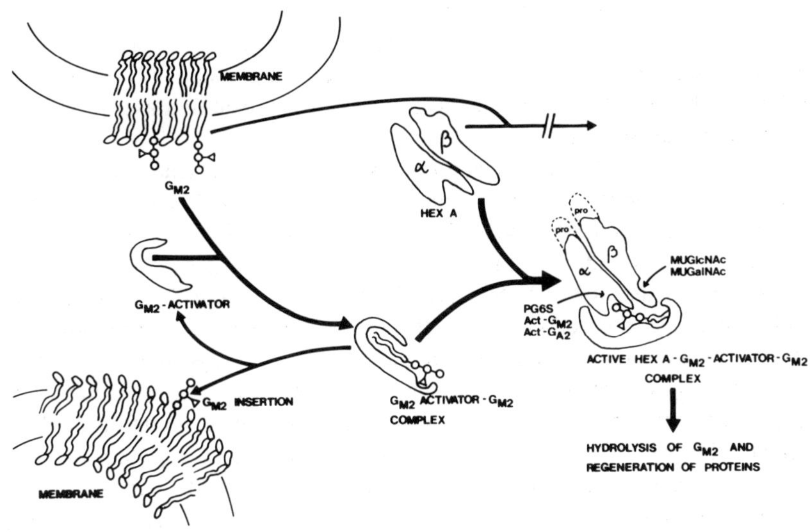

Fig. 4. Model for the lysosomal catabolism of ganglioside G_{M2}. Hexosaminidase A cannot attack membrane-bound ganglioside G_{M2}. Instead, the ganglioside is extracted from the membrane by the activator protein and the (water-soluble) activator/lipid complex is the substrate for the enzymic reaction. Of the two catalytic sites on hexosaminidase A, only the one on the α-subunit cleaves ganglioside G_{M2}. The hexosaminidase precursor ("prohex A") is also active on the activator/G_{M2} complex (Hasilik et al., 1982). After the reaction, the product, ganglioside G_{M3}, is reinserted into the membrane and the activator protein is available for another round of catalysis (Conzelmann and Sandhoff, 1979; Sandhoff and Conzelmann, 1984; Conzelmann et al., 1982).

dialysis techniques as well as affinity labelling of the activator protein by a suitable ganglioside derivative (Neuenhofer and Sandhoff, 1985) suggest a hydrophobic binding site of the G_{M2}-activator protein for ganglioside G_{M2} besides a carbohydrate binding site. Experiments with labelled liposomes showed that the G_{M2} activator protein transfers gangliosides G_{M1}, G_{M2}, G_{D1a}, G_{M3} and glycolipid G_{A2} from donor to acceptor liposomes (Conzelmann et al., 1982). In enzymatic studies the G_{M2} activator stimulates the degradation of glycolipids G_{M2}, G_{A2} and globoside by hexosaminidase A up to 100-fold and more but has no significant effect on the degradation of water-soluble substrates. On the other hand almost no ganglioside G_{M2}-degrading activity was found with hexosaminidase B in the presence of G_{M2} activator protein. This strict isoenzyme specificity contrasts sharply with the isoenzyme specificity observed in the presence of detergents. Kinetic studies indicate that the ganglioside G_{M2} activator complex is the Michaelis-Menten substrate for hexosaminidase A (Fig. 4) (Conzelmann and Sandhoff, 1979).

Human lysosomal hexosaminidase A is a heteropolymer composed of two different subunits, α- and β-subunits, each of which carries an active site (Kytzia and Sandhoff, 1985). Inhibition studies with different lipid and synthetic substrates indicate that the active site on the β-subunit hydrolyzes predominantly water-soluble $\underline{N}$-acetylglucosaminides and $\underline{N}$-acetylgalactos-aminides, whereas the active site on the α-subunit hydrolyzes predominantly a sulfated $\underline{N}$-acetylglucosaminide and the glycolipid G_{M2} and G_{A2} in the presence of G_{M2} activator protein. This view is supported by studies on a mutated hexosaminidase A obtained from patients with a mutation allelic to Tay-Sachs disease (α^-) (Kytzia et al., 1983; Sonderfeld et al., 1985 a).

The mutated hexosaminidase A carries an active β- and an inactive α-site. Kinetic studies indicated that the lipid-free G_{M2} activator competes with a water-soluble substrate for the α-site of hexosaminidases A and S (Hex S consisting of two α-subunits) but not with the water-soluble substrate for the β-site of hexosaminidases A and B, respectively. Subcellular fractionation studies of cultured human fibroblasts demonstrate a lysosomal localization of the G_{M2} activator, which has a molecular weight of 22,000. It is a glycoprotein which is stable up to $60^{\circ}C$ and has an isoelectric point at pH 4.8. Besides this lysosomal form, a 24,000 molecular form exists which is exported from cultured cells in the presence of NH_4Cl. The 24,000 molecular form is the only identifiable G_{M2} activator species in fibroblast obtained from patients with I cell disease. Both molecular forms cannot be detected by immunoprecipitation in fibroblasts of patients with variant AB of GM_2 gangliosidosis (Hirabayashi et al., 1983; Banerjee et al., 1984; Burg et al., 1985). Feeding of variant AB fibroblasts with G_{M2}

activator restores G_{M2} degrading activity in these cells (Fig. 5) (Sonderfeld et al., 1985 b).

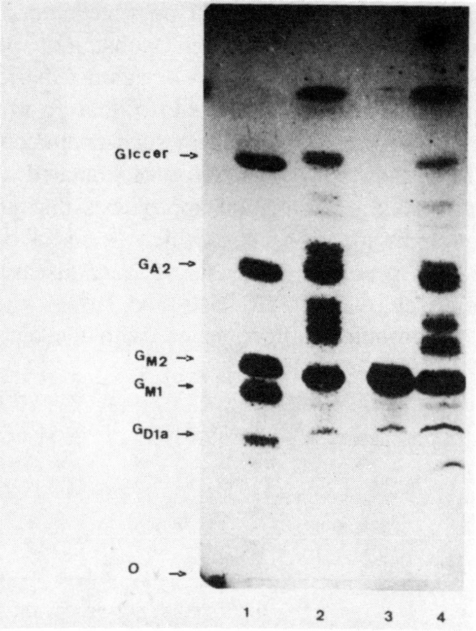

Fig. 5. Metabolism of $[^3H]G_{M2}$ in G_{M2} gangliosidosis, variant AB, cells as function of exogenously added G_{M2}-activator protein. The fibroblasts were fed with $[^3H]G_{M2}$ (5×10^{-5}M) and the indicated amounts of purified G_{M2}-activator protein. After 70 h, the cells were harvested and processed as described (Sonderfeld et al., 1985 b). TLC separation of total lipid extracts: Lane 1 = standards G_{D1a}, G_{M1}, G_{M2}, G_{A2}, GlcCer; Lane 2 = total lipid extracts of normal cells; Lane 3 = total lipid extracts of G_{M2}-gangliosidosis, variant AB, cells; Lane 4 = total lipid extracts of G_{M2}-gangliosidosis variant AB, cells, after feeding 30 μg G_{M2}-activator protein for 70 h; 0 = origin.

RESIDUAL ENZYME ACTIVITIES IN GANGLIOSIDE STORAGE DISEASES

Different allelic mutations in one gene locus (e.g., in that of the α- or β-chain of hexosaminidase A or of the activator protein) may lead to extremely variable clinical and neuropathological forms of the same biochemical variant of gangliosidosis. A crucial difference observed between the different clinical courses, e.g., infantile and adult forms, are different residual activities of the mutated hexosaminidase A against its natural substrate. For a complete understanding of the pathogenesis of the gangliosidoses, it would be very important to determine exactly the residual activity of a mutated enzyme against all possible natural substrates *in vivo*. Mutated enzymes might have altered substrate specificities, changed stabilities against proteolytic degradation or against thermal inactivation, and might be inhibited differently compared to their normal counterparts by components of the lysosomal environment such as mucopolysaccharides.

Unfortunately, the *in vivo* activity of mutated enzymes cannot be determined directly. The experimental approaches that are so far considered to yield the best approximations and which seem to be the most suitable methods for pre- and postnatal diagnosis of these diseases are (a) feeding of natural substrate to cultivated fibroblasts and (b) *in vitro* determination of enzyme activity in fibroblast homogenates, with the natural substrate in the

<u>Table 1</u>: DEGRADATION OF GANGLIOSIDE G_{M2} BY FIBROBLAST

EXTRACTS OF PATIENTS WITH G_{M2}-GANGLIOSIDOSIS

(CONZELMANN ET AL., 1983)

PROBANDS	GANGLIOSIDE G_{M2}-DEGRADATION (pmol/h.mg.AU*)		
	MEAN	RANGE	
NORMAL	535	296 - 762;	$\underline{n}$ = 9
HETEROZYGOTES (DIFFERENT GENOTYPE)	285	121 - 395;	$\underline{n}$ = 4
VAR. B			
INFANTILE	2.4	0.8 - 3.8;	$\underline{n}$ = 5
JUVENILE	15.8	13.6 - 18.0;	$\underline{n}$ = 5
ADULT	19.1	13.1 - 32.8;	$\underline{n}$ = 9
VAR. O			
INFANTILE	6.6	3.6 - 9.5;	$\underline{n}$ = 2
JUVENILE	23.6	9.5 - 39.4;	$\underline{n}$ = 3
HEALTHY	105	75 - 143;	$\underline{n}$ = 2

* AU Activator Units

presence of the respective activator protein. Data obtained with the latter method indicate some residual activity in adult patients of variant B, which is small but significantly higher than that of infantile forms of G_{M2}-gangliosidosis (Tab. 1) (Conzelmann et al., 1983). Residual activites in the range of 10 - 20% of mean control value already appear to be compatible with normal life. These findings stress the importance of small variations of the low residual enzyme activities, that are found in the patients, for the development of different clinical courses of a disease.

Observations that moderate residual activities (e.g., 10 % of normal) may still suffice to sustain normal catabolism of a substrate and that small variations within the low residual activities observed in patients (below 5 % of normal) greatly influence the clinical course of the disease can be understood on the basis of a greatly simplified kinetic model (Fig. 6)

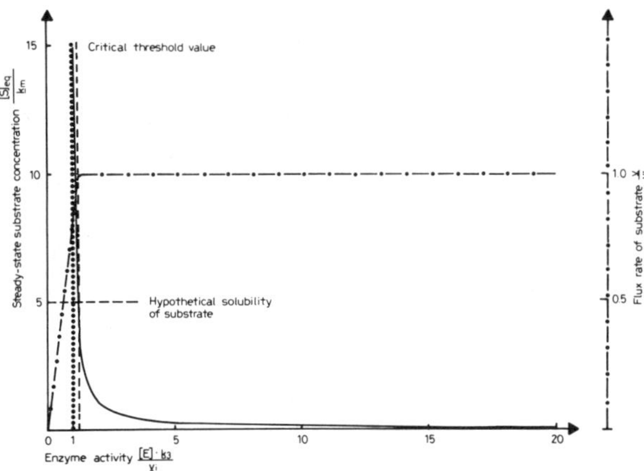

Fig. 6. Steady-state substrate concentrations as a function of enzyme concentration and activity (Conzelmann and Sandhoff 1983/84). The model underlying this theoretical calculation assumes influx of the substrate into a compartment at a constant rate (v_i) and its subsequent utilization, by the enzyme (for details, see text). —— = $[S]_{eq}$, steady-state substrate concentration; ····· = theoretical threshold of enzyme activity; – – – = critical threshold value, taking limited solubility of substrate into account; – · – · – = turnover rate of substrate (flux rate).

(Conzelmann and Sandhoff, 1983, 1984). Assuming a constant influx rate of the substrate, e.g., into the lysosomal compartment of an individual cell, and a degradation rate proportional to the degree of saturation of the available enzyme, the steady-state substrate concentration can be calculated as a function of the residual enzyme actitivty. For lysosomal enzymes, which are probably not regulated, this steady-state concentration should be far below the K_m value. It is then evident that even substantial reductions of the amount or turnover number of the enzyme will only lead to a moderate increase in the steady-state concentration of the substrate, but not in an accumulation, since the remaining activity is still sufficient to cope with the substrate's influx rate. Only when the residual activity falls below a critical threshold, will the overall turnover rate be more or less reduced. The residual turnover rate, and hence the rate at which the substrate accumulates, will then depend on the difference between threshold activity and actual residual enzyme activity in the lysosome. Since influx rate of the substrate as well as enzyme activity are different in different organs and cell types and may even vary between individual cells (e.g., neurons), the consequence of a mutation on different organs and cell types may be quite variable.

These calculations were performed as a first-order approximation on the basis of simple Michaelis-Menten kinetics without considering any complications which may arise, e.g., from regulatory properties of the enzyme considered or from the interaction of accumulating substrates with other components within the lysosomes. Still, they give an impression of how critical small variations of low residual enzyme activities (e.g., between 1 % and 5 % of normal) may be for the genesis, or the avoidance, or for the severity of a lysosomal storage disorder. It is to be expected that ganglioside metabolism is different in different types of neurons. Indeed, studies of the developmental profiles of individual gangliosides clearly show different ganglioside patterns for different brain regions (Vanier et al., 1971; Seyfried et al., 1983; Rösner, 1982). Therefore it must be assumed that individual neurons will be affected differently by the same degree of deficiency in a catabolizing enzyme. More detailed knowledge on the glycolipid metabolism of the various brain regions and its regulation is needed in order to formulate these questions more precisely.

REFERENCES

Banerjee, A., Burg, J., Conzelmann, E., Carroll, M., Sandhoff, K. (1984) Hoppe-Seyler's Z. Physiol. Chem. 365: 347-356.

Basu, S., Basu, M., Kyle, J. W. and Chon, H.-C. (1984) in Ganglioside Structure, Function, and Biomedical Potential, eds. Ledeen, R. W., Yu, R. K., Rapport, M. H. and Suzuki, K., Plenum Press, New York, pp. 249-261.

Burg, J., Banerjee, A., Sandhoff, K. (1985) Biol.Chem. Hoppe-Seyler 366: 887-891.

Christomanou, H. (1980) Hoppe-Seyler's Z. Physiol. Chem. 361: 1489-1502.

Conzelmann, E., Burg, J., Stephan, G., Sandhoff, K. (1982) Eur. J. Biochem. 123: 455-464.

Conzelmann, E., Kytzia, H.-J., Navon, R., Sandhoff, K. (1983) Am. J. Hum. Genet. 35: 900-913.

Conzelmann, E., Sandhoff, K. (1978) Proc. Natl. Acad. Sci. USA 75: 3979-3983.

Conzelmann, E. and Sandhoff, K. (1979) Hoppe-Seyler's Z. Physiol. Chem. 360: 1837-1849.

Conzelmann, E. and Sandhoff, K. (1983/84) Dev. Neurosci. 6: 58-71.

Conzelmann, E., Sandhoff, K. (1986) Advances in Enzymology, in press.

Conzelmann, E. and Sandhoff, K. (1986) "Methods in Enzymology" in press.

D'Azzo, A., Hoogeveen, A. T., Reuser, A. J. J., Robinson, D., Galjaard, H. (1982) Proc. Natl. Acad. Sci. USA 79: 4535-4539.

Di Gregorio, F., Ferrari, G., Marine, P., Siliprandri, R., Gorio, A. (1984) Neuropediatrics 15: 93-96.

Feizi, T. (1985) Nature 314: 53-57.

Gatt, S., Bahrenholz, Y. (1973) Ann. Rev. Biochem. 42: 61-90.

Hakomori, S.I. (1975) Biochim. Biophys. Acta 417: 55-89.

Hasilik, A., von Figura, K., Conzelmann, E., Nehrkorn, H., Sandhoff, K. (1982) Eur. J. Biochem. 125: 317-321.

Hirabayashi, Y., Li, Y.-T., Li, S.-C. (1983) J. Neurochem. 40: 168-175.

Ho, M. W., O'Brien, J. S. (1971) Proc. Natl. Acad. Sci. USA 68: 2810-2813.

Kytzia, H.-J., Hinrichs, U., Maire, I., Suzuki, K., Sandhoff, K. (1983) EMBO J. 2: 1201-1205.

Kytzia, H.-J., Sandhoff, K. (1985) J. Biol. Chem. 260: 7568-7572.

Ledeen, R. (1985) Trends in Neuro. Sciences 8: 169-174.

Li, Y.-T., Li, S.-C. (1983) in Boyer, P.D. (ed.) The Enzymes, Vol. XVI, Academic Press, New York, pp. 427-445.

Markwell, M. A. K., Svennerholm, L., Paulson, J.C. (1981) Proc. Natl. Acad. Sci. USA 78: 5406-54.

Neuenhofer, S., Sandhoff, K. (1985) FEBS Lett. 185: 112-114.

Neufeld, E. F., D'Azzo, A., Proia, R. L. (1984) in Barranger, J. A., Brady, R. O. (eds.) Molecular basis of lysosomal storage disorders. Academic Press, London/New York, pp. 251-256.

Rösner H. (1982) Brain Res. 236: 49-61.

Sandhoff, K. (1977) Angew. Chem. Int. Ed. 16: 273-285.

Sandhoff, K. (1984 a) Funkt. Biol. Med. 3: 141-151.

Sandhoff, K. (1984 b) in Barranger, J. A., Brady, R. O. (eds.) Molecular basis of lysosomal storage disorders. Academic Press, London/New York, pp. 19-49.

Sandhoff, K. and Pallmann, B. (1978) Proc. Natl. Acad. Sci. USA 75: 122-126.

Sandhoff, K., Christomanou, H. (1979) Hum. Genet. 50: 107-143.

Sandhoff, K. and Conzelmann, E. (1984) Neuropediatrics Suppl. 15: 85-92.

Sarmientos, F., Schwarzmann, G. and Sandhoff, K. (1986) Eur. J. Biochem. submitted.

Schachter, H., Roseman, S. (1980) in Lennarz, W.J. (ed.) The biochemistry of glycoproteins and proteoglycans.Plenum Press, New York, pp. 85-160.

Scheel, G., Acevedo, E., Conzelmann, E., Nehrkorn, H. and Sandhoff, K. (1982) Eur. J. Biochem. 127: 245-253.

Scheel, G., Schwarzmann, G., Hoffmann-Bleihayer, P. and Sandhoff, K. (1985) Eur. J. Biochem. 153: 29-35.

Seyfried, T. N., Miyazawa, N., Yu, R. K. (1983) J. Neurochem. 41: 491-505.

Sonderfeld, S., Brendler, S., Sandhoff, N., Galjaard, H. and Hoogeveen, A. T. (1985a) Hum. Genet. 71: 196-200.

Sonderfeld, S., Conzelmann, E., Schwarzmann, G., Burg, J., Hinrichs, U., Sandhoff, K. (1985 b) Eur. J. Biochem. 149: 247-255.

Stanbury, J. B., Wyngaarden, J. B., Fredrickson, D.S., Goldstein, J. L., Brown, M. S. (1983) The metabolic basis of inherited disease, 5th edn. McGraw Hill, New York.

Stevens, R. L., Fluharty, A. L., Kihara, H., Kaback, M. M. Shapiro, L. J., Marsh, B., Sandhoff, K., Fischer, G. (1981) Am. J. Genet. 33: 900-906.

Svennerholm, L. (1984) in Cellular and pathological aspects of glycoconjugate metabolism. Inserm Vol. 26: 21-44.

Vanier, M. T., Holm, M., Öhmann, R., Svennerholm, L. (1971) J. Neurochem. 18: 581-592.

von Figura, K., Hasilik, A. (1984) Trends Biochem. Sci. 9: 29-31.

Wenger, D. A., Sattler, M., Roch, S. (1982) Biochim. Biophys. Acta 712: 639-649.

Yamakawa, T., Nagai Y. (1978) Trends Biochem. Sci. 3: 128-131.

A STRUCTURAL BASIS FOR THE SPECIFICITY OF THE ACTIVATOR PROTEINS OF SPHINGOLIPID HYDROLASES

Colin H. Wynn and Martyn N. Banks, Glycosphingolipid Research Group,

Department of Biochemistry, University of Manchester, Manchester, U.K.

INTRODUCTION

The physiological functions of glycosphingolipids that have now been described are many and various, including the binding of specific ligands such as toxins, hormones and cell growth and differentiation factors (see Wiegandt, 1985) and their role in cell transformation and the development of malignancy (see Hakomori, 1984). For many of these functions there is an implicit need for recognition by the glycosphingolipid molecule of some other macromolecule such as a peptide, glycoprotein or another glycosphingolipid. Since the lipid moiety of the molecule is usually part of a membrane, it must be assumed that the specificity of such interactions must lie in the sugar components. Most of the oligosaccharides involved are relatively simple and, in order to accommodate the variety and specificity of the interactions observed, it is necessary to assume that the conformation of the oligosaccharide is the determining factor.

One of the simplest examples of protein-glycosphingolipid interaction is that of the activator proteins for sphingolipid hydrolysis (see Sandhoff & Conzelmann, 1979). In order for the lipid to be in the correct physical state for degradation by the water-soluble sphingolipid hydrolases, it is necessary for some detergent-like molecule to be present. In vitro bile salts allow the formation of small mixed micelles where the oligosaccharide chains on the surface of the micelle are accessible to the hydrolases. In vivo hydrolysis of the glycosphingolipids takes place in the lysosomes, which contain neither detergent nor bile salts. Instead the solubilisation of the lipid is brought about by a series of proteins, termed activator proteins, which appear to act by extracting the lipid from membranes or lipid aggregates and, by surrounding the hydrophobic regions of the molecule with protein, make it soluble. Although fulfilling a common function there is considerable specificity within the activator proteins and this specificity must depend on the nature of the protein-oligosaccharide interaction. The two glycosphingolipids G_{M1}- and G_{M2}-ganglioside differ only in the absence of the terminal galactose residue in the latter and yet they require different activator proteins for their hydrolysis (Conzelmann & Sandhoff, 1979). The activator protein for sulphatide (cerebroside 3-sulphate) first described by Mehl and Jatzkewitz (1964) has been shown to be identical with that for G_{M1}-ganglioside (Inui et al., 1983; Banks et al., 1985), a finding which would appear to suggest that the terminal galactose is not a major factor in determining the specificity and that there is a major conformational difference between G_{M1}- and G_{M2}-ganglioside.

Although the use of techniques such as mass spectrometry in conjuction with direct probes, negative ion fast atom bombardment or gas chromatography have facilitated the elucidation of the structure and linkage of glycosphingolipids (Hanfland & Egge, 1975; Arita et al., 1983; Ando et al., 1977), the conformation of these molecules is a more intractable problem. The simplest and most direct approach is the use of X-ray crystallography but unfortunately it has not been possible to crystallise most glycosphingolipids in a suitable form. Even where crystals are available, for example galactosyl-ceramide which was reported by Pascher & Sundell (1977), some caution is obviously necessary in extrapolating the results in crystals to structure in a membrane or lipid aggregate. In the absence of suitable crystals, indirect methods such as nuclear magnetic resonance have suggested possible hydrogen bonding and cation binding sites (Czarniecki & Thornton, 1976; Sillerud et al., 1978) although such techniques are often difficult to interpret in more specific conformational terms.

The applicability of computational methods to the prediction of biomolecular conformation has been the subject of increased attention during recent years (see Platt & Robson, 1983). The potential energy of a molecule is calculated from empirical functions representing the interatomic interactions, such as van der Waals repulsion and attraction and electrostatic forces, and this potential energy is then minimised as a function of conformation. The parameters of the interactions are derived from experimental data on simple model compounds. These methods, which were developed for the study of peptides and proteins, have been modified and their applicability to glycosphingolipids tested on molecules, such as glucosyl-ceramide and $\underline{N}$-acetyl-neuraminyl-galactose, where evidence from X-ray analysis is available (Wynn et al., 1986). The similarity between the predicted structures and those given by X-ray analysis is striking. There was also a close parallel between the predicted orientation of the $\underline{N}$-acetylneuraminic acid residue and that suggested from ^{13}C spin relaxation studies by Czarniecki & Thornton (1976) although the glyceryl sidechain was suggested by the latter workers to be rotated through 180° compared to the predicted structure. Their data were equally consistent with the computer prediction and this demonstrates the importance of the multidisciplinary approach to the problems of conformation and the refinement of computer methods and parameters in the light of new experimental evidence.

RESULTS AND DISCUSSION

These predictive techniques have now been applied to G_{M1}- and G_{M2}-ganglioside (Wynn, 1986). The geometry of the individual monosaccharide residues was determined by comparison with known pyranose structures and the rings were fixed in the $^4C_1$ conformation. The extended conformation of the hydrocarbon chains was held constant throughout in the extended conformation found for glucosyl-ceramide (Wynn et al., 1986). All other bonds, i.e. the glycosidic and sidechain links, were allowed to rotate freely. In the energy minimisation procedure the GLOBEX routine of Robson et al. (1982) extending the SIMPLEX procedure of Nelder & Mead (1965) was implemented. This ensured that the minima found were not local, kinetically facile minima but were of the deep meta-stable type. In this way a variety of possible conformers could be investigated en route to the global minimum, which was defined by the fact that no further escape from the minima occurred within the simulation time and that the simplex points had converged to within 0.05 kcal/mole. The three-dimensional cartesian coordinates and the values of the variable angles for the minimum energy conformer were stored and plotted as stereo diagrams by the PLUTO program.

The predicted conformations of G_{M1}- and G_{M2}-ganglioside are shown in Figures 1 and 2. The gross conformational differences in the structures predicted for these two gangliosides are obvious. A major feature is the position of the charged carboxylate anion of the sialic acid residue, which in the G_{M1}-ganglioside lies towards the ceramide moiety and in a cleft formed by the sialic acid, galactose and glucose rings, whereas in the G_{M2}-ganglioside it lies away from the ceramide moiety in a cleft formed by the sialic acid, galactose and N-acetylgalactosamine rings. This is achieved by rotation of the glycosidic link of the sialic acid through approximately 180° between the two structures and is stabilised in the G_{M2}-ganglioside by interaction with the galactosamine ring which is also rotated. Since the charge on the sulphate in sulphatide would reside in a similar orientation to the ceramide moiety as in G_{M1}-ganglioside, it is possible that the specific interaction with its activator protein is also through this charged area and that this would explain the lack of significant interaction with G_{M2}-

GM1 GANGLIOSIDE

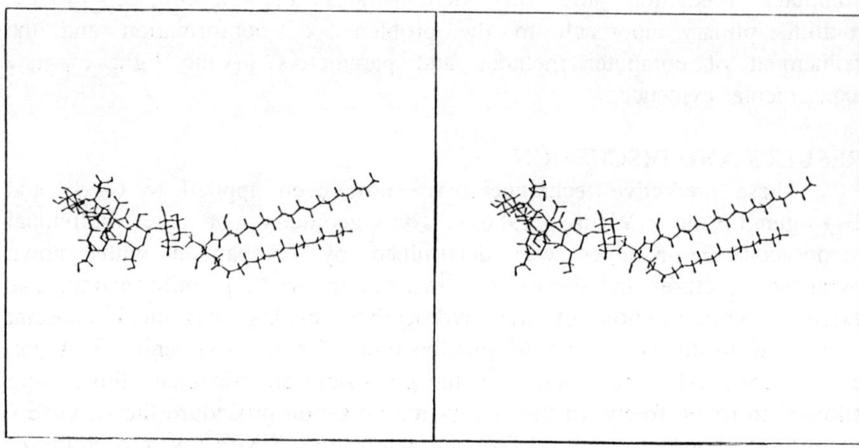

Figure 1. Stereo-diagram of G_{M1}-ganglioside after energy minimisation.

GM2-GANGLIOSIDE

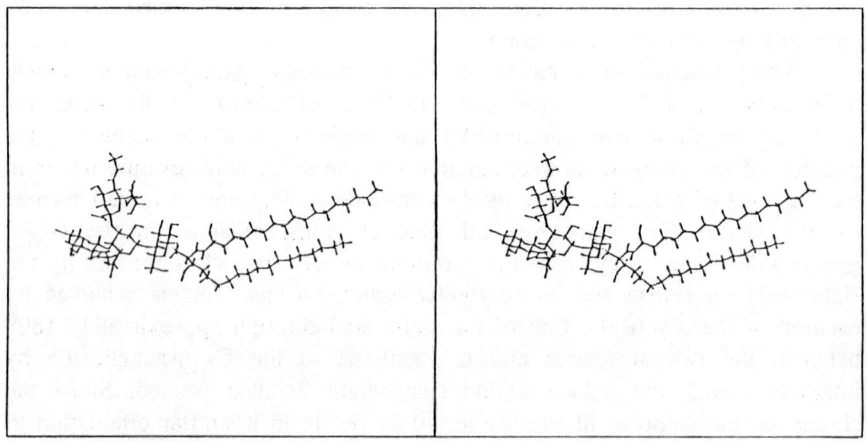

Figure 2. Stereo-diagram of G_{M2}-ganglioside after energy minimisation.

ganglioside. However, since the same activator protein will also function for globotriaosylceramide, some recognition features other than the charge must exist. A close inspection of the G_{M1}-ganglioside structure reveals a nearly planar surface of hydoxyl groups formed from the C_6 hydroxyl groups of glucose, galactose and galactosamine, this feature being absent from the predicted G_{M2}-ganglioside structure. Work is at present in progress on the conformation of globotriaosylceramide and it would be interesting if it were to show a similar planar hydrophilic surface with potential hydrogen-bonding to the activator protein, possibly via the oligosaccharide moiety of this glycoprotein.

The gene for the activator protein for sulphatide and G_{M1}-ganglioside has recently been cloned and the amino acid sequence of virtually the complete mature protein determined by base sequence analysis (Dewji et al., 1986). The sequenced portion contains 67 amino acids and there is a glycosylation site 21 residues from the N-terminus. A preliminary analysis of the precursor of this protein shows that this sequence of 67 amino acids is followed by a lysine residue and then an arginine residue (Dewji, personal communication) and it is possible that this represents a site for proteolytic processing and that the C-terminus of the mature protein is a lysine residue. Thus the activator protein contains both features, i.e. an oligosaccharide moiety and an exposed area of positive charge, that would be predicted to be necessary from the study of the conformation of G_{M1}-ganglioside. It is hoped that studies on the prediction of the secondary and tertiary structure of this protein, at present in progress in this laboratory, will clarify the nature of the protein-glycosphingolipid interaction.

REFERENCES

Ando, S., Kon, K., Nagai, Y. and Murata, T. (1977) J. Biochem. (Tokyo) 82: 1623-1631.

Arita, M., Iwamori, M., Higuchi, T. and Nagai, Y. (1983) J. Biochem. (Tokyo) 93: 319-322.

Banks, M.N., Gutteridge, C.L. and Wynn, C.H. (1985) Biochem. Soc. Trans. 13: 1150-1151.

Conzelmann, E. and Sandhoff, K. (1979) Hoppe Seyler's Z. Physiol. Chem. 360: 1837-1849.

Czarniecki, M.F. and Thornton, E.R. (1976) J. Am. Chem. Soc. 98: 1023-1025.

Dewji, N., Wenger, D., Fujibayashi, S., Donoviel, M., Esch, F., Hill, F. and O'Brien, J.S. (1986) Biochem. Biophys. Res. Commun. 134: 989-994.

Hakomori, S-I. (1984) Trends Biochem. Sci. 9: 453-458.

Hanfland, P. and Egge, H. (1975) Chem. Phys. Lipids 15: 243-247.

Inui, K., Emmett, M. and Wenger, D.A. (1983) Proc. Natl. Acad. Sci. U.S.A. 80: 3074-3077.

Mehl, E. and Jatzkewitz, H. (1964) Hoppe Seyler's Z. Physiol. Chem. 339: 260-276.

Nelder, J.A. and Mead, R. (1965) Comput. J. 7: 308-313.

Pascher, I. and Sundell, S. (1977) Chem. Phys. Lipids 20: 175-191.

Platt, E. and Robson, B. (1983) In "Computing in Biological Science" (Geisow, M.J. and Barrett, A.N., eds) 91-131. Amsterdam: Elsevier Biomedical Press.

Robson, B., Douglas, G.M. and Platt, E. (1982) Biochem. Soc. Trans. 10: 388- 389.

Sandhoff, K. and Conzelmann, E. (1979) Trends Biochem. Sci. 4: 231-233.

Sillerud, L.O., Prestegard, J.H., Yu, R.K., Schafer, D.E. and Konigsberg, W.H. (1978) Biochemistry 17: 2619-2628.

Wiegandt, H. (1985) In "New Comprehensive Biochemistry: Vol. 10, Glycolipids" (Wiegandt, H., ed) 199-260. Amsterdam: Elsevier Science Publishers B.V. (Biomedical Division).

Wynn, C.H. with Robson, B. (1986) J. Theor. Biol., submitted.

Wynn, C.H., Marsden, A. and Robson, B. (1986) J. Theor. Biol. 119: 81-87.

<u>Section</u> <u>4</u>

OF PROBES AND ANAESTHETICS

OF PROBES AND ANAESTHETICS : a round table discussion

Alec Bangham, Ingolf Bernhardt, Charles Brooks, Dennis Chapman, Clive Ellory, Roger Klein, Alister Macdonald, Tim Rink, Konrad Sandhoff, John Tyman and Karel Wirtz

"It is a capital mistake to theorize before one has data. It has long been an axiom of mine that the little things are infinitely the most important"
> (Sherlock Holmes) Sir Arthur Conan Doyle, 1859-1930.

MEMBRANE PROBES

Klein: As I said to some of you, I thought that I would act as the Devil's Advocate and try to instil in you a sort of general sceptical attitude to some of the membrane dogma. We have talked about administering ALEC to cure respiratory distress syndrome; my sceptical attitude was the result of having Alec administered as a research student for two or three years followed by aggressive resuscitation and ever after that I have been very sceptical.

What I thought I would talk about is fluidity, or rather I would stimulate you to talk about fluidity. Two broad areas: what can probes tell us about membrane fluidity and what does this actually mean in terms of enzyme functioning? The second broad topic, how do anaesthetics work at membrane level? - do we actually know of a mechanism by which we can explain their behaviour? I thought my contribution might be to start the discussion about probes. Most of you are aware that probes are supposed to tell you about the micro-environment in the membrane; they are supposed to tell you about phase transitions and phase separations, about molecular rotational dynamics, and yet there are some very serious problems about the use of probes. The physicists, and particularly the physical chemists, will tell you that probes only tell you where the probe is and about the way the probe behaves, but not about the rest of the membrane, and that in practice you are perturbing the normal behaviour of the membrane.

Now, my contribution goes back some years (Klein, 1982). I was interested and have been interested for some time in Arrhenius, who would probably be extremely embarrassed that his name is used by membrane biologists when they convert rate data plotted against temperature into straight line plots to produce something that they call break points, which are supposed to tell us about phase separations or phase transitions in membranes.

It might be worth saying that he was a physical chemist who was interested in immunochemistry about the stage the Arrhenius plot paper was published in 1889. He realised its limitations and in the 100 years since that was published we have persistently mis-interpreted what he said.

Svante Arrhenius

Coming back to something much more serious, most of you will recognise this equation as the free energy equation and the temperature dependent part of it as the enthalpy:

$$\Delta G = \Delta E + P.\Delta V + T.\Delta S \tag{1}$$

Now the problem is that if you have a single-state equation and you plot it in the form of rate, or fluorescence of a probe, or signal intensity against one over the absolute temperature you do get a slope that is the enthalpy of the reaction. However, the situation is far more complex if you have two phases. The simplest possible model is a crystalline phase and a liquid phase in equilibrium; you start with the supposition, which I think is quite reasonable, that your reaction rate or fluorescence intensity is finite in both phases. It is quite unreasonable I think to start with the assumption that it is zero in one of the phases. It is a very convenient mathematical trick but probably not true. You end up with a rather complex expression like this. We do not need to worry too much at this stage about the analysis of this expression:

$$\text{Rate} = \lambda^2 \cdot \frac{kT}{h} \cdot \exp\frac{(T\Delta S_2^* - \Delta H_1^*)}{RT} \left\{ P.N. \exp\frac{(\Delta S_1^* - \Delta S_2^*)}{R} + \right.$$

$$\left. \exp\frac{(\Delta H_1^* - \Delta H_2^*)}{RT} \right\} \bigg/ (PN + 1) \qquad (2)$$

$$N = \exp\left\{ \frac{\Delta E}{R} \left(\frac{1}{T_m} - \frac{1}{T} \right) \right\}$$

We have an expression for the slope which is actually fairly straight forward when you plug in some numbers, and you have an expression for the point of maximum slope.

$$\text{slope} = -T_m - \frac{(\Delta H_2^*)}{R} - \frac{P\Delta E}{R}, \left\{ \frac{e^\alpha}{Pe^\alpha + 1} - \frac{1}{P+1} \right\} \qquad (3)$$

where

$$\alpha = (\Delta S_1^* - \Delta S_2^*)/R$$

since at T_m, $N = 1$

The point of maximum slope in a triphasic Arrhenius plot, *even for a gel/liquid-crystalline partition coefficient of unity*, does not necessarily correspond with the phase transition temperature T_m. This is because of the logarithmic transformation of the ordinate. The distance $(1/T - 1/T_m)$ is given, for $P = 1$, by the approximate expression

$$\frac{(\Delta S_1^* - \Delta S_2^*)}{2\Delta E} - \frac{R \ln 2}{\Delta E} + \frac{R}{\Delta E} \cdot \ln\left(1 + \frac{1}{\exp(\Delta S_1^* - \Delta S_2^*)/R} \right)$$

for values of ΔE very much greater than either ΔH_1^* or ΔH_2^*. Which further simplifies, for large differences in $(\Delta S_1^* - \Delta S_2^*)$, to

$$\left(\frac{1}{T} - \frac{1}{T_m} \right) = \frac{(\Delta S_1^* - \Delta S_2^*)}{2\Delta E} - \frac{R \ln 2}{\Delta E} \qquad (4)$$

Now I think the important feature about the point of the maximum slope is that it is not where you think it is; it is not at the phase transition. Most people when thinking of this change in rate of reaction or change in fluorescence intensity assume that the plot, which is a $1/T$ versus log(parameter) plot or Arrhenius plot gives you a phase transition which is shown as a break of slope (Figure 1). That is not true. Simply because there is a mathematical transformation of both axes and it turns out that the phase transition is separated from the point of maximum slope by a certain amount which is calculable. Now the rate expression brings in an important point and that is the parameter $\underline{P}$, which is a partition coefficient; if your probe does not partition equally between one phase and another, you cannot assume anything about the Arrhenius plot of this form.

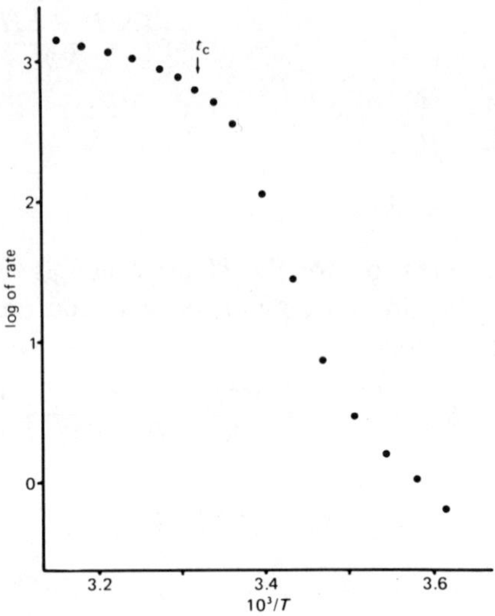

Figure 1. Arrhenius plot for the Na^+,Mg^{2+}-ATPase from Acholeplasma
laidlawii (T_m = 28.8°C).
(from Silvius and McElhaney, 1980)

It turns out that the simplified analysis shows that if you use two
probes with known partition coefficients, such as the cis and trans-parinaric
acids, you can show that they should report a difference in phase transition
temperature which relates to their partitioning behaviour. This is just an
approximation for that temperature difference and the next equation shows
the calculations.

$$\delta T \cong \frac{T_m^{\frac{1}{2}}}{4} \cdot \ln \frac{(P_1)}{P_2} \tag{5}$$

This was done for cis and trans-parinaric acid.

$$\delta T = (T_1 - T_2) = \frac{T_1 \cdot T_2 \cdot R}{\Delta E} \ln (\sim 5) \cong 1\cdot3 \ °C \tag{6}$$

I did this the correct way round; I knew what the partition coefficients were
and I quite purposely did not look up the reported transition temperatures

for the two probes until afterwards and was delighted to find that the calculation came out to 1.3°C, which was within 10% of the actual experimental difference that the two probes reported (Sklar et al., (1979)). So you have to take account of the partition coefficient.

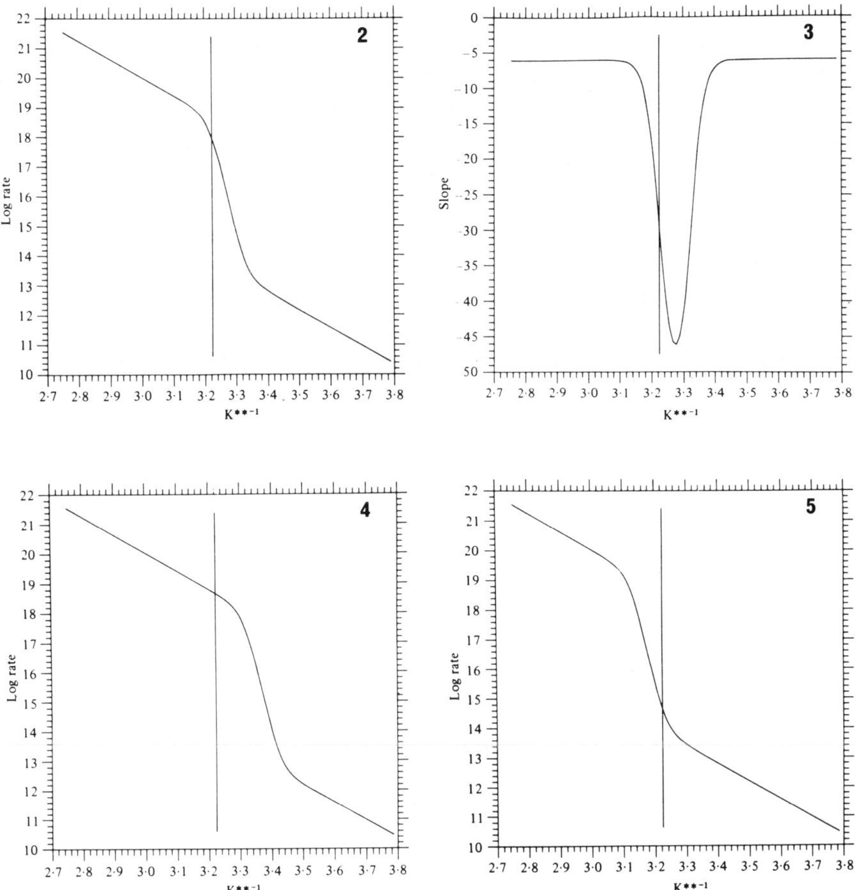

Figure 2. Arrhenius plot of the logarithm of the reaction rate against the reciprocal of the absolute temperature for the computer-simulated gel-to-liquid crystalline partition model with the following parameters: $\Delta \underline{H}^{*}_1 = \Delta \underline{H}^{*}_2 = 48$ kJ/mole; $\Delta \underline{S}^{*}_1 = 10$ J/mole/ K; $\Delta \underline{S}^{*}_2 = -30$ J/mole/ K; $\underline{T}_m = 310.15$ K; half width = 2 K; $\underline{P} = 1$ (from Klein, 1982).

Figure 3. Derivative plot of data shown in Fig. 2.

Figure 4. As Fig. 2 but $\underline{P} = 100$. Figure 5. As Fig. 2 but $\underline{P} = 0.01$.

This is just to show you what actually happens if you take the first derivative of such an Arrhenius plot; Professor Chapman has already mentioned the use of derivatives. They are extremely useful for showing either maxima, minima or points of inflection. You find with a partition coefficient of one, i.e., for probes which partition equally between both phases, that the known transition temperature and the point of maximum slope (the point of inflection on the curve) are different. Figure 3 is for a partition coefficient of 1:1 and the Figure 2 shows the plot from which this figure was derived. The others (Figs. 4 and 5) are for partition coefficients of 100:1, and 100:1 in the other direction. Now really what I show this for, is to make the point that unless you know how your probe partitions between your two phases, you have great difficulty in saying just what the information means. Figure 3 is just an example of what happens if you put a realistic phase transition width into the equation. We have been talking about phase transitions here that are abnormally sharp by biological standards. To get this sort of Arrhenius plot you require a phase transition of between 1 and 2 K. On the other hand if it is 10 K you see something like Figure 6. Even that is fairly sharp by biological standards.

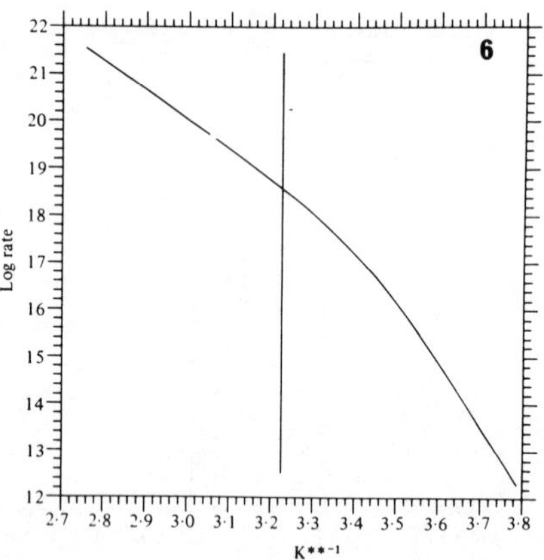

Figure 6. As Fig. 2 but $\underline{P}$ = 10 and half width = 10 K.

The point of showing it is to raise another problem that one has with such plots: they are not two straight lines. When you get to a transition width that is reasonably biological, your two straight lines become a curve. Whether or not you choose to draw them as two straight lines with a ruler is your business, but if you apply proper statistics to the problem you really should not be drawing straight lines at all. So I think the point that I would make by way of trying to stimulate you to tell me that I have been talking absolute nonsense, is that you must be extremely careful how you use probes to investigate the properties of the membrane. You must know how the probes distribute themselves within the membrane and you may, in fact, be in the situation of being forced into a corner and saying 'well, actually the probe tells us about the probe; it does not tell us about the membrane, neither does it tell us about membrane proteins'. Indeed you must be aware that if you attach a probe to a membrane protein or membrane lipid it may no longer behave in the way that it behaved without the probe. So at that point I throw it open to discussion.

Chapman: We agree with you. You do not want any more comments than that, do you? We agree with you! I do not want to do all the talking - I have just given a lecture. I have always been very cautious about probes and that is why I try to avoid them as far as possible. In some cases, however, one has to use probes and be aware of the difficulties that are associated with them. For example, let us consider spin-labelled probes and fluorescent probes. There is at the moment a discord between the interpretations placed upon the results which one obtains with a lipid protein system when one compares fluorescent probes, spin-labelled probes, deuterium probes and infra-red spectroscopic studies. The direct interpretation of the fluorescent probe data says that the protein orders the lipid, i.e., it stiffens up the lipid. There is more order produced. The ESR technique was at one time interpreted in a similar way. The NMR and infra-red techniques show that this is absolutely incorrect. So that is one situation in the literature where there is a discord between the interpretation of results from different probes on the same lipid-protein system.

Now there is another problem, if I can just mention this, about probes as well. If I put a probe into a lipid bilayer, say a liposome system, then I may deduce a measure corresponding to what one may call micro-viscosity which can be related to fluidity - or if you want to be more sophisticated you can talk about order parameters. So probes in the lipid bilayer are sensing the lipid's environment. Now if we put an intrinsic protein into that lipid system, we have an immediate problem. Is the probe sensing the change in fluidity of the lipid caused by the protein, or is the probe itself also feeling the influence of the protein? Is there an effect on the probe due to colliding with the protein within the protein-lipid system? We can see that when we try to interpret the probe data from a real biomembrane where proteins are present, as distinct from simple lipid bilayers, we have problems.

Bangham: Perhaps you can just remind me of what sort of molar concentration these probes are used at.

Chapman: It is not relevant what concentration is used.

Bangham: But it is relevant because if you were putting a molecule into a membrane, one knows very well then that the colligative properties of that membrane are altered by the molecule.

Chapman: Yes that is right, but I mean -

Bangham: I am entitled to ask you whether your putting one mole percent into that membrane gives you an answer which is different.

Chapman: This is an old argument which is misleading. When I used to argue with scientists using spin probes about perturbation effects and that they should be careful with their interpretation, the answer was always that we only need tiny concentrations of the spin label probe, around 10^{-4} M; this is a very small concentration, therefore it is not perturbing the system. Roger Klein rightly points out that the probe is providing information about its <u>local</u> environment but it is the local environment which it is perturbing. So that is what I mean when I said it really does not matter what the concentration is. Even if the probe is fantastically sensitive, if the probe perturbs the environment that is what is being detected.

Klein: Exactly, but I would have said there are two completely separate effects that one must distinguish. One must distinguish the colligative effects, which on extrapolation to zero concentration should then give you whatever effect you are looking for. However, the sort of effect I was pointing at earlier was partitioning, which upon extrapolation to zero still gives you error. I think it is important to realise that, as well as Professor Chapman's point about the concentration being 10^{-4} molar; it could be 10^{-7} molar, those results would still be wrong. But whereas if it were purely colligative and you were depressing freezing point, which is essentially what we would be talking about, at 10^{-7} molar the difference from zero molar would be infinitesimal. So it is very important to distinguish the two effects.

Chapman: At the same time you have to be realistic about the way we work. When we look at the intercalated Ca^{2+}ATPase protein in reconstituted systems, we see in the literature that the enzymatic assays show the presence of breaks in their Arrhenius plots: I think for example with Ca^{2+}ATPase in DPPC there is a break at 30°C. Now this is interesting because the break occurs at 30°C and yet the lipid transition temperature is 41°C. Now we used fluorescent probes (I did not like to use them!) and with these probes you can detect a break at 30°C. We did X-ray diffraction, IR spectroscopy, almost every technique we could find to study this system including a measurement on protein rotation. The latter method uses a triplet probe attached to the protein. There is a consistent pattern observed with all the physical techniques showing that a physical change occurs at 30°C related to the enzymatic change at this temperature. The next question was how to interpret these data. We interpreted these data (supported by freeze fracture electron microscopy) that in these systems the proteins do not remain isolated as we lower the lipid temperature, instead they aggregate to form patches. And the reason is that there are two melting processes, one at $30^{\circ}C$ for the protein-lipid patch and one at $41^{\circ}C$ corresponding to the melting of the remainder of the <u>pure</u> lipid.

In this case because we have supported the study with the use of a range of physical techniques, it seems clear cut that there is full agreement of the physical technique data including probes with the enzymatic activity data.

Klein: In a strange sort of way one might interpret that to say that the protein is the probe, and that the difference between $41^{\circ}C$ and $30^{\circ}C$ was in fact based on this sort of thing. I think that when one is measuring enzyme activity, one has got to be aware that you are not just measuring the lipid environment, but you are measuring all the thermodynamic parameters to do with the catalytic rate constant and Michaelis constant within the enzyme.

Chapman: My message, as far as probes are concerned, is try to keep away from them. Supposing you are faced with this problem - "I am interested in measuring protein rotation; how do I do it?". In the case of bacterial rhodopsin one has an internal probe "retinal". When we examine the Ca^{2+}ATPase protein there is, however, no convenient chromophore. What do I do? Do I say well, probes are difficult therefore I should not do anything? As long as the scientist is aware of the difficulties which the probe can introduce and points them out, then I say fair enough. It is when scientists carry out experiments with a probe and pretend that it is not a probe. Let us consider the cholesterol molecule; if we attach a nitroxide group instead of the OH-group, or we place a nitroxide group instead of the alkyl chain, is it fair to consider that the new molecule is still cholesterol? Can I say that I am studying cholesterol-lipid interactions and it just happens to be a cholesterol with a nitroxide group on top of it?

Klein: Perhaps I ought to broaden this and say to what extent do you believe in the fluidity measurement that comes out of such probes?

Chapman: Well that depends. I think that we can have a working approach and a pure physical chemical approach - and this will depend on what you want to do. If you have a working hypothesis, let us say Shinitzky uses fluorescent probes and finds some correlations between the movement of the probe and some particular function. I think that this approach is great! But the pure physical chemist may say "What is it that you are measuring?" and adopt a very rigorous approach; then the arguments begin involving the meaning of viscosity in two-dimensions, the meaning of the order parameters, the location of the probe, the perturbation due to the probe etc. All quite fair questions - some of which we have already discussed. What does one do then? Shinitzky says, look I have got useful empirical relationships, I am going to use the word fluidity as a name for what I measure with this fluorescent probe. Then I think that I could accept this as a useful approach.

Bernhardt: You would disturb the lipid phase and change the transition temperature if you put the protein into the bilayer.

Chapman: Of course this happens! I mean if you take the protein and put it into the lipid bilayer, for example the Ca^{2+}ATPase, the transition temperature changes; yes, it broadens out. In that particular case at increasing protein:lipid ratios the protein-lipid patch becomes larger and there is less lipid available for cooperative melting and the main endothermic peak begins to broaden. So of course an intrinsic protein can shift the lipid transition temperature. Furthermore if we have a lipid bilayer and we interact with an intrinsic protein, we can also affect the lipid transition temperature.

Sandhoff: I would like to ask you a question and to introduce the Nicholson Singer model which assumes that the entire membrane lipid bilayer is almost homogeneous. What is the proof of this, or what is the proof that on the contrary we have lateral domains in biological membranes? What are you measuring with these probes?

Chapman: What are we measuring with the probes? - we actually did experiments of this type with Dr Oldfield, many years ago. The probe goes to the most fluid part of the heterogeneous lipid system. So it is always measuring the most fluid region in my opinion. The question you raise, of course, about heterogeneity I actually touched on in my lecture with regard to sphingolipids. I said supposing that there is hydrogen bonding? Would those lipids tend to cluster even in the fluid state of a lipid system?

Sandhoff: There is evidence for this from the work of Tom Thompson. He has shown in the erythrocyte membrane that the neutral glycolipids actually form domains, but gangliosides do not. What other evidence do we have about lateral domains? I ask very specifically since this may be a point of attack for general anaesthetics in the membrane.

Chapman: Well I think Alec Bangham will talk about anaesthetics. When I put up the biomembrane diagram, I said it was a consensus summary. It was a drawing, but I made the point very clearly to you that we have to find information about the lipid, about the protein and about lipid-protein interactions. If you want an example of a heterogeneous lipid system I would suggest that you take the <u>Acholeplasma</u> <u>laidlawii</u> system. There is no doubt about it that in this system we can have both fluid and solid lipid present in the membrane. It depends what you mean by heterogeneity. If you mean is every part of a biomembrane identical? I would say unlikely.

Klein: Perhaps Karel, you would say something about doing measurements of fluorescence lifetimes in heterogeneous systems.

Wirtz: Yes, but on a very specific system where we used phosphatidylcholine carrying parinaric acid at the 1- or at the 2-position. We were interested in a transfer protein that specifically bound phosphatidylcholine. In this particular instance, we were investigating a one-to-one parinaroylphosphatidylcholine-transfer protein complex by measuring the decay of the fluorescence anisotropy of the 1-parinaroyl or 2-parinaroyl chain. As for this complex we observed that the rotational correlation times were different for the 2-acyl chain as compared with the 1-acyl chain. In fact these correlation times indicated that both acyl chains were immobilized on the protein in binding sites which were specific for each chain. I think that this applicaton is a nice example of using a probe molecule without it disturbing the system.

In another approach using parinaroyl-labelled lipids we have compared the bilayer of phosphatidylcholine vesicles with the plasma membranes prepared from the electric organ of <u>Torpedo</u>. Here we used both 2-parinaroyl phosphatidylcholine and 2-parinaroyl phosphatidylinositol. Determination of the fluorescence lifetimes indicated that the two lipid probes experienced different environments in those membrane systems. In addition, we could estimate from the fluorescence anisotropy decay that the rotational diffusion times of these two lipids were different. These kinds of experiment enable one to determine whether two lipids introduced into the same membrane system experience different environments and interactions.

Klein: The other point of fact that either Clive Ellory or Tim Rink might like to comment on is that in many membranes one obtains a break in some activity or some probe parameter, for example, in rat liver typically around 20°C or so. And yet if you look for the phase transition it may be -20°C and when people start saying that these are phase transitions, and yet they are so far away from the true measurable phase transition, what does it mean?

Ellory: We tend to make the comparison between different transport systems expressed in the same cell membrane. For example, in the human red cell you can compare glucose transport, amino acid transport, the sodium pump, and $Na^{+}:K^{+}:Cl^{-}$ co-transport as a function of temperature. If you construct Arrhenius plots you find widely differing slopes and break points for the individual systems. All that tells you is that the bulk lipid phase is not influencing each transport system in the same way. However, in these experiments you are measuring two things, the effect of temperature on the lipid environment, but of course there will also be direct effects of temperature on the transport protein <u>per</u> <u>se</u>. Nevertheless, the conclusion that intrinsic membrane proteins do behave differently in response to changing temperature makes it unlikely that the bulk lipid environment is modulating their function.

That is the point I would really like to ask. We are talking here about these probes going to different patches, but do proteins accumulate particular phospholipids? I think we heard that thylakoid membranes perhaps hold PS more closely to them, but how much specific lipid interaction with proteins is there anyway? I mean we do not believe in annulus lipid, or do we?

Wirtz: We were interested in the plasma membrane of the electric organ because of the claim that the acetylcholine receptor has a preference for phosphatidylinositol. Determination of the rotational mobility of parinaroyl-phosphatidylinositol and parinaroyl-phosphatidylcholine in this membrane did not corroborate the claim that phosphatidylinositol interacts with the receptor.

Bernhardt: In 1978 Sanderman et al. discussed this problem and it was again discussed in 1984. They could demonstrate that the proteins can be influenced in some cases by the fluidity as we understand this to be a parameter of the membrane. But some other proteins are influenced by specific phospholipids and in the last two years there is some evidence in the literature that some proteins are influenced by the head groups: they work better with negatively charged phospholipids and other proteins need lipids with specific acyl chains or double bonds, and such proteins are now discussed in the literature.

Chapman: I think the situation is this if I can just say a word about it. It was very attractive when you looked at the lipid analysis of the membrane to try to say, well perhaps some of these lipids have specific interactions with the proteins and then to relate this to pictures of boundary lipid, annulus lipids and so on. To begin with, the interaction was supposed to be a hydrocarbon chain interaction with the protein. Unfortunately in those cases where evidence was put forward to suggest that there was a long-lived boundary lipid, the actual evidence does not really support this conclusion. But of course we must not swing to the other extreme and say there is no specific lipid-protein interaction anywhere. What we have to do is to look at each system carefully and we will find that in some cases there may be, and other cases there may not be, particular lipid-protein interactions, and just take it step by step. The overall generalization that came through some years ago confused the field.

Rink: I would just like to pick up a point about probes. Perhaps there are two general classes. There are probes that you use when you think you understand the physical chemistry or chemistry, and you put them in a system and try to interpret their signals mechanistically in terms of what probes are telling you, whether they are spin labels, fluorescent labels, certain modified fatty acids or cholesterols. If you think you can interpret those signals, you may do if you are brash, or brave, in terms of structural or functional information about the lipids, or the proteins or both. There are other probes which you know do things when things happen to the membrane. I am in the business of trying to understand stimulus-response coupling and for example it was quite an advance when people produced derivatives of certain fluorescent dyes which would somehow report membrane potential. It turns out that there are at least two ways in which that can be done. One is that the dyes are lipid-soluble ions which distribute between the two bulk phases, separated by the plasma membrane or organelle membrane, depending on the potential. Not surprisingly some of these have turned out to be remarkably toxic and extremely potent; and we are not talking about 10^{-4} molar, we are talking about 10^{-8} molar for some of these probes like the thiodicarbocyanines. There are others that sit in the membranes and give a fluorescence or an absorbance signal which is remarkably linearly related to the potential gradient between the two sides of the membrane. We actually do not know how they work. But they do, they are very linear and they are reproducible. You can calibrate them against another reliable method such as the standard electrode method. Then they will report the potential; but you did not know that before you started, for they were found more by trial and error. Nonetheless, they do not seem to be very perturbing. But you have to do the experiment to find out what they are telling you and validate the approach; so you come at it from a different angle.

Chapman: I think they do not have to be too pure but you have to be careful. I always think of Heaviside, when he produced operational mathematics; all the pure mathematicians said what he was doing could not be done because it was illogical. And he said 'I do not care because I am doing it'. And a hundred years later they worked out the reasons why it could be done logically after all.

Bernhardt: Could I raise another point about specific lipids near to proteins? Proteins do not have smooth surfaces.

Chapman: How do you know?

Bernhardt: I will quote experiments. If one takes bilayers made from one lipid and introduces a protein there is a large increase in permeability. If one takes different lipids and introduces a protein, then there is not such a large increase in permeability, and the interpretation of these results in the literature is that in these cases there is a better fit between lipid and protein because the surface is not smooth. This is why we need specific lipids associated with proteins to fill these areas.

Wirtz: May I follow up on this? We have had some experience with the parinaroyl-labelled lipids. If you introduce these probe lipids into a bilayer structure you will find that the fluorescence anisotropy does not decay to zero. One reaches a finite residual value. One interpretation is that this value reflects the local order experienced by the lipid probe. We have compared the physical behaviour of parinaroyl-phosphatidylcholine and parinaroyl-phosphatidylinositol in the plasma membrane from the electric organ of <u>Torpedo</u>, and found that the residual anisotropy was lower for phosphatidylinositol than for phosphatidylcholine. This could mean that phosphatidylinositol interacts with membrane proteins to the extent that the 2-parinaroyl chain experiences less local order than in the surrounding bilayer structure.

Chapman: Can I make a comment? I have always been amazed that they do not check that the conformation of the protein is unaltered. When scientists reconstitute a membrane protein with dimyristoyl or dipalmitoyl lecithin, the first thing that one might expect is that there will be a change of conformation of the protein. Some of the experiments that we have done showed reasonable agreement between reconstituted structure as seen by infrared spectroscopy and the protein structure in the sarcoplasmic reticulum. Another point which is relevant concerns the length of the hydrocarbon chain of the lipid compared with the hydrophobic section of the protein. I would not, therefore, talk about complexes or specific lipids in that sense, but there obviously has to be architectural agreement between that part of the protein that sticks into the lipid bilayer and roughly the length of the lipid in which it is now positioned.

Klein: I think that there is another point too, something that we have been interested in is the use of parinaroyl PC; in fact we have attempted to measure the phase transition for these compounds, and could not find a phase transition between -30 and +90°C. Again underlining this point, here you have a fatty acid that is minimally perturbing in the sense of the more usual fluorescent probes. You assume that it will behave like a fatty acid, a C18 fatty acid. Yet you cannot find a phase transition and this begins to make one wonder what is going on. I think Clive Ellory has some experience with measuring rate processes in lipid systems that show really very paradoxical behaviour and this ties in with what Dennis Chapman was saying earlier, that certain infrared measurements show "transitions" in the same region for rat tissue as for enzyme processes, whereas the true lipid transition is way, way below.

Ellory: There is a paradoxical increase in permeability as you lower the temperature and it is pretty phenomenal. I am not sure that we should go into an absolute explanantion.

I think that you have to be realistic; I think that if you are looking at a real biological system, even as simple as a red cell, you can get into very deep water quickly. What Roger Klein is referring to is the fact that if you cool a red cell and look at its permeability to potassium, in a human red cell you go through a minimum around $10^{\circ}C$ and then the permeability increases enormously.

Chapman: Let me give you an example. Some workers suggested that there was a phase transition at $18^{\circ}C$ in a certain cell system as deduced from enzymatic measurements. I knew that that particular membrane was full of cholesterol, on a mole for mole ratio. I found it difficult to believe that it corresponded to the lipid phase transition; I thought it was more likely to be a small change of the protein structure. But they said, well of course what we did, we extracted the lipid from the membrane and we looked, using a spin labelled probe, to see if there was a phase transition at eighteen degrees. I said "Was there a transition?" They said "No". I said there was not a lipid phase transition - but was there a phase transition in the membrane resulting from the lipid? "Yes". I asked how do you rationalize that? They said "What we are suggesting is that the bilayer is made up of two lipid layers, and it is the phase transition of the top lipid and not the bottom lipid layer which is occuring. When we extract the lipid, we mix all the lipids together". If you follow this logic without further experimental proof then I believe that trouble will occur.

Klein: I think Professor Chapman's point is that whereas lipid phase transitions are co-operative, intermolecular events, they were saying well we will strip one half off the bilayer and say that it has a phase transition, but that has not anything to do with the other half of the bilayer. And the other half of the bilayer has a transition which has nothing to do with the first half. That is so plainly crazy. The physical chemistry of co-operative transitions in membranes and indeed in polymers is fairly well understood, and so that this explanation is just interpretative nonsense.

Tyman: Speaking as a maverick outside the bilayer, it does occur to me that there are two extremes in the probe situation. The probe is either part of the molecules which comprise the bilayer, or it is a foreign molecule. Fluorescent molecules generally, with the exception of the diphenyl hexatriene carboxylic acids, bear very little relationship to the components of the whole system. Some of the fluorescent probes, for example the anthraquinones, seem to be colossal molecules which must perturb the system. Surely more skill should be used by organic chemists in devising small additions to molecules which will enhance, for example, the use of NMR which is, I believe, one of the most notable techniques. Why should one not have a fluorine atom in the chain. Then there are the nitroso-labels for ESR which must perturb very little. These are some of the groups that would perturb the situation least. I also believe that protein structures have been found largely by chiro-optical methods. Why should one not have other types of artificial helical structures which would inter-relate with the bilayer? Since it is to the protein part, it seems to me, that probes have contributed least of all.

Klein: I think your point about the diphenyl hexatriene derivatives is interesting because I remember going to a meeting in London where Yehudi Levine had a poster. I think it should have struck terror into the users of DPH and derivatives. He showed that depending on lipid composition the absolute axiomatic dogma of using a fluorescent probe could be violated. That is you have to assume that the angle between the absorption and the emission dipole does not change. If it changes with lipid composition, you do not know where you are. And in fact Yehudi put up this information and there were lots of very worried people, saying is it really true because if it is we cannot really do this sort of experiment. So even with DPH one should not pretend that it looks like a membrane hydrocarbon.

Tyman: Perhaps one could devise molecules which perturb the system least?

Klein: I think the nearest molecules we have to that are probably the parinaric acids; because they are fluorescent and more than any other fluorescent probe they look like a fatty acid. And they have been very useful.

Brooks: Does Dr Chapman feel that the modern methods of infra-red spectroscopy and Fourier transform spectroscopy would allow more use to be made of the inherent cholesterol in the membrane, as a probe of the surrounding structure?

Chapman: Yes if you wanted to. I mean for example what you could do, if you are interested, to look at cholesterol itself and its environment. The advantage, as I tried to put over, was that here there is no probe, it is just a natural material itself. We do not have to do any organic synthesis; if you want to, of course, you can deuteriate part of the chains and if you want to you could deuteriate part of the cholesterol nucleus.

Brooks: Or perhaps introduce ^{13}C-labelling and do parallel NMR and infrared studies.

Chapman: There are advantages of this particular technique of NMR. You know that with the NMR technique, if the vessel is not small they rotate very fast, then you get a very narrow signal. There is no problem like that with infrared spectroscopy. Another feature is the comparison of the time scale, 10^{-15} per second, an instant snapshot of the system, as compared to the slower rate of the NMR. So the answer is yes.

Brooks: But you did also mention computer accumulation which would slow things down in as much as you would be observing the summation of a series of snapshots.

Chapman: Well the particular feature about it is that you can actually get something like seventy whole spectra per second, so you can think in terms of kinetics.

Klein: Whereas the individual snapshot is 10^{-15} seconds, when you do a time average you are actually not 10^{-15} seconds at all; at 70 per second, you are down to a sampling time of around 10 milliseconds.

ANAESTHETICS

Klein: I think that perhaps we had better move on and I would like to ask Alec Bangham to introduce the subject of anaesthetics.

Bangham: First of all, how did you know that I have just written a somewhat provocative review on the subject? Look out for it (Bangham and Hill, 1986).

There are a few explanations which account for the manner in which the catastrophic physiological consequences of anaesthesia, cold narcosis or, for that matter, a short, sharp upper-cut, come about. Most studies terminate with the presentation of ever-better correlations between an end-point in a model system (dough consistency, rubber elasticity, bacterial, protozoal or animal mobility, liposome permeability, luciferase activity etc.) and oil-water partition coefficients or with some arbitrary biological end-point.

It is my thinking that (general) anaesthetics such as halothane, N_2O or xenon (if you can afford it) ultimately cause unconsciousness by facilitating the collapse of the pH gradients that are required to sustain a concentration gradient of certain neuro-transmitters, notably the catecholamines. As weak bases they can be shown to accumulate in intra-cellular vesicles by a mechanism known as ion-trapping in which a pH gradient is sustained by a balance of pump and leak of protons and potassium ions.

From what is currently known about (a) the permeating pathways for non-electrolytes, ions and protons across membranes, e.g., liposomes, (b) the effect of anaesthetics on such pathways and (c) the effect of temperature and pressure on both liposomes and whole animals, it is possible to develop a

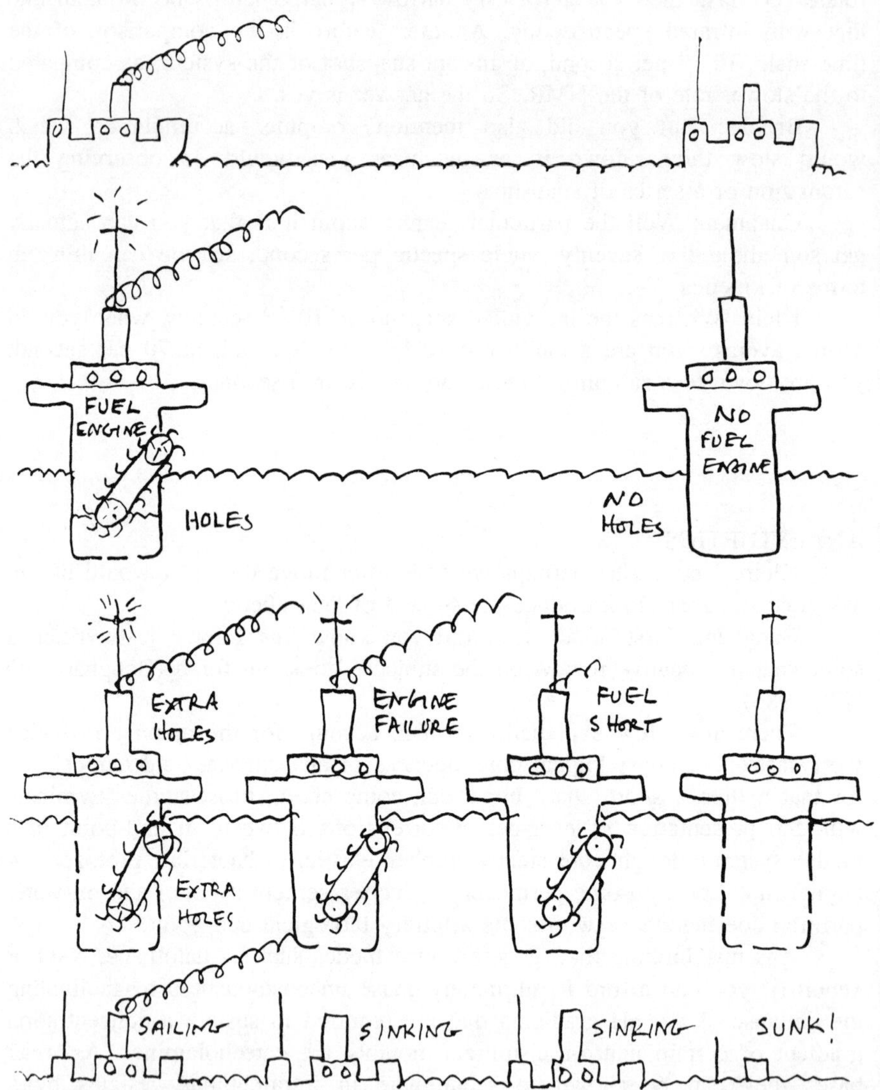
FUEL
ENGINE
HOLES
NO
FUEL
ENGINE
NO
HOLES
EXTRA
HOLES
EXTRA
HOLES
ENGINE
FAILURE
FUEL
SHORT
SAILING
SINKING
SINKING
SUNK!

testable hypothesis. It is called the 'proton pump-leak' hypothesis and involves a number of linked biophysical and biochemical processes. (d) It assumes that a living animal or plant is in a steady state regarding all concentration gradients; passive leaks across membranes are balanced by temperature-, pressure- and energy-dependent ion/ion and/or proton/ion pumps (enzymes), working within an aqueous phase. (e) Consciousness is thus dependent upon inter-neuronal communication via release of transmitter substances. (f) Transmitter substances, characteristically either weak bases or weak acids, e.g., catecholamines, accumulate passively by the ion-trap mechanism in vesicles rich in acid buffer, held to a low pH by the activity of H^+/K^+ energy-driven pumps. Interference with this finely balanced system, either by changing the chemical potential of the hydrophobic (membrane) phase at normal temperature and pressure (NTP) (with anaesthetics), or by changing the chemical potential of both hydrophobic and aqueous (pump) phases by hyperbaric, hypothermic or anoxic conditions imposed (inevitably) on the whole animal, would result in the re-setting of the steady-state parameters.

I have a little cartoon which I think illustrates this point. Perhaps you cannot see it, but imagine yourself on the seashore, and you are looking out to sea, and you see two boats and they are both afloat. The one on the left here has actually got smoke coming out of the chimney. The one on the right has not. You come a little closer to these boats, and you realise the one on the right is a liposome, it is afloat because it has got a virtually impermeable bottom. It is sinking very slowly all the time, but it seems afloat because of the impermeable bottom. You and I, and everything living, is more represented by the left hand side here, where the bottom has got a lot of holes in it, but there are bilge pumps. We are afloat because we are in a steady state, where a leak is precisely matched by a pump. Now the moment you realise the implications of the situation, there are all kinds of ways in which you can become unconscious. The anaesthetist comes along at normal temperature and pressure and he punches a few extra holes in your bottom and your concentration gradients are depleted by a small amount. For example, if in the cartoon, the radio operator has his office just by the water line, the few extra holes in bottom will sink the boat and he will never send out the message. That is important at normal temperature and pressure. But we can widen this whole concept of consciousness, and that is if for example the engineer goes to sleep as he is very inclined to do in lecture theatres, he fails to keep the bilge pump going. So once again you slip into unconsciousness. Now that maybe a little frivolous in a meeting like this, but I have seen it happen very often in my experience. More important in this concept is if you fall into the North Sea at this time of year, your core temperature will decrease and your pumps will become throttled back.

Now it is not a question of unconsciousness because your membranes are affected. It is that the pumps are not working, and enzymes have Q_{10}'s. Although my teachers always told me that the enzymes work faster if they are warmed up they never told me that they work more slowly if you cool them down. Fuel constraint would be a further method of embarrassing the bilge pump from working, for example if I give Roger Klein a short sharp uppercut or wring Tim Rink round the neck, they would become unconscious quickly, and that is simply because the pumps have now been deprived acutely of energy. Returning then to the question of anaesthesia, when the anaesthetist approaches you in the operating theatre, he has chosen a substance, a chemically inert substance which has a rather nice oil-water partition coefficient, i.e., about 1000:1 or 2000:1, which ensures that sufficient molecules, of those that make it into your aqueous phase, will end up in the membranes, in all your membranes at an appropriate concentration. As a general statement, by the time your membranes have dissolved four moles percent, four molecules of xenon, for example, and halothane, gin or whatever, you can achieve thermodynamic disordering of those membranes by an amount which by common experience amongst anaesthetists throughout the world, renders you unconscious. In other words it just causes the boat to sink to a different waterline. So now the question is what is the amplification mechanism? It is no good saying that the leakiness of a liposome doubles when you have four anaesthetic moles percent, that does not explain why you become unconscious, nevertheless you will find the whole history of anaesthesia is filled with such correlations. People go out, they take a rubber band and they find it stretches by a certain amount when they are exposed to 50 different anaesthetics, at anaesthetic concentrations; that correlation does not tell you how you become unconscious.

We too correlated the permeability of liposomes, as models for membranes, to solutes, with and without anaesthetics, measuring among other things the enthalpy of activation of permeation. We plotted the enthalpies for our solutes against the number amongst hydrogen bonds which the solute might have to break, and found that the apparent energy of activation for diffusion increased with the number of hydrogen bonds. That is very reasonable, because if a molecule can jump out of water, it does not know whether it is jumping into a membrane or a vacuum, and you need more energy as you increase the number of hydrogen bonds that have to be broken. Now that's very reasonable.

When we measure ions which have a lot of water molecules, very strangely you find that the activation energy is low, 13-15 kilocalories per mole, no more than is found for four hydrogen bonds, which is quite unrealistic. However, the flux for ions is miniscule, and for these groups of molecules (e.g., sucrose and glucose) the permeability can be 10 to 12 times greater than for the ions. In other words there are clearly two mechanisms for these solutes crossing the membrane. This group which, if they can dehydrate, can diffuse and rehydrate. We understand that. And that event takes place through all the membranes - the total surface area of the membrane. If we have another event which is related to ions, it is totally improbable that they should dehydrate. The energy of activation for dehydration of ions is 30 kilocalories. We know the enthalpy is low, but we know that the probability of dehydration is vanishingly small. That adds up to another mechanism, to the fact that the membrane is only opening in small areas, the ions which are crossing the membrane are only crossing over minute areas of the membrane. So we think that the ions can cross aqueous slits, not pores; I do think that they are aqueous slits, line dislocations which occupy very small areas of the membrane. And the point about these line dislocations is that their number increases - it is a probability event - as you add anaesthetic the probability of permeation as an event increases. If you do a plot against temperature, $1/\underline{T}$, the probability of the number of slits occurring increases with temperature - very reasonable. And lastly you can reverse this depression. In other words, the event in the membrane which allows ions to cross is related to the chemical potential, or the Gibbs free energy of the membrane system. Then finally, we still have not offered you an explanation of the amplification mechanism. This is really quite a difficult one to swallow, but we find we think it is related to the collapse of pH gradients in certain transmitter vesicles.

Sandhoff: That is a very fascinating model you are proposing here. But these slits have to be shown experimentally; that is the problem. From the literature it is known that general anaesthetics do have quite a variety of effects on membranes. It is known that they can affect carrier proteins. For example it has been shown in this country that the sodium inward current in the squid giant axon is reduced (Haydon, 1984). It is known that they modulate receptor function. For example, the opening time of the nicotinic acetylcholine receptor is reduced in the presence of general anaesthetics. And this effect again can be modulated by varying the cholesterol content of the lipid phase (Lechleiter and Gruener, 1984). NMR studies in the States indicate that general anaesthetics may concentrate at the hydrophilic-hydrophobic interface of the lipid bilayer between the water and lipid phase.

Maybe that they concentrate on the protein surface as a result of hydrophilic-hydrophobic interactions too. We do not know. It is known that they inhibit enzymes. So they may not only work on lipid bilayer structures but on membrane-bound proteins throughout the entire system. By modulating receptors and carriers they can change the efficiency of the transmission in the nervous system. We are interested not in anaesthesia per se but in an observation we made several years ago and I am just asking for an explanation. The observation was that hydrolysis by a membrane-bound enzyme is stimulated by general anaesthetics: by xenon gas, by nitrous oxide, and by halothane. If we assay the enzyme with a water-soluble substrate, which of course is not part of the same membrane, we do not see any stimulation. The kinetic studies that we did on the degradation of the lipid substrate suggested some kind of two-dimensional kinetics within the plane of the membrane (Scheel et al., 1982).

One possible explanation for the stimulating effect of general anaesthetics is that the membrane may have a lateral domain structure and only a few molecules may be freely mobile and available to the enzyme. General anaesthetics may dissolve these clusters and thus increase the availability of the substrate.

Klein: I think perhaps the point that is worth making to clarify this issue, is that the anaesthetics that fall into this class are generally called general anaesthetics which tend to be non-polar compounds, and there are the local anaesthetics too which tend to be quarternary nitrogen compounds, with some form of alkyl chain on them. I think what we are really talking about is in the old dogma, 15 or so years ago, that general anaesthetics were always assumed to act as fluidising agents for the membrane hydrocarbon core; that was never of course thought for the local anaesthetics, it was always thought that the local anaesthetics with the charged nitrogen could interfere with either sodium or particularly potassium permeability. I think that we ought to be clear what we are talking about here; we are talking about general anaesthetics: non-polar, non-charged compounds which may on the other hand have a considerable polarizability as has halothane. Professor Sandhoff's point is about it concentrating at the interface rather than diffusing completely freely through the hydrocarbon core, finding it in higher concentration near the interface; we discussed this yesterday evening and I said, was it in fact known whether the system was at equilibrium? Because in fact if the measurements were not made at thermodynamic equilibrium, that is when the system was still reaching equilibrium, a concentration maximum near the interfacial region is exactly where you would expect there to be a maximum, because after all the anaesthetic came from the outside, and to get into the inside it has got to pass through this region and molecules only pass down concentration gradients; they do not magically equilibrate throughout a membrane without a driving force.

Sandhoff: We did not do this experiment so you had better ask the authors (Yokono et al., 1981; Kaneshina et al., 1981; Kaneshina et al., 1982). I think that these were NMR studies on liposomes and on micelles.

Chapman: What does it show and how does it show it?

Sandhoff: Ueda and coworkers measured the line width of proton NMR signals from the head group and from the fatty acyl chains of phosphatidylcholine in liposomes. They found that the choline head groups are mobilized by inhalational anaesthetics at about half the concentrations necessary to mobilize the methylene groups of the fatty acids. Studies from the same group on detergent micelles indicated that inhalation anaesthetics preferentially bind to the amphiphilic interface.

Chapman: Could it be - as another speculation, since these are all speculations anyway - that the anaesthetics interact with some clathrate-type arrangement with the carbohydrate groups, for example the gangliosides?

Sandhoff: No, if you just take liposomes as models you can see the changes.

Chapman: You have the old Pauling idea. Could there be clathrate-type structures or particularly associated carbohydrates or glycoproteins?

Sandhoff: But these measurements I mentioned were done in the absence of carbohydrates.

Chapman: Sure, I am asking beyond that. Have you considered that?

Sandhoff: That is a possibility of course.

Rink: This is one of the things that has always interested me. First of all let us just deal with the local anaesthetics and then come back to the general anaesthetics. Local anaesthetics basically block sodium channels. Fairly selectively, and that is a good anaesthetic system. The thing about general anaesthetics is that they are highly specific: so that is a lousy name! You give them to an animal, and you get an animal that becomes analgaesic and unconscious; yet the whole of pharmacology and physiology is based on the fact that most of the systems in a tolerably anaesthetised animal are almost fully functional including their reflexes. What is it that causes unconsciousness?

Is it really a membrane mechanism that is special to those regions, or is the fact that the thinking, pain, and feeling processes require so many steps in neural circuitry that a small change throughout is amplified and you cannot get complex function?

Chapman: Could it be that there is some parallel to the morphine-type molecules where tiny, tiny amounts can actually knock an elephant over, make an elephant unconscious? Is there some analogue that again links up with the carbohydrate which is interacting with this?

Rink: With ether and nitrous oxide, and other general anaesthetics? I do not know; I mean, I sit on the sidelines of all these arguments as to whether it's even a lipid effect. Is this all epi-phenomenal? Are there hydrophobic regions in proteins which really are not specifically, but rather preferentially prevented from operating by these things?

Macdonald: May I just make the observation that this sort of reductionism, seeking to provide a simple explanation for something as complicated as central analgesia let alone anaesthesia, may be a waste of time. I think there will be a whole galaxy of molecular pertubations, but even if we accept Alec's model, even if we say that is fine, it does not explain anything. It does not explain the extremely complex signal processing and integration that is going on in a very peculiarly perturbed way which clinicians can handle. That is the explanation of anaesthesia we are looking for. Holes and leaks and so on, I really do not think are at the right level of organisation.

Bangham: I am sorry, I really have to take issue very strongly with you, because although I have already discussed liposomes we have also taken the trouble of isolating and studying synaptic vesicles. These we believe are responsible for the integration of consciousness in the CNS, believing them to be responsible for passing on messages. You can take the synaptic vesicles, load them with isotopically labelled catecholamines and you can expose them to anaesthetics. You then measure the efflux of protons and <u>pari</u> <u>passu</u>, mole for mole, as the protons come down, catecholamines come out. Now is that not a specific and amplified mechanism orchestrating the consciousness of the CNS? I am not saying that it is not very complicated, of course it is. Naturally when you become unconscious on the anaesthetist's table, it is not going to be the same as when you go to sleep in a lecture theatre, when the engineer turns the burners down, nor will it be the same as when you fall into the North Sea. All these situations are going to be different. But it suggests an amplification mechanism. Tim Rink talked about anaesthetic partitioning into proteins, and of course Lieb and Franks are postulating this (Franks and Lieb, 1984, 1985). But are you going to tell me categorically that there is luciferase sitting in the appropriate parts of our CNS which cares to flash when we are unconscious? I do not believe it.

Macdonald: That is all part of the reductionist mythology too you see. It is a great way of doing nice experiments, but it is not pitched at a level which is appropriate to explain anaesthesia.

Rink: Well, how do you account for the fact that agents which act at very specific endorphin or morphine receptors are effective analgesics? The success of neuro-pharmacology never ceases to amaze me. You can do amazingly simple things to the brain and get the answers you want and the therapeutic effects you want. It could not be much simpler than blocking one class of receptors and you end up with phenomenal analgesia.

Klein: To take Alister's point, the simple mechanisms really belong to the local anaesthetics, where you can actually say you block a sodium channel and a peripheral nerve shows loss of action potential conduction. You can do that by cooling it. That is simple. I think what Alister is getting at is in the CNS we can measure lots of things, we can say so and so happened, but the overall result in terms of neuronal integration is really exceedingly complex.

Rink: But is it not marvellous that you can achieve these highly desirable states with such extremely simple, as far as we know, chemical interventions like receptor blockade?

Klein: Can I ask Alec and perhaps also Alister, what do you think happens if I hit you very hard on the head?

Bangham: I think the enzymes are suddenly deprived of oxygen. These pumps which are keeping you in a steady state and your pumps are suddenly embarrassed.

Klein: I do not agree with you, you only get deprived of oxygen if your blood pressure drops.

Bangham: If I hit you very hard? You experience an acute ischaemia. The brain is notoriously sensitive to deprivation of oxygen concentration by whatever means. These enzymes require this oxygen as a fuel. You only have to drop the steady state by a very small amount. As I have already said, if the radio operator happens to be at the water line he is going to be embarrassed. I do not see the problem.

Macdonald: I do not see the question. You are talking about two different things. I do not think one observation supports or eliminates the other.

Klein: I think your earlier point, that of the problem of general anaesthetics, is that it is nothing simple like an action potential running down a single peripheral nerve; it is something very much more complex.

Macdonald: There are many ways of dislocating a complex system.

Rink: I would not bank on general anaesthetics being complex. In fact I have to bank on things like that not being complex. We think that we can manipulate the brain quite simply.

Macdonald: I am saying that the complexity is in the integration.

Rink: Nobody imagines that using an analgaesic or an anaesthetic is going to help you understand the brain.

Bangham: I do not agree. If your drug has the right p$\underline{K}$ and the right partition coefficient and is able, by purely physical means such as diffusion, to end up in one of these transmitter vesicles, it is going to kick out what was in there before. Am I right?

Rink: That is one mechanism for doing it.

Bangham: The orchestration is going to depend on the p$\underline{K}$ of the drug and its permeability and how fast it gets there. One of the most toxic compounds is, of course, ammonia. How does that work? It collapses pH gradients.

Rink: I think that ammonia acts on one or other receptors in the spinal cord.

Bangham: Ammonia will cross the membrane as fast as water.

Rink: Sure but it works, you do need a fair bit of ammonia to collapse pH gradients, and it is very toxic at very low concentration - sub-micromolar.

Macdonald: One of the interesting things about the so-called explanation, at a molecular level, for general anaesthetics is the extremely low concentration. You mentioned Haydon's work; he would be the first to point out that his effects are occuring at above clinical doses of the general anaesthetics that he was using for his voltage clamp experiments.

Rink: They are doing things in those experiments which are quite illuminating on the lipid protein interactions in a nerve membrane. But they are looking at things that actually do not happen in anaesthetised patients; you do not lose the action potential.

Macdonald: And Haydon always makes that very clear.

Klein: I think you will probably find that it is basically Alec Bangham's point that the average neuronal membrane depends on the passive leak in one direction and active pumping in the other. If you tamper with either of them and effect the overall balance the resting state then changes. If you change the resting state far enough the thing will not fire. Now Alister's point was of course that this is far too simple. I am not sure that I would agree with Alister on that, I think that one has to think of a simple mechanism that may effect one particular area, let us say, of the mesolimbic system, and that changes the level of consciousness, whereas you may have 99% of the apparatus still fully functional. The fact that your one percent is sub-functional, whether it is your fuse or your bath plug or your hole in the bottom of your boat, the 1% non-functional bit of the membrane may be the critical bit.

Macdonald: In the physiological world it is well known that general anaesthetics have presynaptic and postsynaptic effects.

Bangham: What effects? You have made a general statement there. What effects do general anaesthetics have on presynaptic and postsynaptic membranes?

Macdonald: If you take a preparation like a brain slice, where you can from field effects determine where different anaesthetics like halothane and methoxyfluorane have their main inhibitory block, then they are found to differ. For example halothane might have its most effective blockade, under certain conditions, presynaptically. It can also have a postsynaptic blockade which will only become more important if you increase the dosage. You have got a whole spectrum of points, where inhibition or perturbation can occur. The point I really want to make is that neurologically it may not make very much difference at what point in the throughput of the synapse you perturb it. On the other hand, it may. We simply do not know.

Bangham: Where is the weakest link? What I want to hear from you, because this is very exciting news to me, is that people have actually demonstrated that anaesthetics, general anaesthetics at anaesthetic concentrations, block synaptic transmission. Is that a fact?

Macdonald: Yes.

Bangham: What is the mechanism for that block? What I am suggesting is that it comes about as the result of depletion of transmitter substances.

Macdonald: Well if it is postsynaptic then it cannot.

Bangham: How do you know it is not, how do you eliminate the presynaptic depletion of synaptic vesicle contents?

Macdonald: You can do that by applying your transmitter by iontophoresis, so you have got that as a constant. So you can show under certain defined conditions, that you can get an anaesthetic to work by a mechanism which would not involve your particular idea. This I suspect does play a part, but it will be one of dozens of perturbations.

Klein: If I can bring the two together and say that the general anaesthetic postsynaptically is affecting your hyperpolarization-depolarization phenomenon. And that is almost certainly at pump leak level. I would guess that it is likely to be at receptor level; someone may contradict that, whereas Alec's mechanism may well apply presynaptically.

Sandhoff: It has been shown for the nicotinic acetylcholine receptor, which is postsynaptic, that the opening time of the receptor, if you add acetylcholine, is reduced. And thereby with general anaesthetics like halothane, for example, you have less influx of sodium and less efflux of potassium and thereby the probability that you get enough postsynaptic excitatory potential in order to trigger an action potential, is reduced. And you have amplification in the nervous system since there are so many neurones eventually involved, that this may have an effect.

Brooks: Could I make a comment about general anaesthetics being generally non-polar. These are not all small molecules. The steroid anaesthetics, which were in clinical use until a year or two ago, are fairly non-polar, and one of the first to show activity was, I believe, 21-hydroxy-5-β-pregnane-3,20-dione hemisuccinate sodium salt.

Bangham: The steroid anaesthetics, the Glaxo series, have been shown to act exactly as halothane in terms of lowering the melting temperature of DPPC - there is nothing special. It is a colligative effect.

Klein: So I think on that note we should end the discussion and thank all those speakers who took part and made it so lively.

REFERENCES
Bangham, A.D. and Hill, M.W. (1986) Chem. Phys. Lipids in press
Franks, N.P. and Lieb, W.R. (1984) Nature Lond. 310: 599
Franks, N.P. and Lieb, W.R. (1985) Chem. Brit. 919-921
Haydon, D.A., Elliott, J.R. and Hendry, B.M. (1984) in "Current Topics in Membranes and Transport", Vol. 22, pp. 445-482
Kaneshina, S., Lin, H.C. and Ueda, I. (1981) Biochim. Biophys. Acta 647: 223-226
Kaneshina, S., Kamaya, H. and Ueda, I. (1982) Biochim. Biophys. Acta 685: 307-314
Klein, R.A. (1982) Quart. Rev. Biophys. 15: 667-757
Lechleiter, J. and Gruener, R. (1984) Proc. Natl. Acad. Sci. U.S.A. 81: 2929-2933
Scheel, G., Acevedo, E., Conzelmann, E., Nehrkorn, H. and Sandhoff, K. (1982) Eur. J. Biochem. 127: 245-253
Silvius, J.R. and McElhaney, R.N. (1980) Proc. Natl. Acad. Sci. U.S.A. 77:1255-1259
Sklar, L.A., Miljanich, G.P. and Dratz, E.A. (1979) Biochemistry 18: 1707-1716
Yokono, S., Shieh, D.D. and Ueda, I. (1981) Biochim. Biophys. Acta 645: 237-242

Section <u>5</u>

MEMBRANE STRUCTURE AND FUNCTION

RECENT STUDIES OF LIPIDS, BIOMEMBRANES AND MEMBRANE MIMETIC SURFACES

D. Chapman, D.C. Lee and A.A. Durrani

Department of Biochemistry and Chemistry, Royal Free Hospital School of Medicine, Rowland Hill Street, London NW3 2PF

INTRODUCTION

Scientists interested in the study of biomembrane structure have been concerned to understand and characterise the lipid matrix of biomembranes. A variety of techniques such as calorimetry, ESR, NMR spectroscopy and fluorescence spectroscopy have now been applied to model and natural biomembranes. Questions concerning the lipid disorder, lipid fluidity, lipid phase transitions, lipid diffusion and the orientation of different portions of the lipid molecules have been examined. With many of the techniques used, a probe is required to obtain the appropriate information e.g., a nitroxide group, a deuterium probe or a fluorescent probe molecule such as diphenyl hexatriene. Much discussion has taken place about the reliability of the interpretation of some of these probe measurements involving doubts about the perturbation effects that the probes may have upon the lipid system which at the same time they are measuring. In some instances there are clear divergences in conclusion, e.g., simple interpretations of probe methods give conflicting conclusions concerning the ordering of the lipid chains in a bilayer matrix when intrinsic proteins are present (Restall and Chapman, 1986).

Recently an old instrumental technique has been given a new potential for the examination of biomembrane systems following the introduction of microcomputers and Fourier transform methods. This technique is infrared spectroscopy. It enables useful information to be obtained about lipid phase transitions and order but in addition to this can also provide new information concerning the secondary structures of membrane proteins. In many cases the molecules themselves, or portions of the molecules, act as intrinsic probes thereby eliminating the necessity for external probes to be added.

In this paper we would like to point out recent studies made in our laboratory using this infrared spectroscopic technique. We will also indicate some new inventions which we have developed where we mimic biomembrane structures, particularly the lipid portion, to produce new haemocompatible surfaces for biomaterial design.

INFRARED SPECTROSCOPY

Vibrational spectroscopy has several advantages for membrane studies. Firstly, variations in frequency, linewidth and intensity are sensitive to structural transitions of both lipid and protein components. Secondly, the vibrations of individual groups provide structural information on highly localised regions of the bilayer. Thus, the C-H stretching absorptions of the lipid acyl chains are readily distinguished from the carbonyl stretchings of the interfacial region and the phosphate stretchings of the polar headgroup. Of particular importance in the study of lipid-protein interactions is the non-perturbing nature of the technique. The addition of an external probe molecule is not required and the absorptions of the lipid and protein groupings reflect their genuine environments. Other techniques, such as ESR and fluorescence, are limited by the perturbations which the added reporter groups may induce. Finally, the time-scale of the molecular vibrations is of the order of 10^{13} s^{-1} which ideally complements the ESR and NMR timescales of 10^{8} s^{-1} and 10^{5} s^{-1} , respectively.

INSTRUMENTATION

The application of the traditional, dispersive IR instrument in the biological field expanded with the widespread use of minicomputers. Computer aquisition of IR data allows the operator to store spectra in undegraded form. Spectral subtractions, enhancements and expansions may then be carried out at a later date (Chapman et al., 1980). Following the advent of the fast Fourier transform algorithm, spectrometers based on the Michelson interferometer are now widely used (Fourier transform infrared). An FT-IR spectrometer is comprised of two parts: an optical bench containing an interferometer and a computer which controls all aspects of spectral scanning and analysis. The interferogram of a scan, or the sum of many such scans, is converted by means of a fast Fourier transform into the conventional form of transmittance (or absorbance) versus wavenumber spectrum. A short description of the mathematical treatments involved is that given by Griffiths (1980). The principal advantage of FT-IR compared with dispersive IR is that the former is able to obtain spectra of higher signal to noise ratio in a given scanning time. This arises through the multiplex or Fellgett's advantage, the detector examines all of the scanned spectrum almost simultaneously, and the Jacquinot advantage, the high optical throughput of the interferometer owing to the absence of slits. A further advantage of FT-IR is the much greater abscissa (wavenumber) accuracy which is achieved via laser referencing.

MODEL AND NATURAL BIOMEMBRANES

An early study of films of phosphatidylcholine by Chapman et al.(1967) provided assignments of the C-H stretching and bending modes, the carbonyl band and a variety of phosphate vibrations. These results showed that the spectra, particularly in the region below 1400 cm^{-1}, were remarkably dependent on temperature and the method of sample preparation. Considerable fine structure was observed in spectra recorded of samples after dehydration or at low temperatures. The band progression due to CH_2 wagging modes and the CH_2 rocking band near 720 cm^{-1} were found to be particularly sensitive. For anhydrous samples, the single 720 cm^{-1} band present at room temperature split into several components at -186°C. This transition coincides with a change in acyl chain packing from hexagonal to orthorhombic according to X-ray diffraction of the same samples.

The sharp main endothermic phase transition of aqueous phospholipid bilayers results in pronounced alterations in the methylene band parameters (Asher and Levin, 1977; Cameron and Mantsch, 1978; Chapman et al., 1980; Cortijo and Chapman, 1981). The band maximum frequencies of the CH_2 asymmetric and symmetric stretching bands (3000 cm^{-1} to 2800 cm^{-1}) are sensitive to the static order of the acyl chains. The introduction of an increased proportion of <u>gauche</u> conformers above the phase transition causes a shift in these bands to higher frequencies. The width of the IR absorptions is determined by rotational, translational and/or collisional effects (Casal et al., 1980). Thus, the CH_2 bandwidths are sensitive to the degree of motional freedom of the CH_2 groups. They are sensitive, therefore, to the phase transition but they also reflect changes which do not result in an alteration in the proportion of <u>gauche</u> conformers, such as the extent of librational or torsional motion. As part of their extensive studies of lipid phase behaviour by FT-IR, Mantsch et al. (1982, 1983) have also studied the C-H modes of aqueous dispersions of phosphatidylsulphocholines and phosphatidyl-ethanolamines.

THE INTERFACIAL AND HEADGROUP REGIONS

The structure of the interfacial region of lipid assemblies may be examined via the ester group vibrations. The most intense of these bands are the C=O stretching frequencies between 1750 cm^{-1} and 1700 cm^{-1} . Two absorption bands in this region are associated with the two ester groupings in diacyl lipids. The $\underline{sn}$-1 carbonyl gives rise to a band near 1740 cm^{-1} and a band near 1725 cm^{-1} is associated with the carbonyl at the $\underline{sn}$-2 position (Bush et al., 1980; Mushayakarara and Levin, 1980; Levin et al., 1982). This splitting arises, in part, because of the conformational inequivalence about the C1-C2 bonds of the $\underline{sn}$-1 and $\underline{sn}$-2 chains which adopt $\underline{trans}$ and $\underline{gauche}$ conformations, respectively, and through possible differences in the extent of hydration. Difference IR spectra of hydrated samples normally reveal only a single broad C=O band contour. The application of either spectral deconvolution (Casal and Mantsch, 1984) or second-derivative (Lee et al., 1985b) calculations usually reveals two or more components.

The phosphate moiety of the headgroup gives rise to several strong vibrations in the infrared region. Asymmetric and symmetric stretching modes for the PO_2^- group are found near 1250 cm^{-1} and 1085 cm^{-1}, respectively (Casal and Mantsch, 1984; Arrondo et al., 1984). Weaker single bond P-O stretching modes are found in the region 900-800 cm^{-1} (Casal and

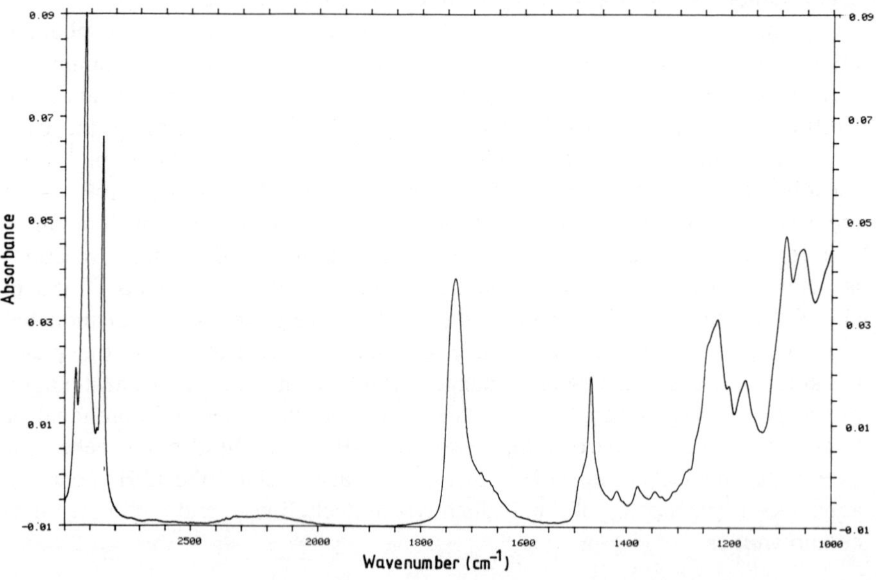

Figure 1. Fourier transform infrared spectrum at 4 cm^{-1} resolution of a suspension of dimyristoylphosphatidylcholine in water at 10°C after subtraction of the background water spectrum.

Mantsch, 1984). A shoulder near 1060 cm^{-1} on the PO$_2^-$ symmetric stretching band in the spectrum of DPPC (dipalmitoylphosphatidylcholine) is attributed to a R-O-P-O-R' stretching mode (Arrondo et al., 1984). The IR spectrum showing the absorption bands from a phospholipid dispersion in water is shown in Figure 1.

PHOSPHOLIPID-CHOLESTEROL INTERACTIONS

There have been several IR studies on the effect of the incorporation of cholesterol on the static order of the acyl chains of aqueous phospholipids (Asher and Levin, 1977; Umemura et al., 1980; Cortijo and Chapman, 1981; Cortijo et al., 1982). These studies have shown that cholesterol causes an increase in the proportion of <u>gauche</u> conformers below the phase transition temperature (T_m) and a decrease in these conformers above T_m compared with a pure lipid bilayer. This is manifested as an increase in both the CH$_2$ asymmetric and symmetric stretching frequencies below T_m and a decrease in these parameters above T_m (Cortijo and Chapman, 1981; Cortijo et al., 1982). The width of the phospholipid phase transition, as detected by IR, is increased by the presence of cholesterol in the bilayer but the midpoint of the transition is unaffected (Asher and Levin, 1977). At very high cholesterol concentrations (e.g. 1.5:1 phospholipid:cholesterol molar ratio) almost no change in the relative proportions of <u>trans</u> and <u>gauche</u> conformers occurs with temperature (Cortijo et al., 1982). These observations are in accord with the results obtained by calorimetric, NMR and ESR techniques.

There is little IR evidence concerning the perturbation of the interfacial and headgroup regions which may be exerted by cholesterol. A combined IR and Raman investigation of the C=O stretching modes of DPPC (Bush et al., 1980) demonstrated that no hydrogen bonding occurs between the fatty acyl carbonyl and the 3β-OH group of cholesterol in anhydrous samples. In hydrated samples, cholesterol reduces the conformational inequivalence between the <u>sn</u>-1 and <u>sn</u>-2 carbonyls by perturbing the latter. This was revealed as a decrease in the relative intensity in the 1720 cm^{-1} region. A study of the PO$_2^-$ stretching modes and the asymmetric N$^+$(CH$_3$)$_3$ stretching mode of fully hydrated DPPC showed no variation in frequency on the introduction of cholesterol into the bilayers (Umemura et al., 1980). This suggests that there is little interaction between cholesterol and the phosphorylcholine head-group.

CEREBROSIDES

Fourier transform infrared spectroscopy has been used to study the stable and metastable forms of a range of cerebrosides in aqueous systems. The spectra provide evidence for different degrees of inter- and intramolecular hydrogen bonding, involving principally the amide group, in these different states. A comparison has been made with the spectra of a cerebroside containing an α-hydroxyl group in the fatty acyl chain. This cerebroside does not show metastability and its hydrogen bonding characteristics are shown to be different (Lee et al., 1986). Fourier transform infrared spectra of glucocerebrosides from Gaucher's spleen and type II galactocerebroside from bovine brain in their metastable and stable forms are presented in Figure 2. The spectral region shown reveals amide I (1670-1600 cm^{-1}) and amide II (1575-1510 cm^{-1}) absorptions associated with the amide linkage between the fatty acyl chain and the sphingosine base. A reduction in both the frequency of the amide I shoulder and the bandwidth of both components indicates increased hydrogen bonding involving the amide group in a more rigid stable state. An increase in amide II frequency also indicates greater hydrogen bonding in the stable state.

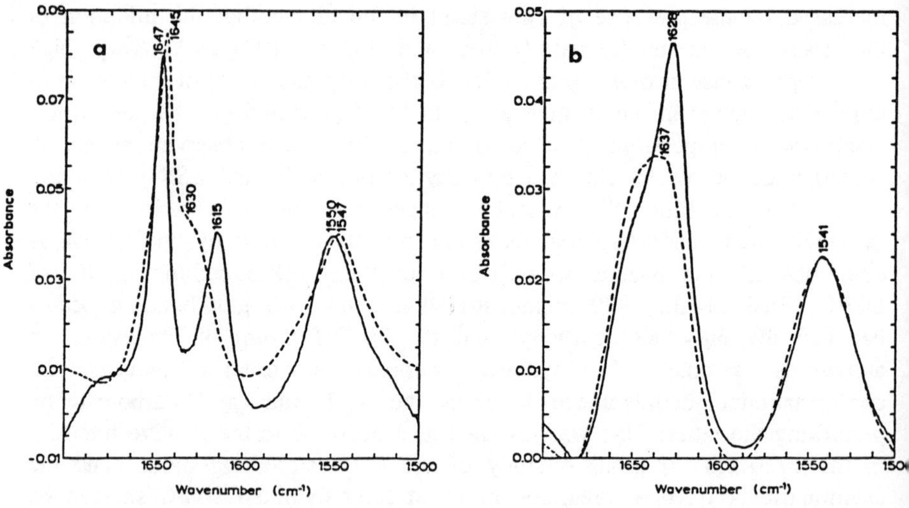

Figure 2. Fourier transform infrared spectra after subtraction of the background water spectrum of (a) glucocerebrosides from Gaucher's spleen and (b) type II galactocerebrosides from bovine brain at 20°C. Spectra were recorded after quenching in ice water from 90-95°C (broken lines) to produce the metastable states, and after reheating and incubation at 70°C for 5 min (solid lines) to produce the stable states.

LIPID-PROTEIN INTERACTIONS

Perturbations to the hydrophobic acyl chain region may be studied via the C-H asymmetric and symmetric stretching frequencies. These parameters are sensitive to the static order (trans/gauche isomerisation) of the acyl chains. Cortijo et al. (1982) have reported the effects of incorporation of the intrinsic proteins Ca^{2+}-ATPase and bacteriorhodopsin, and the intrinsic polypeptide gramicidin A on the acyl chain order of (dimyristoyl-phosphatidylcholine) DMPC and DPPC bilayers. They found that below T_m, these molecules behaved in a similar manner to cholesterol; that is they caused an increase in the proportion of gauche isomers. The presence of the intrinsic molecules within the bilayer prevents the acyl chains from attaining the all-trans conformation at low temperatures. Above T_m, the perturbation of the acyl chains differs from that observed with cholesterol. At high lipid:protein molar ratios (e.g. DMPC/Ca^{2+}-ATPase of 150:1) a reduction in the proportion of gauche conformers with respect to the pure lipid bilayer was observed. However, when the concentration of the intrinsic protein or polypeptide was increased this effect was removed and the static order of the acyl chains was essentially the same as in the pure lipid system. More recently, we have investigated lipid-protein interactions using diper-deuteriomyristoylphosphatidylcholine, an analogue of DMPC in which the acyl chain hydrogen atoms are completely substituted by deuterium atoms (Lee et al., 1984). It was shown that above the lipid phase transition temperature, low concentrations of gramicidin A and alamethicin caused a small ordering of the lipid chains while bacteriorhodopsin had no effect. At high concentrations each intrinsic molecule caused a disordering of the lipid chains above T_m.

MEMBRANE PROTEINS
POLYPEPTIDES AND CONFORMATIONAL ASSIGNMENTS

Information on the conformation of polypeptides and proteins may be derived from their IR spectra (Krimm, 1962; Susi, 1969; Fraser and MacRae, 1973; Fawcett and Long, 1973; Thomas and Kyogoku, 1977). The assignment of the characteristic absorptions (the so-called 'amide' bands) has been assisted by the analysis of model compounds such as N-methyl-acetamide and is summarised in these reviews. For many years there was a great deal of controversy concerning the molecular description of these vibrations and this is summarised by Susi (1969). It is now generally accepted that the amide I to amide VII vibrations cannot be described adequately by any single displacement coordinate. It should be noted that these amide bands are not the only absorptions observed. The IR spectra of polypeptides and proteins also include contributions from the amino acid side-chains.

The amide I and II bands are sensitive to the secondary structure of polypeptides and proteins and are the most frequently used modes in

TABLE I

Characteristic amide I and II frequencies for various polypeptide conformations

Conformation	Amide I (cm^{-1})		Amide II (cm^{-1})
	H$_2$O	^{2}H$_2$O	
α-helix	1652 (s)	1650 (s)	1546 (s)
	1646 (w)	1644 (w)	1516 (w)
random coil	1656	1643	1520
anti-parallel	1632 (s)	1632 (s)	1530 (s)
chain β-sheet	1690 (w)	1675 (w)	1510 (w)
parallel chain	1632 (s)	1630 (s)	1530 (s)
β-sheet	1648 (w)	1645 (w)	1550 (w)

s: strong. w: weak. Suspension of soluble proteins in ^{2}H$_2$O shifts the amide II band (principally N-H bending) to frequencies near 1450 cm^{-1} (amide II').

conformational analysis. The fundamental theory has been described by Miyazawa (1960). Some early applications of this approach are those of Miyazawa and Blout (1961), Krimm (1962), Susi et al. (1967), Timasheff and Susi (1966) and Timasheff et al. (1967). These and other studies, both experimental and theoretical (Jakes and Krimm, 1971; Abe and Krimm, 1972; Bandekar and Krimm, 1980; Krimm and Bandekar, 1980; Dwivedi and Krimm, 1984; Chirgadze and Nevskaya, 1976 a,b; Nevskaya and Chirgadze, 1976; Chirgadze et al., 1973; Chirgadze and Brazhnikov, 1974; Kawai and Fasman, 1978) have enabled confident correlations between amide I and II band frequencies and secondary conformation. These assignments are presented in Table I.

The structure of the polypeptide antibiotic gramicidin A has been investigated by IR spectroscopy in solution and after insertion into phospholipid bilayers (Cortijo et al., 1982; Sychev and Ivanov, 1982; Buchet et al., 1985). Gramicidin A is a linear polypeptide antibiotic that renders biological membranes and lipid bilayers passively permeable to small monovalent cations by forming a channel structure (Chappell and Crofts, 1965; Hladky and Haydon, 1970 and 1972). It consists of fifteen alternating D- and L-amino acids (Sarges and Witkop, 1965). The structure of the polypeptide in lipid bilayers is a matter of some dispute and is dependent on temperature, lipid:protein ratio and lipid class (Sychev and Ivanov, 1982). Although an N-terminal to N-terminal linked dimeric structure has been proposed on the basis of NMR evidence (Weinstein et al., 1979), IR and CD data are consistent with the presence of an anti-parallel β-sheet structure in DMPC liposomes at lipid:protein ratios between 350:1 and 35:1 (Sychev and Ivanov, 1982). In DPPC bilayers an anti-parallel double-helical structure (Veatch et al., 1974) is found at high lipid:protein ratios (Sychev and Ivanov, 1982).

STUDIES ON BACTERIORHODOPSIN AND Ca^{2+}-ATPase

The structure of bacteriorhodopsin in the purple membrane of <u>Halobacterium</u> <u>halobium</u> has been studied by IR spectroscopy (Rothschild and Clark, 1979b; Cortijo et al., 1982; Krimm and Dwivedi, 1982; Lee et al., 1985b). These studies have confirmed the existence of a predominantly α-helical structure for this protein. The observed amide I frequency of 1660-1662 cm^{-1} is unusually high for α-helices and has been attributed to 'distorted' helices (Rothschild and Clark, 1979 b) or α_{II}-helices (Krimm and Dwivedi, 1982; Dwivedi and Krimm, 1984). In the α_{II}-helix the N-H bonds are tilted inward toward the helix axis whereas in the α_{I}-structure they are essentially parallel to the axis. Polarised IR spectroscopy has been applied to determine the angle of tilt of the α-helices in dry, oriented specimens (Rothschild and Clark, 1979 a). In agreement with the electron microscope data, the dichroism of the amide A, I and II bands indicated an average spatial orientation for the helices of less than 26° from the membrane normal.

Difference, second-derivative and fourth-derivative spectra of an aqueous suspension of purple membrane at $20^{\circ}C$ after subtraction of the absorption of the water have been presented elsewhere (Lee and Chapman, 1986). The amide I maximum is found at 1660 cm^{-1} and may be attributed to distorted or α_{II}-helices. A predominantly α-helical structure is also indicated by the amide II maximum at 1545 cm^{-1}, in agreement with the electron diffraction data of Henderson and Unwin (1975). However, an amide I band shoulder at 1630-1640 cm^{-1} suggests the presence of some β-structure. This absorption is rendered more clear by the second- and fourth-derivative treatments. The appearance of a band at 1684 cm^{-1} suggests that a proporton of this β-structure is anti-parallel pleated sheet. Bands at 1618 cm^{-1} and 1517 cm^{-1} may be tentatively assigned to tyrosine side-chain absorptions (Chirgadze et al., 1975). Bands in the region 1469-1439 cm^{-1} are principally due to C-H deformation modes from the lipid acyl chains.

The structure of the sarcoplasmic reticulum (SR) of rabbit striated muscle has also been investigated by IR spectroscopy (Cortijo et al., 1982; Mendelsohn et al., 1984; Lee et al., 1985 b; Arrondo et al., 1985). Examination of the amide I and II maxima indicated that the major protein present in this membrane, the Ca^{2+}-ATPase, contains unordered as well as α-helical conformation (Cortijo et al., 1982; Lee et al., 1985 b). Fourier deconvolution (Mendelsohn et al., 1984) and second-derivative analysis (Lee et al., 1985 b) of the amide I region revealed a minor amount of β-sheet

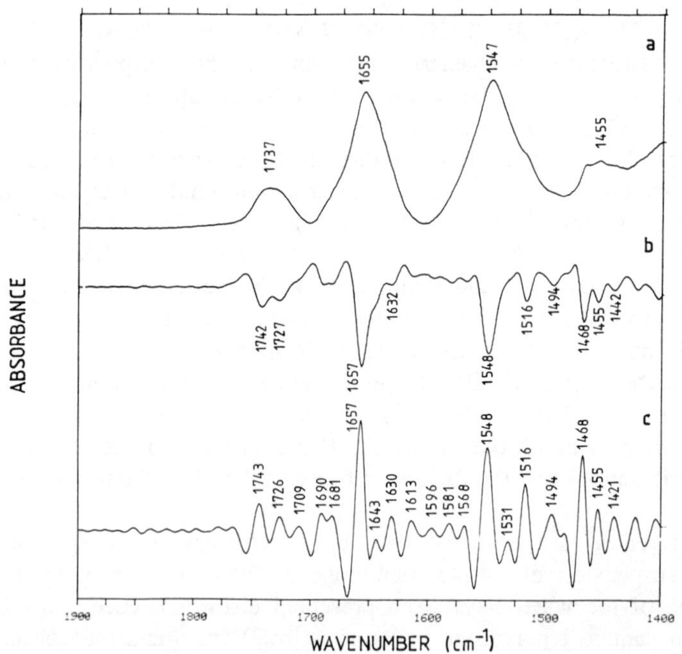

Figure 3. Fourier transform infrared spectra of sarcoplasmic reticulum in 10mM HEPES, 50mM sucrose, 1M KCl, pH 7.4 at 20°C. (a) Difference spectrum generated by the subtraction of the spectrum of the buffer, (b) second-derivative of (a), (c) fourth-derivative of (a).

structure. Temperature-dependent changes in the amide II region revealed by Fourier self-deconvolution may have been due to a reduction in the proportion of α-helical structure as the temperature was increased although the influence of increased ^{1}H-^{2}H exchange was not eliminated (Mendelsohn et al., 1984).

Second-derivative (Lee et al., 1985b) and fourth-derivative (Lee and Chapman, in preparation) analysis of the amide I and II bands has revealed a remarkable stability to temperature changes over the range 8° to 50°C. Figure 3 shows difference (a), second-derivative (b) and fourth-derivative (c) FT-IR spectra of sarcoplasmic reticulum membranes at 20°C after subtraction of the absorption of the aqueous buffer. Three main absorption bands are seen in the difference spectrum: the amide I and amide II bands from the protein at 1655 cm^{-1} and 1547 cm^{-1}, respectively, and a C=O stretching band from the lipid at 1737 cm^{-1}. The frequency of the amide I maximum of 1655 cm^{-1} may be assigned to the presence of a predominantly α-helical structure for Ca^{2+}-ATPase. However analysis of the second-derivative reveals the presence of β-sheet structure, with absorptions at 1632 cm^{-1} and 1680-1690 cm^{-1}, and which has been predicted from the primary sequence (Allen et al., 1980; MacLennan et al., 1985). Further analysis using the fourth-derivative reveals components in the amide I region which may be assigned to α-helical structure (1657 cm^{-1}), β-sheet structure (1681 cm^{-1} and 1630 cm^{-1}) and β-turns (1690 cm^{-1}) (Susi and Byler, 1983; Lee et al., 1985b). The band at 1531 cm^{-1} in the amide II region may also be due to β-structure.

We have also obtained the first infrared spectra of the purified glucose transporter from the human erythrocyte membrane (Lee et al., 1985a). Second-derivative analysis of the spectra obtained from preparation in H$_2$O and ^{2}H$_2$O revealed a predominantly α-helical protein with a minor proportion of β-sheet. Further studies are in progress to characterise conformational changes in this protein and the Ca^{2+}-ATPase which occur as part of their transport cycles.

The future of infrared spectroscopy as a tool for the investigation of lipids and biomembranes seems assured. Its nature as a non-destructive and non-perturbing technique combined with the fact that it can be applied to all phases of virtually any substance means that the field is open for the study of many new systems. The greater sensitivity and spectral resolution afforded by modern instrumentation have now made it possible to begin looking at the membranes of living cells. Casal and co-workers (1979, 1980) have studied the intact and deproteinated plasma membranes of <u>Acholeplasma</u> <u>laidlawii</u> and recently studies of the gel to liquid crystalline phase transition in live <u>A.</u> <u>laidlawii</u> cells have been reported (Mendelsohn and Mantsch, 1985). It seems likely that the study of living systems will continue to develop as will the use of novel methods of data manipulation, such as second-derivative infrared analysis. Quantitative estimates of secondary structure should soon be available for various membrane protein systems which further our understanding of membrane structure and function.

BIOMIMETIC SURFACES

Asymmetries in the distribution of phospholipid head groups have been found in a variety of cells (Zwaal and Hemker, 1982). The accumulated evidence suggests strongly that in the case of blood cells, the observed lipid asymmetry serves a biological purpose in the maintenance of the delicate balance between haemostasis and thrombosis. The extracellular surfaces of the plasma membranes of blood cells are thromboresistant; in strong contrast, their cytoplasmic surfaces are highly thrombogenic. The simplest common feature of the blood-compatible surfaces is the presence of large quantities (up to approximately 90% of the total surface lipids) of phosphorylcholine-containing phospholipids.

We have previously investigated the haemostatic potential of polymeric phosphorylcholine surfaces in dispersion (Hayward and Chapman, 1984). Polymerised liposomes of diacetylenic phosphatidylcholines did not alter the recalcification clotting times of citrated, pooled normal plasma regardless of lipid concentration. This non-thrombogenic nature of polymeric phosphatidylcholine was not altered after incubation in human plasma <u>in vitro</u> for up to 1 week at 37°C.

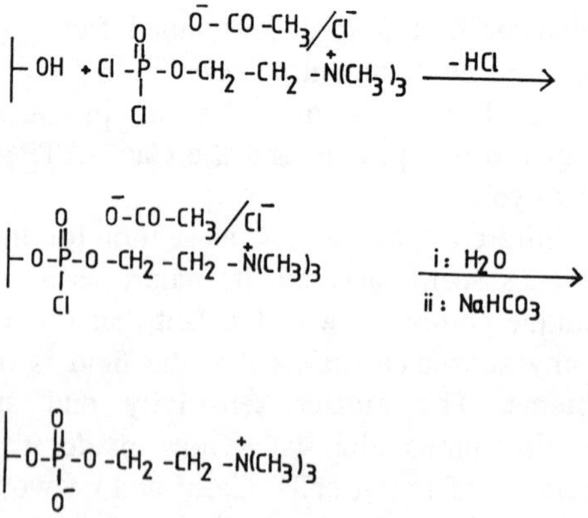

$$\text{—OH} + \text{Cl} - \overset{\overset{\displaystyle O}{\|}}{\underset{\underset{\displaystyle Cl}{|}}{P}} - \text{O-CH}_2 - \text{CH}_2 - \overset{+}{N}(\text{CH}_3)_3 \quad \overset{\bar{O} - CO - CH_3 / Cl^-}{\xrightarrow{\quad -HCl \quad}}$$

$$\text{—O} - \overset{\overset{\displaystyle O}{\|}}{\underset{\underset{\displaystyle Cl}{|}}{P}} - \text{O-CH}_2 - \text{CH}_2 - \overset{+}{N}(\text{CH}_3)_3 \quad \overset{\bar{O} - CO - CH_3 / Cl^-}{\underset{\text{ii : NaHCO}_3}{\xrightarrow{\text{i : H}_2\text{O}}}}$$

$$\text{—O} - \overset{\overset{\displaystyle O}{\|}}{\underset{\underset{\displaystyle O^-}{|}}{P}} - \text{O-CH}_2 - \text{CH}_2 - \overset{+}{N}(\text{CH}_3)_3$$

Figure 4. The reaction of choline acetate dichlorophosphate with hydroxylated surfaces.

We have now developed (Durrani et al., 1986) a panel of reactive species which should result in the covalent deposition of phosphorylcholine on a variety of solid substrates (see Figure 4). The resultant monomolecular layer provides a self-assembling surface which, in theory, should resemble the lipid interface presented by blood cells. The hydrophilic, membrane-mimetic character of this surface provides opportunities for the generation of new, hybrid biomaterials. The modified surfaces retain the chemical, physical and topological properties of the substrates, while simultaneously exhibiting characteristics of biomembranes (extreme thinness, hydrophilicity, low antigenic potential and increased potential for haemo- and biocompatibility).

Evidence for the covalent deposition of a self-assembling phosphorylcholine layer has been provided for the hydroxylated surfaces of glass and silica (Hayward et al., 1986). Structural integrity of the deposited group is supported by the equimolar association of phosphorus and choline with the reacted surfaces. Modified glass surfaces yield contact angles which are consistent with those found previously for other models of biological membranes. Covalent modification of the treated surfaces is demonstrated by IR spectroscopy. The modified surfaces are thermostable at temperatures up to 375°C for extended periods. These biomimetic surfaces which mimic the exterior of the lipid bilayer of red blood cells may be useful to provide new haemocompatible materials and also for interesting tissue culture applications.

ACKNOWLEDGEMENTS

We wish to acknowledge financial support from the Wellcome Trust and the Science and Engineering Research Council.

REFERENCES

Abe, Y. and Krimm, S. (1972) Biopolymers 11, 1817-1839

Allen, G., Trinnaman, B.J. and Green, N.M. (1980) Biochem. J. 187, 591-616

Arrondo, J.L.R., Goni, F.M. and Macarulla, J.M. (1984) Biochim. Biophys. Acta 794, 165-168

Arrondo, J.L.R., Urbaneja, M.A., Goni, F.M., Macarulla, J.M. and Sarzala, G. (1985) Biochem. Biophys. Res. Commun. 128, 1159-1163.

Asher, J.M. and Levin, I.W. (1977) Biochim. Biophys. Acta 468, 63-72

Bandekar, J. and Krimm, S. (1980) Biopolymers 19, 31-36

Buchet, R., Sandorfy, C., Trapane, T.L. and Urry, D.W. (1985) Biochim. Biophys. Acta 821, 8-16

Bush, S.F., Levin, H. and Levin, I.W. (1980) Chem. Phys. Lipids 27, 101-111

Cameron, D.G. and Mantsch, H.H. (1978) Biochem. Biophys. Res. Comm. 83, 886-892

Casal, H.L., Cameron, D.G., Smith, I.C.P. and Mantsch, H.H. (1980) Biochemistry 19, 444-451

Casal, H.L. and Mantsch, H.H. (1984) Biochim. Biophys. Acta 779, 381-401

Casal, H.L., Smith, I.C.P., Cameron, D.G. and Mantsch, H.H. (1979) Biochim. Biophys. Acta 550, 145-149

Chapman, D., Gomez-Fernandez, J.C., Goni, F.M. and Barnard, M. (1980) J. Biochem. Biophys. Methods 2, 315-323

Chapman, D., Williams, R.M. and Ladbrooke, B.D. (1967) Chem. Phys. Lipids 1, 445-475

Chappell, J.B. and Crofts, A.R. (1965) Biochem. J. 95, 393-402

Chirgadze, Yu.N. and Brazhnikov, E.V. (1974) Biopolymers 13, 1701-1712

Chirgadze, Yu.N., Fedorov, O.V. and Trushina, N.P. (1975) Biopolymers 14, 679-694

Chirgadze, Yu.N. and Nevskaya, N.A. (1976 a) Biopolymers 15, 607-625

Chirgadze, Yu.N. and Nevskaya, N.A. (1976 b) Biopolymers 15, 627-636

Chirgadze, Yu. N., Shestopalov, B.V. and Venyaminov, S.Yu. (1973) Biopolymers 12, 1337-1351

Cortijo, M., Alonso, A., Gomez-Fernandez, J.C. and Chapman, D. (1982) J. Mol. Biol. 157, 597-618

Cortijo, M. and Chapman, D. (1981) FEBS Lett. 131, 245-248

Durrani, A.A., Hayward, J.A. and Chapman, D. (1986) Biomaterials 7, 121-125

Dwivedi, A.M. and Krimm, S. (1984) Biopolymers 23, 923-943

Fawcett, V. and Long, D.A. (1973) in Molecular Spectroscopy 1, 352-445, Chem. Soc. London

Fraser, R.D.B. and MacRae, T.P. (1973) Conformations in Fibrous Proteins and Related Synthetic Polypeptides Chapter 5, Academic Press, New York

Griffiths, P.R. (1980) in Analytical Applications of FT-IR to Molecular and Biological Systems (Durig, J.R., ed.) pp. 11-24, Reidel, Holland

Hayward, J.A. and Chapman, D. (1984) Biomaterials 5, 135

Hayward, J.A. and Chapman, D. (1985) In: Biocompatibility of Tissue Analogs, Vol. II (Williams, D.F., ed.) CRC Press, Boca Raton, Florida pp. 119-132

Hayward, J.A., Durrani, A.A., Shelton, C.J., Lee, D.C. and Chapman, D. (1986) Biomaterials 7, 126-131

Henderson, R. and Unwin, P.N.T. (1975) Nature 257, 28-32

Hladky, S.B.and Haydon, D.A. (1970) Nature 225, 451-453

Hladky, S.B.and Haydon, D.A. (1972) Biochim. Biophys. Acta 274, 294-312

Jakes, J. and Krimm, S. (1971) Spectrochim. Acta 27A, 19-34

Kawai, M. and Fasman, G. (1978) J. Am. Chem. Soc. 100, 3630-3632

Krimm, S. (1962) J. Mol. Biol. 4, 528-540

Krimm, S. and Bandekar, J. (1980) Biopolymers 19, 1-29

Krimm, S. and Dwivedi, A.M. (1982) Science 216, 407-408

Lee, D.C. and Chapman, D. (1986) Bioscience Rep. 6, 235-256

Lee, D.C., Durrani, A.A. and Chapman, D. (1984) Biochim. Biophys. Acta 769, 49- 56.

Lee, D.C., Elliot, D.A., Baldwin, S.A. and Chapman, D. (1985 a) Biochem. Soc. Trans. 13, 684-685

Lee, D.C., Hayward, J.A., Restall, C.J. and Chapman, D. (1985 b) Biochemistry 24, 4364-4373

Lee, D.C., Miller, I.R. and Chapman, D. (1986) Biochim. Biophys. Acta (in press)

Levin, I.W., Mushayakarara, E. and Bittman, R. (1982) J. Raman Spectr. 13, 231- 234

MacLennan, D.H., Brandl, C.J., Korczak, B. and Green, N.M. (1985) Nature 316, 696-700

Mantsch, H.H., Cameron, D.G., Tremblay, P.A. and Kates, M. (1982) Biochim. Biophys. Acta 689, 63-72

Mantsch, H.H., Hsi, S.C., Butler, K.W. and Cameron, D.G. (1983) Biochim. Biophys. Acta 728, 325-330

Mendelsohn, R., Anderle, G., Jaworsky, M., Mantsch, H.H. and Dluhy, R.A. (1984) Biochim. Biophys. Acta 775, 215-224

Mendelsohn, R. and Mantsch, H.H. (1985) In: Protein-Lipid Interactions, Vol.2 (Watts, J., ed.) Oxford University Press, London

Miyazawa, T. (1960) J. Chem. Phys. 32, 1647-1652

Miyazawa, T. and Blout, E.R. (1961) J. Am. Chem. Soc. 83, 712-719

Mushayakarara, E. and Levin, I.W. (1980) J. Phys. Chem. 86, 2324-2327

Nevskaya, N.A. and Chirgardze, Yu.N. (1976) Biopolymers 15, 637-648

Restall, C.J. and Chapman, D. (1986) In: Lipids and Membranes: Past, Present and Future (J.A.F. Op den Kamp, B. Roelofsen and K.W.A. Wirtz, eds.) Elsevier, Amsterdam, pp. 61-92

Rothschild, K.J. and Clark, N.A. (1979a) Biophys. J. 25, 473-488

Rothschild, K.J. and Clark, N.A. (1979b) Science 204, 311-312

Sarges, R. and Witkop, B. (1965) J. Am. Chem. Soc. 87, 2011-2020

Susi, H. (1969) in Structure and Stability of Biological Macromolecules (Timasheff, S.N. and Fasman, G.D., eds.) pp.575-663, Decker, New York

Susi, H. and Byler, D.M. (1983) Biochem. Biophys. Res. Comm. 115, 391-397

Susi, H., Timasheff, S.N. and Stevens, L. (1967) J. Biol. Chem. 242, 5460-5466

Sychev, S.V. and Ivanov, V.T. (1982) in Membranes and Transport Vol. 2 (Martonosi, A.N., ed.) pp. 310-307, Plenum, New York

Thomas, G.J. and Kyogoku, Y. (1977) in Infrared and Raman Spectroscopy, Part C, Practical Spectroscopy Vol. 1 (Braine, E.G. and Grasselli, J.G., eds.) Marcel Dekker Inc.,N.Y.

Timasheff, S.N. and Susi, H. (1966) J. Biol. Chem. 241, 249-251

Timasheff, S.N., Susi, H. and Stevens, L. (1967) J. Biol. Chem. 242, 5467-5473

Umemura, J., Cameron, D.G. and Mantsch, H.H. (1980) Biochim. Biophys. Acta 602, 32-44

Veatch, W.R., Fossel, E.T. and Blout, E.R. (1974) Biochemistry 13, 5249-5256

Weinstein, S., Wallace, B.A., Blout, E.R., Morrow, J.S. and Veatch, W. (1979) Proc. Natl. Acad. Sci. U.S.A. 76, 4230-4234

Zwaal, R.F.A. and Hemker, H.C. (1982) Haemostasis 11, 12

^{31}P-NMR SPECTROSCOPY OF POLYPHOSPHOINOSITIDES IN MICELLES AND VESICLES

Peter A. van Paridon, Ben de Kruijff, Karel W. A. Wirtz

Department of Biochemistry and Institute of Molecular Biology, State University of Utrecht, Transitorium III Padualaan 8, de Uithof, Utrecht, the Netherlands

INTRODUCTION

Phosphatidylinositol 4,5-bisphosphate (PIP_2) is functionally a very important phospholipid in the plasma membranes of cells. It is generally accepted that its metabolism is linked to transmembrane signal transduction. Stimulus-induced activation of PIP_2-phosphodiesterase yields inositol trisphosphate and diglyceride which function as secondary messengers in the cell (Berridge, 1984; Nishizuka,1984). On the other hand it has been reported that the shape of the erythrocyte depends on the levels of PIP_2 in the membrane (Michell,1975; Allan & Thomas,1981; Ferrel & Huestis, 1984). Recently it was proposed that this dependency originates from a specific interaction between glycophorin and PIP_2 which is required for an effective interaction with the membrane skeletal protein 4.1 (Anderson & Marchesi,1985).

To date few studies have appeared on the physical properties of PIP_2 and its immediate precursor phosphatidylinositol 4-phosphate (PIP) (Sugiura,1981; Shibata et al., 1984; Hayashi et al.,1984). It is generally assumed that the very acidic nature of PIP and PIP_2 is an important factor in dictating the behaviour and function of these phospholipids in the plasma membrane. From measuring calcium binding and electrophoretic properties it was inferred that at pH 5.5, PIP_2 in egg-yolk phosphatidylcholine vesicles contains five negative charges (Hauser & Dawson,1967). This would then imply that the protonation of the phosphomonoesters in PIP_2 is remarkably different from that of phosphatidic acid which has only one negative charge at pH 5.5 (Abramson et al., 1964; Koter et al.,1978). Since the charge state of PIP and PIP_2 at physiological pH may have important implications for their physiological function, we have investigated the ionization as a function of pH by use of ^{31}P-nuclear magnetic resonance (NMR) spectroscopy.

^{31}P-NMR SPECTRA OF PIP AND PIP$_2$

Spectra of PIP and PIP$_2$ isolated from bovine brain (Hendrickson & Ballou,1964; Kiselev,1982) in pure micellar form are presented in Fig. 1. The spectrum for PIP$_2$ consists of three well-separated resonance peaks. From a comparison with other phospholipid spectra we conclude that the high-field peak represents the phosphodiester moiety , and the two low-field peaks the 4- and 5-phosphate residues. In agreement with this interpretation the spectrum for pure PIP gives two resonances, with the low-field peak representing the 4-phosphate moiety. These two spectra do not unambiguously assign the resonance peaks for the 4- and 5-phosphate groups in PIP$_2$. However, from ^{1}H- and ^{31}P-NMR spectra of myo-inositol(1,4,5)trisphosphate one was able to conclude that the 4-phosphate moiety was represented by the resonance signal with the highest chemical shift (Lindon et al., 1986). This is in agreement with our tentative identification. The NMR spectrum for PIP$_2$ (2.5 mole %) in mixed phosphatidylcholine (PC) vesicles (Fig. 2) again shows the two well-separated phosphomonoester resonance peaks, while the phosphodiester resonance is obscured by the PC phosphodiester signal. Under comparable conditions PIP has one low-field resonance peak representing the 4-phosphate residue. The spectra of Fig. 1 were recorded at pH 8.0, and Fig. 2 at pH 5.9. The positions of the phosphomonoester resonances reflect the differences in ionization at these pH values. At identical pH these positions were found to be similar for vesicles and micelles.

DETERMINATION OF pK VALUES FROM CHEMICAL SHIFTS

As seen from Figs. 1 and 2 the chemical shift positions of the 4- and 5-phosphate resonances vary with the pH of the medium. This is in agreement with earlier observations that the ionisation of the phosphate groups of lipids affects the chemical shift (Koter et al.,1978). We have used this property to determine the pK values of the 4- and 5-phosphate moieties in PIP and PIP$_2$. A typical titration curve for PIP$_2$ in micellar form is shown in Fig. 3. The resonance position of the phosphodiester moiety is not shifted over the pH range between 4 and 13. This is in agreement with the pK values of 2.5 and 3.1 measured for the phosphodiester residues of phosphatidylinositol and phosphatidylglycerol, respectively (Abramson et al.,1968; van Dijck et al.,1978). As for the phosphomonoester resonance signals, a clear pH dependency of the chemical shift was observed, yielding

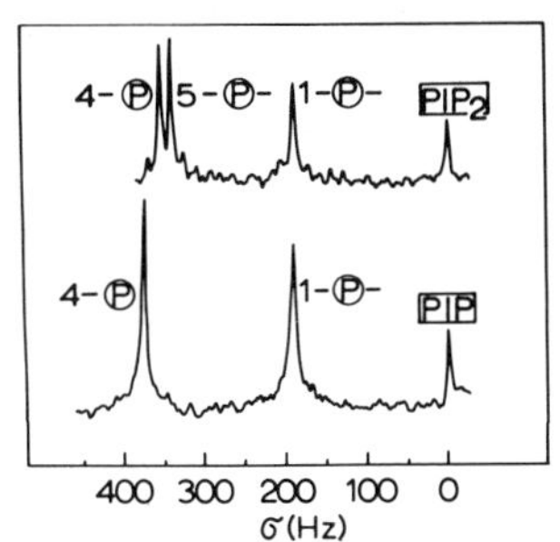

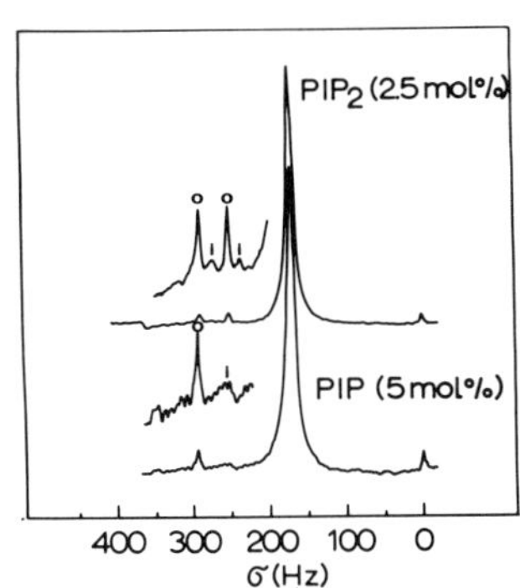

Figure 1. ^{31}P-NMR spectra of micellar PIP and PIP$_2$ in H$_2$O/^{2}H$_2$O (1:1) containing 20 mM Tris-maleate, 5 mM EDTA and 100 mM NaCl, p^{2}H 8.0. Spectra were recorded on a Bruker WH 90 spectrometer operating at 36.4 MHz on ^{31}P-NMR as described before (de Kruijff et al.,1976). Triphenylphosphine was used as external standard. The phospholipids were converted into the sodium form. For all experimental details see van Paridon et al.,1986.

Figure 2. ^{31}P-NMR spectra of PIP and PIP$_2$ in small unilamellar egg-yolk phosphatidylcholine vesicles at p^{2}H 5.9. For experimental details see Fig.1.

a single inflection point at pH 6.7 for the 4-phosphomonoester moiety, and at pH 7.6 for the 5-phosphomonoester moiety. This shows that, in the physiological pH range, the 5-phosphate is more easily protonated than the 4-phosphate residue. For comparison, the first protonation of the phosphomonoester moiety in phosphatidic acid has a pK value of 8.0, and the second protonation a pK of 3.0 (Abramson et al.,1964). The second protonation of the phosphomonoesters in PIP$_2$ could not be measured by ^{31}P-NMR, as at a pH below 3 PIP$_2$ tended to aggregate as a result of complete charge neutralization. Similar titration curves were determined for PIP$_2$ in mixed vesicle form, yielding the pK values as presented in Table 1. It is evident that the pK values both for the 4- and 5-phosphate residues in PIP$_2$ are not affected by whether the measurements are performed on pure PIP$_2$ or on PIP$_2$ present in mixed vesicles. Titration of PIP in pure or mixed-vesicle form yielded pK values of 6.3 and 6.1, respectively. It is apparent from Table 1 that the 4-phosphate on PIP is slightly more acidic

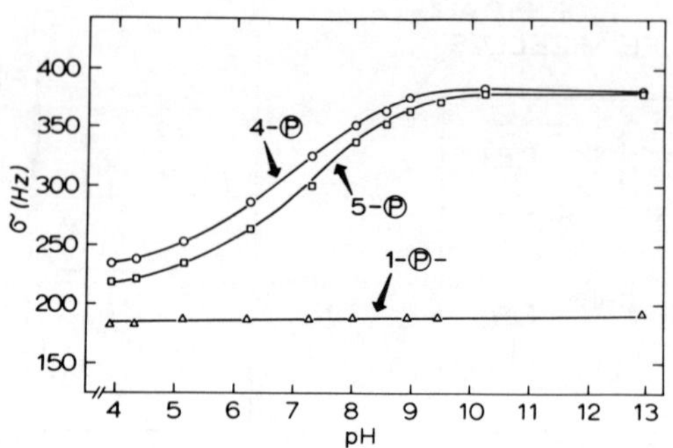

Figure 3. Chemical shift as a function of pH for pure PIP_2 micelles. The chemical shift position of the phosphate resonances relative to the internal standard were determined as a function of the uncorrected p^2H as described in the legend under Fig. 1. The pK values for the protonation of the phosphate residues were derived from the inflection point in the titration curve. For further details see van Paridon et al.,1986.

Table 1. pK values for the first protonation of the phosphomonoester residues in PIP and PIP_2

Phospholipid	pK(4-P)	pK(5-P)
100% PIP_2	6.7	7.6
2.5% PIP_2 in PC vesicles	6.5	7.7
1.0% PIP_2 in PC vesicles	6.7	7.7
100% PIP	6.3	
5.0% PIP in PC vesicles	6.1	

^{31}P-NMR spectra were recorded at various pH of PIP and PIP_2, either pure, or in mixed vesicles with egg-PC.

From the titration curves as shown in Fig.3 the pK values were calculated as described in the legend under Fig.3. The pK values mentioned in Table 1 were corrected for the presence of 50% ^2H_2O (+ 0.2 pH unit ; Wüthrich, 1976) than the 4-phosphate residue on PIP_2. In view of these p$\underline{K}$ values it is clear that both PIP and PIP_2 can undergo charge neutralization in the physiological pH range, resulting in a potential charge of -3 to -5 for PIP_2 and of -2 to -3 for PIP. The p$\underline{K}$ values presented in Table 1 are apparent, as the pH at the interface is different from that of the bulk. The p$\underline{K}$ values of the 4- and 5-phosphate residues in $\underline{myo}$-inositol trisphosphate still remain to be determined. However one would expect a substantial difference between the p$\underline{K}$ values in PIP_2 and $\underline{myo}$-inositol trisphosphate since the phosphates of the latter compound experience the pH of the medium. At the pH of the cytosol of resting cells (i.e. pH 7.1-7.3) PIP_2 has approximately four and PIP three negative charges. At present it is hard to say whether fluctuations in the intra-cellular pH may be sufficient to affect the charge state of PIP and PIP_2. An intra-cellular pH rise has been observed upon activation of the Na^+/H^+ exchanger (Moolenaar et al.,1984). Since this proton pump is present in the plasma membrane one cannot preclude the possibility that at the immediate environment of these pumps the local pH change may be sufficient for changes in the charge state of PIP and PIP_2 to occur, thereby affecting crucial cellular processes.

REFERENCES

Abramson, M.B., Katzman, R., Wilson, C.E. and Gregor, H.P. (1964) J. Biol. Chem. 239: 4066-4072

Abramson, M.B., Colacicco, G., Curci, R. and Rapport, M.M. (1968) Biochemistry 7: 1692-1698

Allan, D. and Thomas, P. (1981) Biochem. J. 198: 433-440

Anderson, R.A. and Marchesi, V.T. (1985) Nature 318: 295-298

Berridge, M.J. (1984) Biochem. J. 220: 345-360

de Kruijff, B., Cullis, P.R. and Radda, G.K. (1976) Biochim. Biophys. Acta 436: 729-740

Ferrel, J.E. and Huestis, W.H. (1984) J. Cell Biol. 98: 1992-1998

Hauser, H. and Dawson, R.M.C. (1967) Eur. J. Biochem. 1: 61-69

Hayashi, K., Mühleisen,M., Probst, W. and Rahman, H.(1984) Chem. Phys. Lipids 34: 317-322

Hendrickson, H.S. and Ballou, C.E. (1964) J. Biol. Chem. 239: 1369-1373

Kiselev, G.V. (1982) Biochim. Biophys. Acta 712: 719-721

Koter, M., de Kruijff, B. and van Deenen, L.L.M. (1978) Biochim. Biophys. Acta 514: 255-263

Lindon, J.C., Baker, D.J., Farrant, R.D. and Williams, J.M. (1986) Biochem. J. 233: 275-277

Michell, R.H. (1975) Biochim. Biophys. Acta 415: 81-147

Moolenaar, W.H., Tertoolen, L.G.J. and de Laat, S.W. (1984) Nature, 312: 371-374

Nishizuka, Y. (1984) Nature 308: 693-697

Shibata, T., Uzawa, J., Sugiura, Y., Hayashi, K. and Takizawa, T. (1984) Chem. Phys. Lipids 34: 107-113

Sugiura, Y. (1981) Biochim. Biophys. Acta 641: 148-159

van Dijck, P.W.M., de Kruijff, B., Verkleij, A.J., van Deenen, L.L.M. and J. de Gier (1978) Biochim. Biophys. Acta 512: 84-96

van Paridon, P.A., de Kruijff, B., Ouwerkerk, R. and Wirtz, K.W.A. (1986) Biochim. Biophys. Acta 877: 216-219

Wüthrich, K. (1976) NMR in biological research: Peptides and Proteins, North-Holland, Amsterdam, p. 31

FLIP-FLOP RATES AND ORIGIN OF PHOSPHOLIPID ASYMMETRY IN THE NATIVE AND MODIFIED RED CELL MEMBRANE CAN BE STUDIED BY PHOSPHOLIPID ANALOGUES

Haest, C.W.M., Kunze, I., Claßen, J., Schneider, E., and Deuticke, B.

Department of Physiology, Medical Faculty, RWTH Aachen, F.R.G.

INTRODUCTION

The transbilayer distribution and reorientation of lipid constituents of biological membranes has considerable structural and functional implications. An asymmetric transbilayer distribution of phospholipids is well established in a number of membrane systems (Op den Kamp, 1979), among which the erythrocyte has been particularly well studied.

The stable asymmetric orientation of amino-phospholipids to the inner membrane surface of the erythrocyte is not the consequence of a slow transbilayer reorientation (flip) of phospholipids (Dressler et al., 1984). This asymmetry has been claimed to be established by a fast ATP-dependent reorientation of the amino-phospholipids from the outer to the inner membrane layer (Seigneuret and Devaux, 1984). However, asymmetry is still preserved after ATP depletion, prolonged incubations and enhancement of flip rates (Dressler et al., 1984). We therefore postulated that a stable orientation of aminophospholipids to the inner membrane surface is maintained by direct or indirect interactions with membrane skeletal proteins (Haest, 1982). In this concept the ATP-driven process highly enhances the establishment of equilibrium distribution of aminophospholipids.

The flip of diacylphospholipids is hard to measure directly at present. Measurements require insertion of labelled diacyl-phospholipids into the membrane by means of purified phospholipid transfer proteins (Op den Kamp et al., 1985; Tilley et al., 1986) or insertion of spin-labelled phospholipid analogues (Seigneuret and Devaux, 1984). An advantage of the use of transfer proteins is the principal possibility of a stoichiometric exchange of an endogenous phospholipid for an exogenous one without perturbation of the lipid domain. The method is, however, time consuming and complicated. Its major disadvantage is the fact that the flip rate has to

Table I: Parameters of the flip process of various lysophospholipids and PAF
in human erythrocytes.

	PAF* (18:0)	LysoPC** (16:0)	LysoPC*** (18:1)	LysoPE*** (18:1)	LysoPS** (18:1)	LysoPI*** (18:0)
k_{in} (10^2 h^{-1})	1.2	1.9±0.3	3.0	7.5	19.2±6.3	5.5
		(n=5) (S.D.,n=17)	(r=0.89,n=6)	(r=0.94,n=5)	(S.D.,n=5)	(r=0.96,n=5)
$t_{1/2}$ (h)	17.3	11.0	7.6	8.0	3.4	5.5
q	0.3	–	0.33	0.87	0.81	0.54
Activation energy (kJ/mol)	(96)	134	71	100	64	96

k_{in} is the rate of unidirectional movement of phospholipid
analogue from the outer to the inner membrane layer

$t_{1/2}$ is the half-time of the flip process calculated using

$$t_{1/2} = q * \ln 2/k_{in}$$

q is the fraction of the phospholipid analogue in the
inner membrane layer under stationary conditions calculated from
the flip data, assuming first-order kinetics, by an iteration
procedure; or from F_i values after prolonged incubations (Bergmann et al., 1984b; Kunze, 1986).

r is the correlation coefficient for the least square fit of the data
used to determine k_{in}

[14]C-labelled LysoPS, LysoPI, LysoPE were prepared from labelled
diacyl-phosphoplipid analogues by removal of the fatty acid at
the position 2 with phospholipase A_2 from bee venom (Bergmann et
al., (1984a).

*   data from Schneider et al. (1986)
**   data from Bergmann et al. (1984a)
*** data from Kunze (1986)

be quantified by the time-dependent increase of the fraction of phospholipid not accessible towards cleavage by exogenous phospholipase A_2.

As an alternative, lysophospholipids and related compounds (platelet activating factor (PAF) and acyl carnitine) may be appropriate probes to investigate the flip process of diacyl phospholipids. The flip of lysophospholipids from the outer to the inner membrane surface can be measured by incorporation of tracer amounts of labelled lysophospholipids into the outer layer and quantification of the time-dependent increase of the fraction of lysophospholipid not extractable by serum albumin (Bergmann et al., 1984a,b). The major advantage of the method is that it is fast and simple with respect to the incorporation of the lipid analogue into the membrane as well as quantification of flip rates. A disadvantage of the method consists in the acylation of lysophospholipids to diacyl analogues after their reorientation to the inner membrane surface (Bergmann et al., 1984a,b). This problem can be overcome by the use of appropriate lipid analogues (acyl carnitine, PAF).

The present contribution deals with the influence of the polar head group and of the fatty acid structure of lysophospholipids on their flip rate and on their steady-state distribution between the inner and outer membrane layer of human erythrocytes. Moreover, data are reported on the enhancement of flip rates of lysophospholipids after chemical modification of membrane proteins, by incorporation of exogenous channel-forming agents and bacterial cytolysins into the membrane, and by addition of local anaesthetics.

RESULTS AND DISCUSSION

Lysophospholipids and PAF reorient (flip) slowly in the membrane of human erythrocytes with half-times between 3 and 17 h at 37°C (Table I). These half-times are comparable to those for the flip of diacyllecithins (Fujii et al., 1985; Op den Kamp et al., 1985). The structure of the polar head group and of the fatty acid chain influences rate and activation energy of the flip process (Table I). Surprisingly, a negatively charged head group does not restrict the flip as compared to a zwitterionic one. Mono-unsaturated lysolecithins reorient faster than fully saturated species (Table I). The passive (ATP-independent) flip of lysophosphatidylethanolamine (lysoPE) and lysophosphatidylserine (lysoPS) is much slower (Bergmann et al., 1984a, Table I) than the ATP-dependent transfer of their diacyl-analogues PE and PS from the outer to the inner membrane surface (Seigneuret and Devaux, 1984; Tilley et al., 1986).

To illustrate the kinetics of the flip of lysophospholipids, data are given for lysoPE and lysophosphatidylinositol (lysoPI) in Figs. 1 and 2.

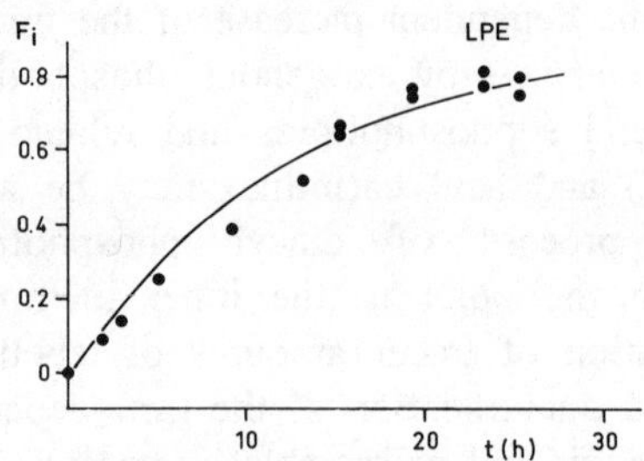

Fig. 1. Time – dependent increase of the fraction (F_i) of ^{14}C-labelled lysophosphatidylethanolamine (lysoPE) reoriented to the inner membrane layer of human erythrocytes after primary insertion into the outer layer. For further methodological details see Bergmann et al. (1984a,b). Acylation of ^{14}C-lysoPE at the inner membrane surface was inhibited by pre-incubation of cells with 100 nmol of non-labelled analogue per ml of packed cells for 5 h at 22°C. This treatment did not influence flip rates of lysoPE. The fraction of lysoPE acylated after 20 h was <5% and could be neglected. Data from 5 experiments were fitted using an iteration procedure and the function (Kunze, 1986):

$$F_i(t) = q \, (1 - \exp(-K_{in}/q * t))$$

where q is the ratio of lysoPE in the inner layer relative to total lysoPE initially inserted and K_{in} the rate constant of unidirectional movement of lysoPE from the outer to the inner membrane layer.

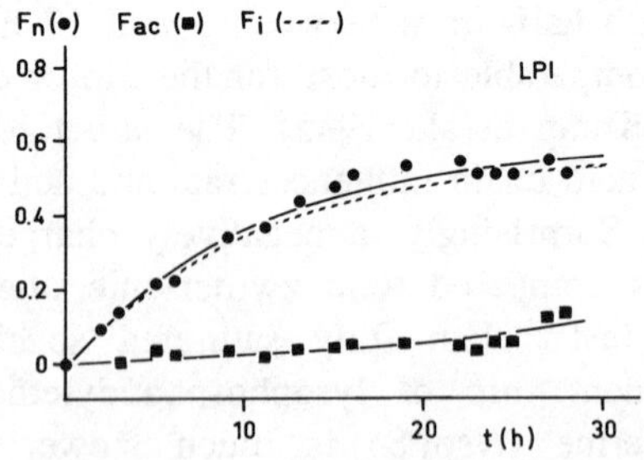

Fig. 2. Time – dependent increase of the fraction (F_n) of ^{14}C-labelled lysophosphatidylinositol (lysoPI) non-extractable by albumin and of the fraction (F_{ac}) of lysoPI acylated to diacyl PI. The fraction of lysoPI in the inner membrane layer (F_i) is represented by the difference F_n - F_{ac}.

From such curves flip rate constants as well as steady-state distributions of the lysophospholipids between the two membrane leaflets can be calculated (Table 1).

Despite the passive (ATP-independent) nature of the slow flip process, lysoPE and lysoPS (Table I) have the same preference for the inner surface of the erythrocyte membrane that was reported by others for the corresponding diacyl-analogues (Verkleij et al., 1973; Seigneuret and Devaux, 1984). LysoPC and PAF also demonstrate the same preference for the outer membrane layer as their diacyl-analogue PC (Table I and Figure3). In contrast, lysoPI does not exhibit the preference for the inner membrane layer reported for its diacyl-analogue PI (Low and Finean, 1977). This may be due to the phosphorylation of the endogenous PI, which occurs at the inner membrane surface. The slow native flip rates of lysophospholipids can be highly enhanced by chemical modification of membrane proteins (Bergmann et al., 1984), by reversible electric breakdown of the membrane barrier (Dressler et al., 1983) or by the insertion of foreign compounds such as local anaesthetics, channel-forming antibiotics or bacterial cytolysins into the membrane (Schneider et al., 1986). Depletion of membrane cholesterol to a cholesterol/phospholipid ratio of 0.40 (0.72 for controls), however, only results in a two-fold increase of flip rate of oleoyllysolecithin. This should be compared with an increase of up to 45-fold in flip rates produced by local anaesthetics. These effects are immediately reversible upon removal of the anaesthetics by washing of the cells.

Chemical modification of membrane proteins by oxidizing agents and the radical-forming agent t-butyl hydroperoxide results in an enhancement of

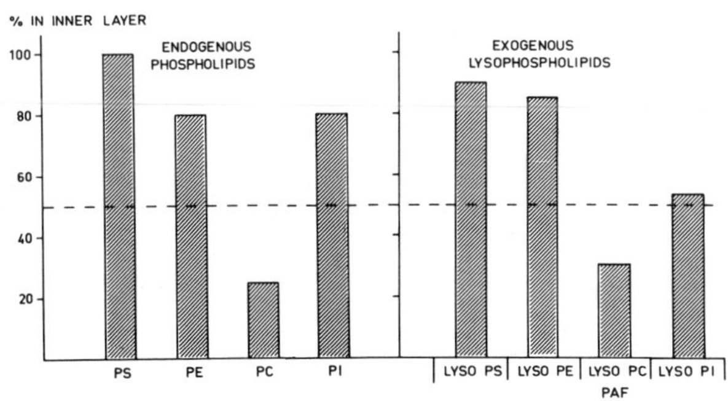

Fig. 3. The fractions (%) of various lysophospholipids and PAF in the inner erythrocyte membrane layer under stationary conditions as compared to those of their native diacyl-analogues. Data for PC, PE and PS are from Verkleij et al., (1973), those for PI from Low and Finean, (1977).

Table II: Modification of membrane proteins and of the lipid phase
         enhances the flip (37°C) of phospholipid analogues in
         human erythrocytes.

Modification	Relative flip rate	Phospholipid analogue tested
None	1	18:1 or 16:0 LysoPC, 16:0 Carnitine
Cholesterol depletion Chol./P-lipid 0.40 (0.72)	2	18:1 LysoPC
Cationic amphiphiles		
3.5 mM tetracaine	45	16:0 LysoPC
1.2 mM nupercaine	35	
Channel-forming antibiotics		
Gramicidin A (2×10^6 copies/cell)*	30	16:0 LysoPC
Amphotericin B (2×10^6 copies/cell)*	30	
Cytolysins α-Toxin from <u>S. aureus</u>* (10^5 copies/cell)	35	16:0 LysoPC
Cytotoxin from <u>Ps. aeruginosa</u> (700 µg/ml packed cells)**	35	18:1 LysoPC
Electric breakdown (0°C, 8kV/cm, 40µs)	>2000 (0°C)	18:1 LysoPC
<u>N</u>-ethylmaleimide (5 mM, 60 min, 37°C, pH 8)	5	18:1 LysoPC
t-Butyl hydroperoxide (2 mM, 30 min, 37°C, pH 8)	7	16:0 Carnitine
Na-periodate (5 mM, 120 min, 37°C, pH 8)	15	18:1 LysoPC
Diamide (5 mM, 120 min, 37°C, pH 8)	100	16:0 LysoPC

Data were obtained as described in the papers from the authors'
laboratory given in the references

  * Numbers refer to membrane-incorporated molecules

** Numbers refer to the total amount of toxin in the cell suspension flip rates which depends in its extent on the exposure time and the concentration of the agent. The most powerful reagent is the disulfide inducing agent diamide, which produces an up to 100-fold increase of the flip rate (Table II).

We suggest that such enhancements of the flip of phospholipids result from the formation of fluctuating local defects in the lipid phase producing aqueous pores by which the polar head groups can bypass the steep hydrophobic barrier normally preventing their transbilayer reorientation. This concept is supported by three lines of evidence. First, studies on reconstituted proteoliposomes (van Hoogevest et al., 1984) demonstrated a lipid-dependent formation of leaks and flip sites upon insertion of protein into lipid vesicles. Second, leaks were shown to occur in erythrocyte membranes after treatments leading to an enhancement of flip rates such as chemical modification (Deuticke et al, 1983; Heller et al., 1984), electric breakdown (Schwister and Deuticke, 1985) and incorporation of local anaesthetics (Deuticke, unpublished data). Third, insertion of membrane-spanning polyenes and peptides into the lipid domain of erythrocytes enhances flip rates (see above), while a polyene antibiotic like pimaricin, unable to form a channel probably because it cannot span the hydrophobic lipid domain, does not enhance the flip. This clearly demonstrates the specificity of the enhancing effect and indicates the requirement of a perturbation of the inner membrane layer to induce flip enhancement.

CONCLUSIONS

Measurements of the flip-flop of phospholipid analogues (lysophospholipids, palmitoylcarnitine, PAF) are convenient tools to (1) detect fluctuating local structural defects in the membrane lipid domain acting as flip sites and as leaks for hydrophilic solutes; (2) investigate the origin of the asymmetric transbilayer distribution of phospholipids in membranes; (3) demonstrate fluctuating defects at the hydrophobic protein-lipid interface after modification of intrinsic membrane proteins.

ACKNOWLEDGEMENT

This work was supported by the Deutsche Forschungsgemeinschaft (SFB 160/C3).

REFERENCES

Bergmann, W.L., V. Dressler, C.W.M. Haest and B. Deuticke (1984a). Biochim. Biophys. Acta 772, 328-336.
Bergmann, W.L., V. Dressler, C.W.M. Haest and B. Deuticke (1984b). Biochim. Biophys. Acta 769, 390-398.

Deuticke, B., B. Poser, P. Lutkemeier and C.W.M. Haest (1983). Biochim. Biophys. Acta 731, 196-210.

Deuticke, B., K.B. Heller and C.W.M. Haest (1986). Biochim. Biophys. Acta 854, 169-183.

Dressler, V., K. Schwister, C.W.M. Haest and B. Deuticke (1983). Biochim. Biophys. Acta 732, 304-307.

Dressler, V., C.W.M. Haest, G. Plasa, B. Deuticke and J.D. Erusalimsky (1984). Biochim. Biophys. Acta 775, 189-196.

Fujii, T., A. Tamura and T. Yamane (1985). J. Biochem. 98, 1221-1227.

Haest, C.W.M. (1982). Biochim. Biophys. Acta 694, 331-352.

Kunze, I. (1986). M. D. Thesis, RWTH Aachen.

Heller, K.B., B. Poser, C.W.M. Haest and B. Deuticke (1984). Biochim. Biophys. Acta 777, 107-116.

Low, M.G. and J. B. Finean (1977). Biochem. J. 162, 235-240.

Op den Kamp (1979). Annu. Rev. Biochem. 48, 47-71.

Op den Kamp, J.A.F., B. Roelofsen and L.L.M. van Deenen (1985). Trends in Biochemical Sci. 320-323.

Schneider, E., C.W.M. Haest and B. Deuticke (1986). FEBS Letters 3535, 198, 311-314.

Schneider, E., C.W.M. Haest, G. Plasa and B. Deuticke (1986). Biochim. Biophys. Acta 855, 325-336.

Schwister, K. and B. Deuticke (1985). Biochim. Biophys. Acta 816, 332-348.

Seigneuret, M. and Ph. F. Devaux (1984). Proc. Natl. Acad. Sci. USA 81, 3751-3755.

Tilley, L., S. Cribier, B. Roelofsen, J.A.F. Op den Kamp and L.L.M. van Deenen (1986). FEBS Letters 194, 21-27.

van Hoogevest, P., A.P.M. Du Maine, B. de Kruijff and J. de Gier (1984). Biochim. Biophys. Acta 771,119-126.

Verkleij, A.J., R.F.A. Zwaal, B. Roelofsen, P. Comfurius, D. Kastelijn, L.L.M. van Deenen (1973). Biochim. Biophys. Acta 323, 178-193.

INFLUENCE OF LIPID COMPOSITION ON PASSIVE ION TRANSPORT OF ERYTHROCYTES

I. Bernhardt, A. Erdmann, R. Glaser, G. Reichmann[+] and R. Bleiber[+]

Bereich Biophysik, Sektion Biologie, Humboldt-Universität Berlin; [+]Institut für Pathologische und Klinische Biochemie, Bereich Medizin, Humboldt-Universität Berlin

INTRODUCTION

Davson (1939) had already reported that erythrocytes suspended in isotonic solutions of low ionic strength lose cations. Later these findings were made more precise and interpreted as a change of K^+ permeability of human erythrocytes dependent on the transmembrane potential (Donlon and Rothstein, 1969). We extended the investigations of the influence of reducing the extracellular ionic strength on Rb^+ (K^+) efflux to erythrocytes from various mammalian species (Table 1).

RESULTS AND DISCUSSION

The osmolarity of the solutions used was kept constant by adding sucrose. The method of continuous efflux measurements employed for all experiments was described in detail by Bernhardt et al. (1984). In this case ^{86}Rb was used as an indicator for K^+. The rate constant of Rb^+ efflux was calculated as the negative slope of the linear regression line obtained by a semilogarithmic plot of counts against time and expressed in min^{-1}.

Up to now all authors who have investigated the change of Rb^+ (K^+) efflux from erythrocytes as a dependence on the ionic strength of the solution, explained this change with differences in the driving force, especially the transmembrane potential. It could be shown that the transmembrane potential cannot be solely responsible for this effect. At least for bovine erythrocytes the transmembrane potential deduced from ^{36}Cl distribution changes in response to the extracellular ionic strength similarly to human erythrocytes (Table 2).

The main differences between erythrocytes of the various mammalian species are the relative contents of the phospholipids (Nelson, 1967) as well as the structure of the acyl chains of their fatty acids (Wessels and Veerkamp, 1973). Differences in the so-called membrane fluidity are an expected result of these compositional differences. So the mean number of double bonds is considered to be a rough indicator of the fluidity (Deuticke, 1977; Farias et al., 1975).

Table 1. Rate constant $\underline{k}$ of Rb^+ efflux of erythrocytes of various mammalian species in solution of high and low ionic strength.

species	NaCl+KCl 147 mM $\underline{k}x10^4(min^{-1})$	NaCl+KCl 5.7 mM $\underline{k}x10^4(min^{-1})$	ratio of $\underline{k}$[*] (%)
cow	7.0±0.6	8.7±1.4	124 n.s.
pig	3.3±1.0	5.0±1.9	152 n.s.
rabbit	8.4±2.8	12.3±1.2	146 s.
horse	3.8±1.0	9.1±3.9	240 s.
rat	13.6±2.6	39.8±9.7	293 s.
cat	6.9±1.1	23.1±3.2	333 s.
man	8.1±0.4	63.6±4.2	785 s.

[*] $\underline{k}$ in solution of (NaCl+KCl)= 147 mM taken as 100%. $\underline{T}$= 310 K; pH= 7.4; osmolarity= 290 mOsM; values are ±1 S.D. "s." and "n.s." denote, respectively, significant or non-significant differences of the rate constants at a level of $\underline{p}$= 0.05 according to Welch's $\underline{t}$-test.

Table 2. Transmembrane potential ($\Delta\Psi$) of human and bovine erythrocytes in solution of high and low ionic strength.

(NaCl+KCl) concentration (mM)	man $\Delta\Psi$ (mV)	cow $\Delta\Psi$ (mV)
147.0	-5.7±4.7	-10.1±3.6
5.7	+45.1±3.7	+40.4±4.8

$\underline{T}$= 293 K; pH= 7.4; osmolarity= 290 mOsM; values are ±1 S.D.

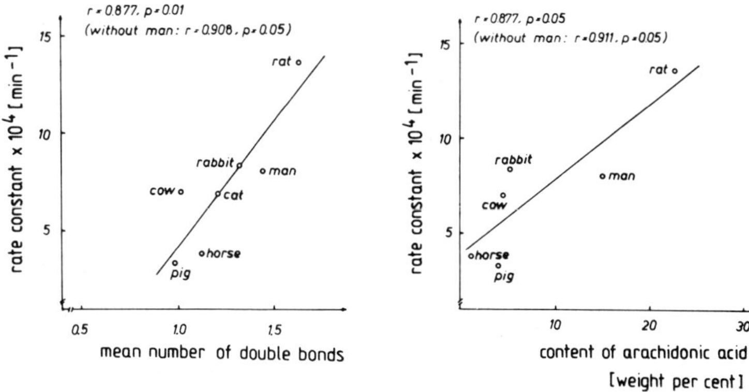

Fig. 1. Rate constant of $Rb^+(K^+)$ efflux of erythrocytes from various mammalian species in solution of high ionic strength $((NaCl+KCl)= 147$ mM) as a function of the mean number of double bonds (taken from Wessels and Veerkamp, 1973; data for cat from Deuticke, 1977) and as a function of the content of arachidonic acid of the phospholipids (taken from Wessels and Veerkamp, 1973): $\underline{T}= 310$ K; pH= 7.4; osmolarity= 290 mOsM.

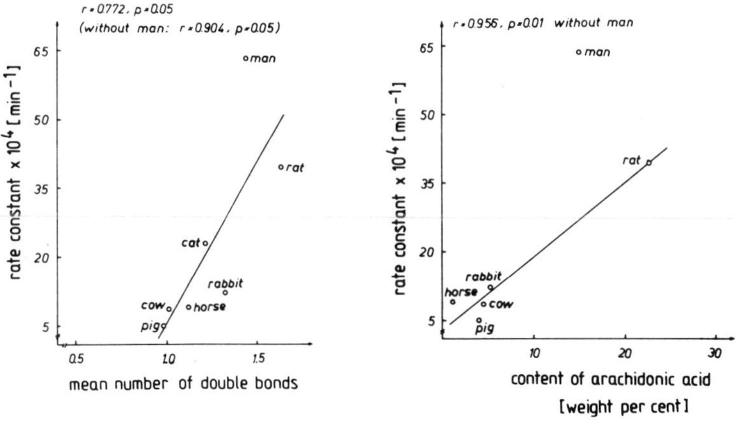

Fig. 2. Rate constant of $Rb^+(K^+)$ efflux from erythrocytes from various mammalian species in solution of low ionic strength $((NaCl+KCl)= 5.7$ mM) as a function of the same parameters as in Fig. 1: $\underline{T}= 310$ K; pH= 7.4; osmolarity= 290 mOsM.

We attempted to correlate the rate constants of Rb$^+$ (K$^+$) efflux in solutions of high and low ionic strength with data for the membrane lipids of various mammalian species. There was no correlation with the relative content of the four main phospholipids (taken from Wessels and Veerkamp, 1973 and from Nelson, 1967). Differences in the asymmetric distribution of the phospholipids between inner and outer leaflet of the membrane, too, are not the reason for the different rate constants, because it is known that all phosphatidylserine is confined to the inner, and almost all sphingomyelin to the outer leaflet of the membrane, at least for the erythrocytes of man, cow and rat (Verkleij et al., 1973; van Dijck et al., 1976; Renooij et al., 1976). In Fig. 1 it can be seen that it was possible to establish a correlation between the rate constant of Rb$^+$ (K$^+$) efflux from erythrocytes of various mammalian species in solutions of high ionic strength and the mean number of double bonds or the relative content of arachidonic acid of the phospholipids. Fig. 2 shows the same results for the rate constant in solution of low ionic strength, if the values of human erythrocytes are exluded. The reason for this exclusion might be the existence of Na$^+$/K$^+$ cotransport only found in human erythrocytes (Ellory et al., 1982; Lauf, personal communication).

The question is if these correlations indicate a special effect of the arachidonic acid, or only the importance of membrane fluidity. It is very difficult to answer this question because the relatively high content of this fatty acid containing four double bonds especially influences the mean number of double bonds used as a rough indicator of membrane fluidity. In contrast to the findings regarding the rate constants for Rb$^+$ (K$^+$) efflux in bovine erythrocytes, the erythrocytes of newborn calves show the same increase of the rate constant in solution of low ionic strength as human erythrocytes (Fig. 3). The increase depends on the age of the calves. Investigations of the lipid composition demonstrate that there is no change concerning the head groups of the membrane phospholipids with the age of the calves, but that there is a change of the content of arachidonic acid (20:4) and of linoleic acid (18:2) (Fig. 4). So it is possible to correlate the rate constant of Rb$^+$(K$^+$) efflux from calf erythrocytes in solution of low ionic strength with the content of arachidonic acid depending on the age of the calves (correlation coefficient r= 0.951). It is not possible to correlate this rate constant with the mean number of double bonds.

The special importance of arachidonic acid for transport phenomena becomes more and more obvious (Gruber and Deuticke, 1973; Lu and Chow, 1982). Kuypers et al. (1984) could show that human erythrocytes will be leaky for K$^+$ if the native phosphatidylcholine is partly replaced by phosphatidylcholine containing arachidonic acid as one of the fatty acids. It

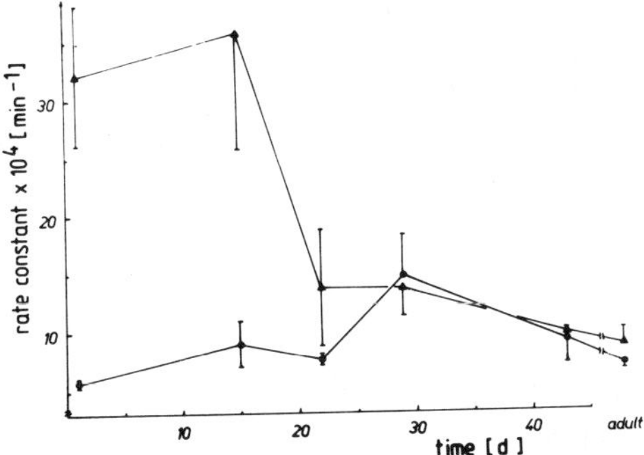

Fig. 3. Dependence of the rate constant for Rb^+ (K^+) efflux from calf erythrocytes in solutions of high (●) ((NaCl+KCl)= 147 mM) and low (▲) ((NaCl+KCl)= 5.7 mM) ionic strength on the age of the calves: $\underline{T}$=310 K; pH=7.4; osmolarity=290 mOsM; vertical bars indicate ±1 S.D.

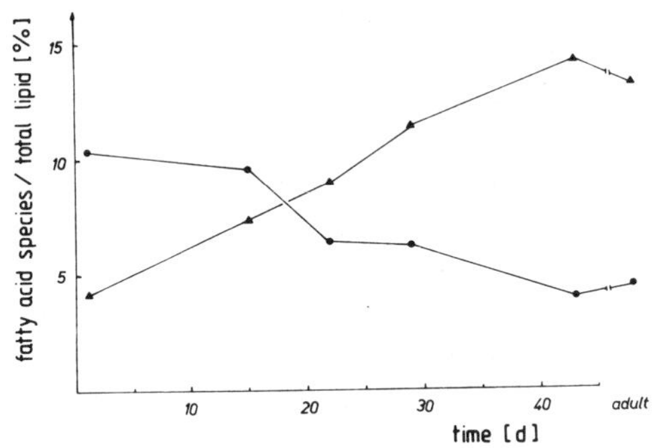

Fig. 4. Relative content of arachidonic acid (●) and linoleic acid (▲) of the phospholipids of calf erythrocytes as a dependence on the age of the calves (fatty acid determination by gas chromatography as described by Preiss et al., 1982; values for adult cows according to Haest - personal communication).

is possible to assume that the ionic strength of the solution alters the properties of the eythrocyte membrane depending on the content of arachidonic acid of the phospholipids. So the value of the rate constant for the $Rb^+(K^+)$ efflux in solutions of low ionic strength seems to depend on the content of arachidonic acid. In connection with the results shown in Fig.

1 it is noteworthy that Bergmann et al. (1984) found a correlation between membrane fluidity (mean number of double bonds) and the flip rate of lysophosphatidylcholine. This could lead to the assumption that the reduction of ionic strength of the solution influences the flip-rates and that the content of arachidonic acid in one leaflet of the membrane is the decisive factor for the $Rb^+(K^+)$ flux.

REFERENCES

Bergmann, W.L., Dressler, V., Haest, C.W.M., Deuticke, B. (1984) Biochim. Biophys. Acta 772: 328-336

Bernhardt, I., Donath, E., Glaser, R. (1984) J. Membrane Biol. 78: 249-255

Davson, H. (1939) Biochem. J. 33: 389-401

Deuticke, B. (1977) Rev. Physiol. Biochem. Pharmacol. 78: 1-97

Donlon, J.A., Rothstein, A. (1969) J. Membrane Biol. 1: 37-52

Ellory, J.C., Dunham, P.B., Logue, P.J., Stewart, G.W. (1982) Phil. Trans. R. Soc. London B299: 483-495

Farias, R.N., Bloj, M., Marero, R.D., Sineriz, F., Trucco, R. (1975) Biochim. Biophys. Acta 415: 231-251

Gruber, W., Deuticke, B. (1973) J. Membrane Biol. 13: 19-36

Kuypers, F.A., Roelofsen, B., Op den Kamp, J.A.F., Van Deenen, L.L.M. (1984) Biochim. Biophys. Acta 769: 337-347

Lu, Y., Chow, E.I. (1982) Biochim. Biophys. Acta 689: 485-489

Nelson, G.J. (1967) Biochim. Biophys. Acta 144: 221-232

Preiss, R., Prümke, H.J., Sohr, R., Müller, E., Schmeck, G., Schmidt, J., Banaschak, H. (1982) J. Clin. Pharmacol. 20: 105-112

Renooij, W., Van Golde, L.M.G., Zwaal, R.F.A., Van Deenen, L.L.M. (1976) Eur. J. Biochem. 61: 53-58

Van Dijck, P.W.M., Van Zoelen, E.J.J., Seldenrijk, R., Van Deenen, L.L.M., De Gier, J. (1976) Chem. Phys. Lipids 17: 336-343

Verkleij, A.J., Zwaal R.F.A., Roelofsen, B., Comfurius, P., Kastelijn, D., Van Deenen, L.L.M. (1973) Biochim. Biophys. Acta 323: 178-193

Wessels, J.M.C., Veerkamp, J.H. (1973) Biochim. Biophys. Acta 291: 190-196

PHYSICAL PROPERTIES OF MEMBRANES FORMED FROM PHOSPHOLIPIDS HAVING MONOMETHYL BRANCHED FATTY ACYL CHAINS

Charles R.A. Earl

Rowett Research Institute, Aberdeen, Scotland

INTRODUCTION

Membrane lipids with methyl-branched acyl chains are generally confined to prokaryotes (Kaneda, 1977). Methyl substitution in these lipids commonly occurs at the penultimate or ante-penultimate carbon atom. Higher organisms for the most part contain saturated and <u>cis</u>-unsaturated acyl and alkyl chains in membrane lipids. Nevertheless animals can be presented with substantial amounts of methyl-branched acyl chains derived from phytol which is the alcohol moiety of chlorophyll. After assimilation phytol is metabolised to phytanic acid, 3,7,11,15-tetramethylhexadecanoic acid. This multi-branched isoprenoid acid is normally catabolised rapidly by α-oxidation followed by the β-oxidation sequence of fatty acid degradation. Exceptional amounts of phytanic acid occur in tissue lipids of humans lacking the capacity for an initial α-oxidation step. This condition is associated with the rare neuropathological disease, namely Refsum's Syndrome (Lough, 1973).

Monomethyl branched fatty acids have been found in subcutaneous adipose tissue of lambs given diets with a high cereal content. The position of methyl substitution in these compounds is always at an even numbered carbon atom with respect to the carboxyl carbon. It has been shown that these acids are derived through fatty acid biosynthesis in lamb tissues. Methyl malonate can replace malonate during chain elongation by the synthetase (Scaife et al., 1978).

A concentrate of fatty acids obtained from triacylglycerols of barley-fed lambs has been included in diets given to rats. This mixture contained several homologous series and positional isomers of methyl branched fatty acids (MBFA). The branched-chain fatty acids were found in tissues of the rats where they were present in neutral lipids and phospholipids. No gross pathology was evident in rats given the MBFA concentrate at levels of 2.5% to 7.5% by weight of the diet (Smith et al., 1982). When the levels of dietary inclusion of the MBFA concentrate were 10% and 15% by weight, overt toxic effects occurred. The mixture of MBFA used in these studies contained many isomers where methyl substitution was near the carboxyl end of the acyl chains. When chemically synthesised 3-methylhexadecanoic (3-MeHA), 4-methylhexadecanoic (4-MeHA), or phytanic acid were given to rats at the 5% level they proved severely toxic. All of the methyl-substituted fatty acids noted above are rapidly catabolised by animal tissues and at levels normally occurring in animal and human diets they are entirely without adverse effect.

Interest in the biological effects of MBFA is centred around their use in investigations of the relationship between cellular membrane structure and function. A wealth of information has been obtained, within the framework of the fluid mosaic model (Singer and Nicholson, 1972), on the structure and physical properties of biomembranes (Quinn, 1981; Stubbs and Smith, 1984). One aspect of this model that requires further attention relates to the hydrophobic region proximal to the polar surfaces of the bilayer. The polymethylene sections of acyl chains in this region have a higher orientational order (Seelig and Seelig, 1977). Alteration of acyl structure by methyl substitution can disrupt packing due to an increase in molecular diameter and decrease in the length of uninterrupted polymethylene segments. Membrane properties that could be sensitive to these changes include incorporation of cholesterol in plasma membranes and adsorption of lipophilic substrates of membrane enzymes.

RESULTS AND DISCUSSION

Four monomethyl hexadecanoic acids were chemically synthesised to obtain the 2, 3, 4 and 5-methyl isomers, by using methyl malonation or a Grignard reaction on a methyl ketone. The purity of the compounds was found to be >98% by TLC and capillary GC-MS analysis. These acids were used to make the corresponding diacylphosphatidylcholines. Diphenyl-hexatriene (DPH) or (2-carboxyethyl)-DPH was embedded in each of the phospholipids which were dispersed in ethanediol/water 40% w/w. The fluorescence anisotropy was measured as a function of temperature.

Contributions to the steady-state fluorescence anisotropy (r_s) of DPH arise from restricted angular rotation of the probe in liposomes. The magnitude of r_s is related empirically to orientational order in the acyl chain region of the membranes (Van der Meer et al., 1986). Temperature dependences of r_s for DPH in liposomes of DPPC and several isomeric di-(n-methylhexadecanoyl)-phosphatidylcholines (n-MeHPC) are shown in

Figure 1. For the sequence DPPC, 2-MeHPC, 3-MeHPC, 4-MeHPC and 5-MeHPC, the phase transition temperature was found at approximately 41°C, 28°C, 20°C, 12°C and 0°C respectively. The positions and shapes of the r_s versus T°C curves show also that as methyl substitution in the phospholipids proceeds towards the middle of the acyl chains there is a convergence in the properties of the gel and liquid crystalline states. This finding is in accordance with the prediction made by Suzuki and Cadenhead, 1985.

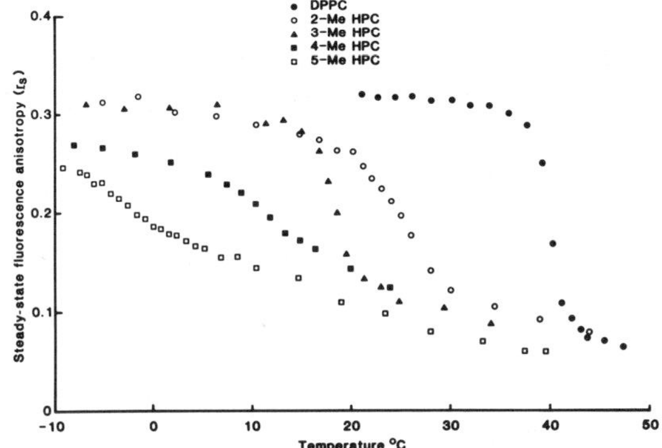

Figure 1. Fluorescence anisotropy of diphenylhexatriene in liposomes of dipalmitoyl, 2-methylhexadecanoyl, 3-methylhexadecanoyl, 4-methylhexadecanoyl and 5-methylhexadecanoyl-phosphatidyl cholines.

When the amphiphilic fluorescence probe (2-carboxyethyl)-DPH (Morgan et al., 1982) was used comparable data for r_s versus T°C were obtained.

In an exploration of the possible association of physical and biochemical effects in membrane structures this study is being extended to include an examination of the influence of _cis_ and _trans_ olefinic groupings in acyl chains (Barton and Gunstone, 1975; Macdonald et al., 1985) for comparison with methyl branching.

REFERENCES

Barton, P.G. and Gunstone, F.D. (1975) J. Biol. Chem. 250: 4470-4476.
Kaneda, T. (1977) Bacteriol. Rev. 41: 391-418
Lough, A.K. (1973) in "Progress in the Chemistry of Fats and Other Lipids", Vol. XIV, part 1, (ed. R.T. Holman), pp. 5-48, Pergamon Press, Oxford.
Macdonald, P.M., Sykes, B.D., McElhaney, R.N. and Gunstone, F.D. (1985) Biochemistry 24: 177-184.

Morgan, C.G., Thomas, E.W., Moras, T.S. and Yianni, Y.P. (1982) Biochim. Biophys. Acta, 692: 196-201.

Quinn, P.J. (1981) Progress in Biophysics and Molecular Biology 38: 1-104.

Scaife, J.R., Wahle, K.W.J. and Garton, G.A. (1978) Biochem. J. 176: 799-804.

Seelig, A. and Seelig, J. (1977) Biochemistry 16: 45-50.

Singer, S.J. and Nicholson, G.L. (1972) Science 175: 720-731.

Smith, A., Lough, A.K. and Earl, C.R.A. (1982) J. Sci. Food Agric. 33: 421-430.

Stubbs, C.D. and Smith, A.D. (1984) Biochim. Biophys. Acta 779: 89-137.

Suzuki, A. and Cadenhead, D.A. (1985) Chem. Phys. Lipids 37: 69-82.

Van der Meer, B.W., Van Hoeven, R.P. and Van Blitterswijk, W.J. (1986) Biochim. Biophys. Acta 854: 38-44.

RECONSTITUTION OF MEMBRANE PROTEINS INTO LIPID VESICLES - A DISCUSSION OF CONDITIONS AND METHODS

Brigitte Schmitz

Molteno Institute, Medical Research Council, University of Cambridge, U.K.

INTRODUCTION

Six years after the description of liposome formation by Bangham (1965), the reconstitution of a membrane protein into vesicles of exogenous phospholipids was first reported (Kagawa and Racker, 1971). Since then a large number of membrane proteins have been isolated and re-incorporated into model membranes, where they often exhibit their normal, or sometimes reduced, biological activity.

An increasing number of techniques for the preparation of such proteoliposomes has been published subsequently, and the reason for this is that the approach to the reconstitution of a new protein is still an empirical one. It has been impossible until now to predict which conditions are necessary for a particular protein to be incorporated into liposomes without loosing its activity. So one just has to try to find out which method works best. This can drive one to despair as the author knows from her own experience, which, according to E. Racker's comment "...if nothing works, invent a new reconstitution procedure..." (Racker, 1985, p.38), may in practice result in the invention of another method (see below).

This article discusses factors which, as far as is known, play a role in the interaction between proteins and lipids during the incorporation process, and gives a brief overview with special emphasis on recent methods which aim at avoiding unphysiological constraints on proteins and lipids. The references cited should be understood as examples; for a more complete reference list as well as further details of the methods see Racker (1979 and 1985), Szoka and Papahadjopoulos (1980) and Darszon (1983).

FACTORS AFFECTING INCORPORATION

One important factor is the physical state of the lipids. Natural membranes are normally fluid, i.e. in the liquid crystalline phase above the phase transition temperature T_m. Many proteins are known to be incorporated more readily into lipid bilayers above T_m and are even sometimes found to be excluded from gel–state lipids (Mateu et al., 1978). Other proteins are incorporated into membranes only over a narrow range around T_m (e.g. glucagon, Epand et al., 1982). A third group of proteins interacts with phospholipids over a wide range of temperature exhibiting preferential interaction with lipids in the gel phase (Epand and Sturtevant, 1984).

The molecular basis for the interaction of proteins with lipids near the phase transition or with gel-state lipids is thought to be due to point or line defects facilitating protein-lipid interactions as well as protein-protein interactions along such defects, while maintaining at the same time lipid-lipid interactions (Epand and Surewicz, 1984). The spontaneous incorporation of membrane proteins into vesicles by freeze-thawing (see methods) probably relies on conditions which induce packing defects.

The importance of the physical state for the incorporation of proteins into liposomes becomes obvious from the observation that M13 virus coat protein shows a maximum for asymmetric incorporation (C-terminus inside and N-terminus outside) near the phase transition. The protein assembles with both termini exposed to the exterior vesicle surface well below T_m; above T_m only little protein is incorporated (Wickner, 1977).

In general three types of lipid–protein interactions are distinguished, having different effects on T_m, transition enthalpy and transition width (McElhaney, 1982): electrostatic, a mixture of electrostatic and hydrophobic, and hydrophobic interactions. Other types of lipid–protein interaction are observed which do not fit into this scheme. Some proteins or peptides (e.g. melittin, Prendergast et al., 1982) have been shown to trigger bilayer to non-bilayer phase transitions, the hexagonal H_{II} phase being of particular interest in recent years (e.g., Taraschi et al., 1982). A Triton-extract of human erythrocyte membrane proteins for example, showed highest incorporation into liposomes composed of PC:PE at a ratio of 25:75, a mixture exhibiting highest bilayer disruption (Hah et al., 1983).

In this context it is also of interest that duramycin, a polypeptide antibiotic, was found to interact only with PE or monogalactosyl diglyceride in artificial and natural membranes. Both lipids are known for their preference for the H_{II} phase (Navarro et al., 1985).

These examples highlight another important aspect, namely the composition of the lipids chosen for the incorporation of a protein. Some of the recently published methods try to avoid unphysiological constraints for proteins and lipids such as detergents, organic solvents, sonication, etc., and instead make use of the potential of certain rather more physiological compounds to enhance the insertion of membrane proteins into liposomes. Charged lipids for example, often have a marked effect on the incorporation. The use of charged lipids certainly reinforces electrostatic interactions, but this is not the only effect. It has been shown that negatively charged lipids (phosphatidic acid, phosphatidylinositol, cardiolipin) also improve the incorporation of acidic proteins (e.g. cytochrome oxidase; Eytan and Racker, 1977). The occurrence of non-bilayer phases may again play a role.

It has been shown that myristic acid and also cholesterol may induce incorporation of proteins into DMPC-vesicles only below the phase transition (Scotto and Zakim, 1985, 1986). Incorporation of the same proteins did not occur in the absence of either myristic acid or cholesterol, or in the presence of these compounds above the phase transition. Although the molecular events are not really known, it is suggested that they are similar for the fusion between unilamellar liposomes and for the incorporation of membrane proteins into vesicles. But these two events do not occur simultaneously, as it could be shown that incorporation precedes fusion.

Lysolecithin, which is known to prefer a micellar phase under certain conditions, enhances incorporation (Eytan et al., 1975, Christiansen and Carlsen, 1983) and again a bilayer disordering effect may be responsible for this; the same applies to the short-chain lecithins used by Dencher (1986).

Some other factors which can affect a protein-lipid reconstitution system deserve consideration as well. Many incorporation studies were, and still are, conveniently performed in the presence of detergents. The interaction between lipids and proteins plus detergents in mixed micelles is different from that without detergents, and points up other aspects which must be considered such as

- the choice of detergent and the detergent/lipid ratio
- the tolerance of proteins towards detergents
- the tolerance of the assay of protein function towards detergents
- the time of exposure to detergents and the time of removal
- the necessity (or not) of the complete removal of the detergent.

Some of these aspects are discussed below with the detergent methods, and in greater detail by Racker (1985).

The ionic strength of the buffer has an influence on lipid-protein interactions, especially when electrostatic interactions are important, and for example on the preparation of large unilamellar liposomes by the freeze-thaw technique (MacDonald et al., 1983). Sometimes cations (Ca^{2+}, Mg^{2+}) enhance incorporation at certain concentrations, higher concentrations often being inhibitory (Racker, 1985). Last but not least, the pH can influence the rate of incorporation.

In many cases proteins are inserted into preformed liposomes. Lipid- or detergent-free membrane proteins tend to aggregate in aqueous solution, or hydrophobic parts of the protein may be sequestered from the aqueous environment slowing down the interaction with lipid vesicles. Dissociation of aggregates or unfolding of the protein by chaotropic agents, detergents, urea or by increasing the ionic strength could yield better incorporation results in such cases.

Other factors which may influence the incorporation are type and size of the liposomes. Some proteins (e.g., cytochrome b_5) favour small vesicles (20 nm) over large vesicles (100 nm) by a factor of about 200 probably due to the looser packing of the polar headgroups of phospholipids in small vesicles (Greenhut et al., 1986). On the other hand, cytochrome c oxidase incorporates readily into multilamellar liposomes, but less or not at all into small unilamellar liposomes. Quantitative biophysical measurements are also sometimes hampered by the fact that different numbers of proteins are inserted into vesicles (Nicholls, 1983).

METHODS
A) DETERGENT METHODS

Incorporation of membrane proteins into lipid vesicles was carried out initially in the presence of detergent, and currently detergent methods are still used with success in many, although not all reconstitution experiments. Detergents with a high critical micelle concentration (e.g., n-octyl-glucoside, cholate) are preferred because they can be removed more completely either by dialysis, by dilution, or using a combination with gel partition chromatography (Sephadex, Sepharose, Biobeads, Extracti-Gel etc.). Mixtures of detergents sometimes show better results (Schmidt and Gräber, 1985).

For the dialysis method (Kagawa and Racker, 1971) phospholipids are sonicated briefly prior to adding protein in the presence of up to 3% detergent. The detergent is then removed by dialysis for at least 20 hours. Vesicles formed during dialysis are unilamellar and sometimes large (up to 200 nm), for example, when prepared with octyl-glucoside (Mimms et al., 1981).

The other procedure for the removal of detergent is the dilution method (Racker et al., 1975), a very simple and rapid method. Phospholipids are sonicated either in the presence or absence of detergent and are mixed with a solution of protein plus detergent to give a final detergent concentration of up to 0.8%. The sample is kept for some time at 4°C (the optimal time, which varies from protein to protein, is from a few minutes onwards); it is then diluted at least 25 fold with cold buffer and left some time (≈1 hr.) at 4°C. If necessary the sample can be concentrated by centrifugation, or using a Millipore filter.

There are some drawbacks, however, although detergent methods are frequently used and often tried first, when a membrane protein is to be reconstituted into vesicles. The first drawback is that detergents are not always well tolerated by membrane proteins. Long exposure to detergents during dialysis may result in low activity, the long dialysis time itself being a disadvantage for labile proteins. Some proteins do not incorporate at all in the presence of detergents, especially when using the dilution method.

It is also extremely difficult to remove detergents completely (own observations), thus making measurements of the physical state of the liposomes, or changes upon incorporation of a protein almost impossible. If one calculates the amount of detergent remaining on a molar basis, for example in the case of the virtually complete removal (99.94%) of octylglucoside, it turns out that there is still enough detergent left to give a molar ratio of phospholipid:detergent of 4:1, because of the large excess of detergent normally used (600:1 in this instance; Petri and Wagner, 1979).

Some of these problems may be solved by combining a detergent method with one of the other methods described below, such as freeze-thawing, sonication, or sonication and freeze-thawing. Removal of detergent can be improved using a combined approach with chromatography on Sephadex, Sepharose or Biobeads, if the protein is not adsorbed on these materials as sometimes happens.

B) SONICATION

Co-sonication of lipids and membrane proteins yields good reconstitution results in some cases (e.g., Racker and Stoeckenius, 1974), but cannot be applied generally because of the sensitivity of many proteins and lipids to sonication. Some proteins may survive sonication for 30 minutes or longer, others lose their activity within the first few minutes. Unsaturated lipids are rapidly autoxidized and degraded during sonication in the presence of even small amounts of oxygen , so sonication must be carried out under oxygen-free nitrogen or better argon, and/or in the presence of an antioxidant.

Another crucial point is that variation in the experimental and technical details may have a great influence on the results of sonication (e.g., probe versus bath sonicator, temperature and local heating effects, type of glassware, volume etc.).

C) FREEZE-THAWING

For this method lipids are briefly sonicated, the protein added and the mixture then rapidly frozen by immersing it in a dry ice/acetone bath or liquid nitrogen. After thawing the mixture at room temperature the sample is sonicated again (Kasahara and Hinkle, 1976). Freezing and thawing induces vesicle fusion, increasing the size of the vesicles up to 500 nm and making the vesicles less leaky (Pick, 1981).

Electrophysiological techniques have been shown to be applicable to proteoliposomes prepared by this method (e.g., single channel recording of a Cl^- channel (Suárez-Isla et al., 1983)).

MacDonald et al. described a variation of this method which produces large unilamellar liposomes with an average size of 15 µm; they used a combination of three methods, sonication, freeze-thawing and dialysis (MacDonald et al., 1983). The freeze-thawing was carried out in saturated KCl solution resulting in large vesicles due to osmotic inflation which occurred during dialysis.

In some instances freeze-thawing plus sonication yields better results in the presence of detergent, but then again there arises the problem of the removal of the detergent (Schenerman et al., 1985).

D) REVERSE PHASE EVAPORATION TECHNIQUE

Based on the method of Szoka and Papahadjopoulos (1978), Darszon et al. (1979) developed a procedure for the incorporation of membrane proteins into large unilamellar liposomes after extraction into a hydrocarbon solvent together with phospholipids. The membrane protein was added to a suspension of lipids in a suitable buffer, ether and/or n-hexane added and the mixture sonicated for a few minutes under argon. Large unilamellar liposomes of about 1 µm are formed after the removal of the organic solvent in a rotary evaporator. Slow hydration of the ether or n-hexane extract composed of lipids and proteins yields liposomes with a size of between 5-100 µm having one or at most a few layers of lipid (Darszon et al., 1980).

Critical aspects are the selection of the organic solvent as well as the volume ratio of aqueous to non-aqueous phase. Ca^{2+} is added as a counter-ion in order to neutralize the overall charge.

This method seems to be applicable to many membrane proteins (Darszon 1983; Rigaud et al., 1983), although organic solvents may not be well tolerated by all proteins, and as with the use of detergents, a potential disadvantage may be that traces of solvents that are difficult to remove may affect vesicle properties.

E) INJECTION METHOD

The major glycoprotein of the human erythrocyte membrane was incorporated into liposomes which were mainly of the unilamellar type as prepared by the method of Batzri and Korn (1973). An ethanolic solution of PC:Chol 2:1 was rapidly injected into a solution of the protein in buffer. After removal of the free glycoprotein by ultrafiltration it was found that $\approx$ 80% of the protein was inserted into the outer half of the bilayer (Juliano and Stamp, 1976; Redwood et al., 1975).

F) UREA METHOD

After having tried several of the methods described in the literature, e.g. differently charged liposomes, detergent dialysis and dilution with various detergents, sonication, and some combinations of these methods as well as incorporation in the presence of diheptanoyl PC (see below, Gabriel et al., 1984), it was obvious that the membrane protein in which we are interested, the variant surface antigen (VSG) from <u>Trypanosoma</u> <u>brucei</u>, could not be incorporated into liposomes using these methods. This protein exists in a water-soluble form (sVSG) and a membrane-bound form (mfVSG). The difference between these two forms is a dimyristoyl diglyceride (for details of the structure see Egge et al. in this book). One would expect the diglyceride to facilitate the incorporation of mfVSG into liposomes, but neither of the methods which we tried resulted in an incorporation of greater than 30%. Our interpretation of these results was that mfVSG exists in aqueous solution in a conformation which prevents the diglyceride from interacting with the environment, probably aided by hydrogen bonding between its oligosaccharide and/or peptide portions, or by forming dimers or oligomers (Gurnett et al., 1986). Detergents seem not to affect the conformation sufficiently for incorporation to occur.

On the other hand urea is known to change the secondary and tertiary structure of proteins. So finally we tried to incorporate mfVSG into vesicles by mixing a PC:PA:Chol (65:5:30) suspension in 8M urea with a solution of mfVSG in 8M urea. This method resulted in 80% incorporation (Schmitz et al., 1986). Bilayers are permeable to urea, and so urea is easily removed by dilution and centrifugation. Liposomes prepared by this technique consist of 5-6 bilayers as shown by E.M. We could also show that mfVSG was incorporated to more than 85-90% into the outer lamella of the liposomes via its diglyceride; after pronase digestion of the incorporated VSG and isolation of the C-terminal glycophospholipopeptide from the liposomes, the terminal diglyceride as well as the dimyristoylphosphatidylinositol were identified by FAB-MS.

This method cannot be applied to proteins that are irreversibly denatured by urea. But it is known that denaturation by urea is often a reversible process (Thomson and Augusteyn, 1984). Work is now in progress to convert the multi-layered liposomes into large unilamellar vesicles by pressure extrusion (Hope et al., 1985).

G) SPONTANEOUS INCORPORATION

A few membrane proteins have been shown to be incorporated spontaneously into vesicles. Cytochrome b_5, for example, an integral membrane protein, binds rapidly below and above T_m to preformed liposomes consisting of PC:PE 95:5 (prepared by sonication, 95% of the vesicles being of approximately 25 nm in diameter; Enoch et al., 1977). Dufourcq et al. (1975) also showed that cytochrome b_5 is readily incorporated into sonicated liposomes containing only egg yolk PC or DPPC. Another protein which is incorporated spontaneously into small unilamellar vesicles of DPPC (20-30 nm, prepared by sonication) above or below the phase transition, is the hepatic asialoglycoprotein receptor (Klausner et al., 1980).

H) FACILITATED INCORPORATION

There are probably not many membrane proteins that can be incorporated spontaneously into liposomes. Different approaches have been developed over the last few years with the aim of avoiding conditions which may have a damaging effect on protein and lipid.

The lipid composition often has a decisive influence on the amount of incorporation as already discussed. Negatively charged lipids especially tend to facilitate incorporation, possibly by disturbing the order of the bilayer. Eytan and Racker (1977) studied the reconstitution of some mitochondrial enzymes, e.g., cytochrome oxidase, into preformed liposomes prepared by sonication. They found that it is necessary, for satisfactory reconstitution, to include up to 30% negatively charged lipids. The concentration of certain cations such as Mn^{2+}, Mg^{2+} (10 mM), or Ca^{2+} (2 mM) seems to be important.

Other lipid compounds which are not normally found in membranes, or only in very small amounts, show a marked effect on the efficiency of incorporation for some proteins. Eytan et al. (1975) described the incorporation of cytochrome oxidase at $0^{o}C$ in the presence of 10% lysolecithin. The activity of the reconstituted system was similar to that found after incorporation using other techniques (e.g., freeze-thawing, sonication, detergent dilution; Racker (1985)). Low concentrations of detergent (0.1%) which do not disrupt vesicles gave similar results.

Amino peptidase, bound to lysophospholipid micelles, was incorporated into liposomes at pH 5. 100% of the protein was incorporated with a ratio of protein:lyso-PL of 1:4 and PL:lyso-PL of 15:1 (Christiansen and Carlsen, 1983).

Scotto and Zakim (1985) reported the spontaneous incorporation of integral membrane proteins (UDP-glucuronyl transferase, cytochrome oxidase and bacteriorhodopsin) into preformed unilamellar DMPC liposomes in the presence of myristic acid as the fusogen (up to 11 mol%) with a maximum at 18°C, i.e., below the T_m for DMPC. There was no incorporation observed at 30°C in the presence of myristic acid, and none at 18°C without myristic acid. They found that cholesterol also induces protein incorporation probably in a similar manner to myristic acid (Scotto and Zakim, 1986). This method, however, produces a mixture of lipid-protein vesicles together with vesicles which are devoid of protein, the vesicles being of rather heterogeneous size distribution.

Another easy and gentle procedure for the incorporation of membrane proteins was described by Dencher (1986). He used short-chain lecithins (C6-C8), which tend to form micelles in aqueous solutions, mixed with long-chain lecithins. Gabriel and Roberts (1984) were the first to describe this technique; although the aqueous suspensions of such mixtures look rather clear, suggesting unilamellar vesicles, we found that under E.M. they form unilamellar "spaghetti-like" structures rather than vesicles (Schmitz and Klein, unpublished observations).

The addition of bacteriorhodopsin to an aqueous mixture of soya bean phospholipids, DMPC, or DPPC with 20 mol% di-heptanoyl-PC, aided by short sonication, leads to the spontaneous formation of lipid-protein vesicles with a vesicle diameter of between 35-100 nm with soya bean, or 100-450 nm with DMPC as the long-chain phospholipid as shown by Dencher using E.M. (Dencher, 1986).

Gould-Fogerite and Mannino (1985) describe the preparation of large proteo-liposomes by rotary dialysis, based on the calcium-EDTA-chelation method of Papahadjopoulos et al. (1975). Cochleates, spirally folded cylinders, are produced from phosphatidylserine:cholesterol 9:1 and the membrane protein (influenza virus glycoprotein) by short sonication and subsequent precipitation during dialysis in the presence of 3 mM $CaCl_2$. Calcium is then removed by dialysis against 10 mM EDTA in a specially devised apparatus; during this procedure large unilamellar liposomes (0.1 - 3 µm) are formed which contain up to 80% of the applied protein.

Although this method is only applicable to negatively charged phospholipids, it represents another gentle procedure for the incorporation of membrane proteins into liposomes.

REFERENCES
Bangham, A.D., Standish, M.M. and Watkins, J.C. (1965) J. Mol. Biol. 13, 238-252

Batzri, S. and Korn, E.D. (1973) Biochim. Biophys. Acta 332, 11-25

Christiansen, K. and Carlsen, J. (1983) Biochim. Biophys. Acta 735, 225-233

Darszon, A., (1983) J. Bionerg. Biomembr. 15, 321-334

Darszon, A., Vandenberg, C.A., Ellisman, M.H. and Montal, M. (1979) J. Cell Biology 81, 446-452

Darszon, A., Vandenberg, C.A., Schönfeld, M., Ellisman, M.H., Spitzer, N. and Montal, M. (1980) Proc. Natl. Acad. Sci. U.S.A. 77, 239-243

Dencher, N.A. (1986) Biochem. 25, 1195-1200

Dufourcq, J., Faucon, J.F., Lussan, C. and Bernon. R. (1975) FEBS Letters 57, 112-116

Enoch, H.J., Flemming, P.G. and Strittmatter, P. (1977) J. Biol. Chem. 252, 5656-5660

Epand, R.M., Boni, L.T. and Hui, S.W. (1982) Biochim. Biophys. Acta 692, 330-338

Epand, R.M. and Sturtevant, J.M. (1984) Biophys. Chem. 19, 355-362

Epand, R.M. and Surewicz, W.K. (1984) Can. J. Biochem. Cell Biol. 62, 1167-1173

Eytan, G.D., Matheson, M.J. and Racker, E. (1975) FEBS Letters 57, 121-233

Eytan, G.D. and Racker, E. (1977) J. Biol. Chem. 252, 3208-3213

Gabriel, N.E., Roberts, M.F. (1984) Biochem. 23, 4011-4015

Gould-Fogerite, S. and Mannino, R.J. (1985) Anal. Biochem. 148, 15-26

Greenhut, S.F., Bourgeois, V.R. and Roseman, M.A. (1986) J. Biol. Chem. 261, 3670-3675

Gurnett, A.M., Raper, J. and Turner, M.J. (1986) Mol. Biochem. Parasitol. 18, 141-153

Juliano, R.L. and Stamp, D. (1976) Nature Lond. 261, 235-237

Hah, J.S., Hui, S.W. and Jung, C.Y. (1983) Biochem. 22, 4763-4769

Hope, M.J., Bally, M.B., Webb, G. and Cullis, P.R. (1985) Biochim. Biophys. Acta 812, 55-65

Kagawa, Y., and Racker, E. (1971) J. Biol. Chem. 246, 5477-5487

Kasahara, M. and Hinkle, P.C. (1976) Proc. Natl. Acad. Sci. U.S.A. 73, 396-400

Klausner, R.D., Bridges, K., Tsunoo, H., Blumenthal, R., Weinstein, J.N. and Ashwell, G. (1980) Proc. Natl. Acad. Sci. U.S.A. 77, 5087-5091

MacDonald, R.I., Oku, N. and MacDonald, R.C. (1983) in Liposome Letters (Bangham, A.D., ed.) pp.63-72. Academic Press, New York

Mateu, L., Caron, F., Luzzati, V. and Billecoq, A., (1978) Biochim. Biophys. Acta 508, 109-121

McElhaney, R.N. (1982) Chem. Lipids 30, 229-259

Mimms, L.T., Zampighi, G., Nozaki, Y., Tanford, C. and Reynolds, J.A. (1981) Biochem. 20, 833-840

Navarro, J., Chabot, J., Sherrill, K., Aneja, R., Zahler, S.A. and Racker, E. (1985) Biochem. 24, 4645-4630

Nicholls, P. (1983) in Liposome Letters (Bangham, A.D., ed.) p.215-229. Academic Press, New York

Papahadjopoulos, D., Vail, W.J., Jacobsen, K., and Poste, G. (1975) Biochim. Biophys. Acta 394, 483-491

Petri, W.A. and Wagner, R.R. (1979) J. Biol. Chem. 254, 4313-4316

Prendergast, F.G., Lu, J., Wei, G.J. and Bloomfield, V.A. (1982) Biochem. 21, 6963-6971)

Pick, U. (1981) Arch. Biochem. Biophys. 212, 186-194

Racker, E. (1979) in Methods in Enzymology (Fleischer, S. and Packer, L., eds.), Vol.55, pp.699-711. Academic Press, New York

Racker, E. (ed.) (1985) Reconstitutions of Transporters, Receptors and Pathological States. Academic Press, London

Racker, E. and Stoeckenius, W. (1974) J. Biol. Chem. 234, 662-663

Racker, E., Chien, T.F. and Kandra, A. (1975) FEBS Letters 57, 14-17

Redwood, W.R., Jansons, V.K. and Patel, B.C. (1975) Biochim. Biophys. Acta 406, 347-361

Rigaud, J.L., Bluzat, A. and Buschlen, S. (1983) Biochem. Biophys. Res. Commun. 111, 363-382

Schenerman, M.A., Racker, E. and Kilberg, M.S. (1985) J. Cell Biology 101, 189a

Schmidt, G. and Gräber, P. (1985) Biochim. Biophys. Acta 808, 46-51

Schmitz, B., Klein, R.A., Egge, H. and Peter-Katalinic, J. (1986) Mol. Biochem. Parasitol., 191-197

Scotto, A.W. and Zakim, D. (1985) Biochem. 24, 4066-4075

Scotto, A.W. and Zakim, D. (1986) Biochem. 25, 1555-1561

Suárez-Isla, B.A., Wan, K., Lindstrom, J. and Mental, M. (1983) Biochem. 22, 2319-2323

Szoka, F., Jr. and Papahadjopoulos, D. (1978) Proc. Natl. Acad. Sci. U.S.A. 75, 4194-4198

Szoka, F., Jr. and Papahadjopoulos, D. (1980) Ann. Rev. Biophys. Bioeng. 9, 467-508

Taraschi, T.F., van der Steen, A.T.M., de Kruiff, B., Tellier, C. and Verkleij, A.J. (1982) Biochem. 21, 5756-5764

Thomson, J.A. and Augusteyn, R.C. (1984) J. Biol. Chem. 259, 4339-4345

Wickner, W.T. (1977) Biochem. 16, 254-258

PHOSPHOLIPID MEMBRANES AS DRUG DELIVERY VEHICLES

L. Michaelis and M.J. Moore

Pharmacology Department, Düsseldorf University, F.R.G. and Biochemistry Department at Hull University, U.K.

INTRODUCTION

Phospholipids such as phosphatidylcholine, phosphatidylethanolamine, phosphatidylserine, and sphingomyelin play an important role in biological tissue and are components of all cell membranes. They have been extracted from a variety of cell types to reconstitute the bimolecular lipid backbone of biomembranes. Today there are a number of techniques available for the production of model membranes which may be summarized in three classes:

1) Monolayers: A small volume of lipid suspension is spread on the aqueous surface of a Langmuir trough to form a mono-molecular layer. 2) Black Lipid Membranes: A drop of lipid suspension is placed on a circular orifice of a cuvette which is submerged in an aqueous electrolyte solution. As the lipid solvent retracts to the edges of the orifice a bimolecular membrane is formed which appears "black" to reflected light. 3) Phospholipid Vesicles: In the most popular production method a lipid film is produced on the walls of a roundbottom flask using a rotary evaporator. Then the aqueous phase is added and the mixture agitated to form large spherical multilayer vesicles, each layer consisting of a bimolecular membrane and being separated from the neighbouring layer by aqueous phase. In a further step these "handshaken" vesicles are subjected to ultrasonic irradiation for any length of time. Thus unilamellar small vesicles are produced with a diameter of around 250 Å.

Soon after Bangham (1) first described these smectic mesophases or liposomes it became obvious that they could be used to deliver material to biological cells. This created considerable excitement. Some enthusiastically thought that it must be possible to target liposomes to special organs or even cell types, to use them as Trojan horses to transport drugs, enzymes, antibodies etc. The ultimate of this "drug targeting" idea was considered to be the destruction of tumor metastatic cells - a fascinating thought indeed. However, today we are far away from such ambitions. Although liposomes are used in many fields of research their clinical application is still in its preliminary stages. An extensive review about liposome technology and medical application has recently been compiled by Gregoriadis (2) and we would like to add some results obtained in collaboration with the ophthalmic hospital at Düsseldorf University.

Small unilamellar liposomes when injected intravenously are preferentially taken up by hepatocytes (3) which contain a large number of transport systems, particularly the one for vitamin A. Via this route it should be possible to deliver carotinoids or vitamin A precursors to the retina.

We are reporting about the use of canthaxanthin as an example. This substance is used in oral suntanning pills and may cause severe irreversible retinopathy. By using liposomes for intravenous injection it could be shown (in rabbits) that concentrations of about 10 mg/kg body weight (e.g. transferred with all reservations to an adult male human being 3 pills of 15 mg per day for about three weeks) may already have negative effects on electroretinograms. Consequently the substance was withdrawn from the German market (September 1985) as an oral suntanning agent.

MATERIALS AND METHODS

Phosphatidylcholine (PC) was extracted from egg yolk by a modified method based on those of Dawson (4), Lea, Rhodes and Stoll (5) and Hanahan, Dittmer and Warashina (6).

Homogeneity of the preparations was checked by TLC on Silicar gel G plates as described by Mangold (7). The extent of oxidation of the phospholipid solutions was determined by the method of Klein (8). Material showing more than 0.2 for oxidation was discarded. Lipid phosphorus was qualitatively tested by using the method of Dittmer and Lester (9) and quantitatively assayed by the determination as ammonium phosphomolybdovanadate at 390 nm.

Purity of canthaxanthin (4,4'dioxo-beta-carotene) was checked on Silicar gel G plates using dichloromethane/diethyl ether 9/1 and assayed quantitatively by spectrophotometry. As the substance is sensitive to oxygen and direct sunlight the necessary precautions were taken.

Liposomes were prepared in the conventional manner i.e. appropriate aliquots of stock solution were mixed, the solvents evaporated, the film resuspended in pyrogen-free isotonic saline and sonicated for any length of time (until laser light scattering gave a value < 1000 Å for diameter) using a Branson B12 water bath at 30°C. The samples were then centrifuged at 1000 g for 10 min to remove any large particles.

Homogeneity of the preparations and liposome diameter were estimated by laser light scattering spectroscopy (10) and transmission electron microscopy.

Liposomes were injected intravenously into anaesthetized chinchilla bastard rabbits. After repeated dosage of liposomes without and with canthaxanthin (16.7 mg PC plus 2.1 mg canthaxanthin) for eight weeks a cumulative dose was injected (200 mg total PC/rabbit resp. 200 mg PC plus 50 mg canthaxanthin). Mydriasis was introduced by tropicamid and after dark adaptation electroretinograms were recorded. Scotopic (dark-adapted) a- and b-waves in the electro-retinogram (ERG) were measured as an estimate of retinal function. Any changes may serve as a sign of possible severe ophthalmic complications. Parameters evaluated were overall amplitude at standard intensity, the maximum a- and b-wave amplitudes and the peak latency periods of scotopic a- and b-waves.

RESULTS AND DISCUSSION

The power spectra of laser light scattering experiments and electron microscopy photographs show that phosphatidylcholine liposome preparations without and with canthaxanthin are < 1000 Å in diameter and relatively homogeneous and reproducible in size. No multilamellar structures can be detected. The choline methyl membrane outside resonance of the NMR-spectrum is shifted toward lower magnetic field when Pr^{3+} is added. This proves that "intact" liposomes are present. That canthaxanthin is incoporated into the liposome membrane can be deduced from the intense colour of the vesicles after centrifugation and also by re-extracting supernatants after varying lengths of sonication followed by centrifugation at about 1000 g and assaying for phospholipid and canthaxanthin. We have incorporated up to 40 mole% of canthaxanthin into the liposome membrane but it was found to be more convenient to use concentrations of around 20 mole%.

The electroretinograms show that canthaxanthin prolongs the peak latencies of scotopic a- and b-waves which is proof that the substance reaches the retina and influences the membrane potential (Figure 1). As canthaxanthin is practically water-insoluble it must reach the retina by a transport mechanism. As up to 80% of small unilamellar liposomes are taken up by hepatocytes it is possible that canthaxanthin uses the retinol binding protein.

On the other hand the substance - or at least a certain percentage of it - may well reach the retina by "direct contact" of the circulating liposomes.

Admittedly there are still a number of questions to be answered but the medical implication to deliver water-insoluble (carotinoid-like) drugs to the retina by incorporating them into small liposomes is quite promising.

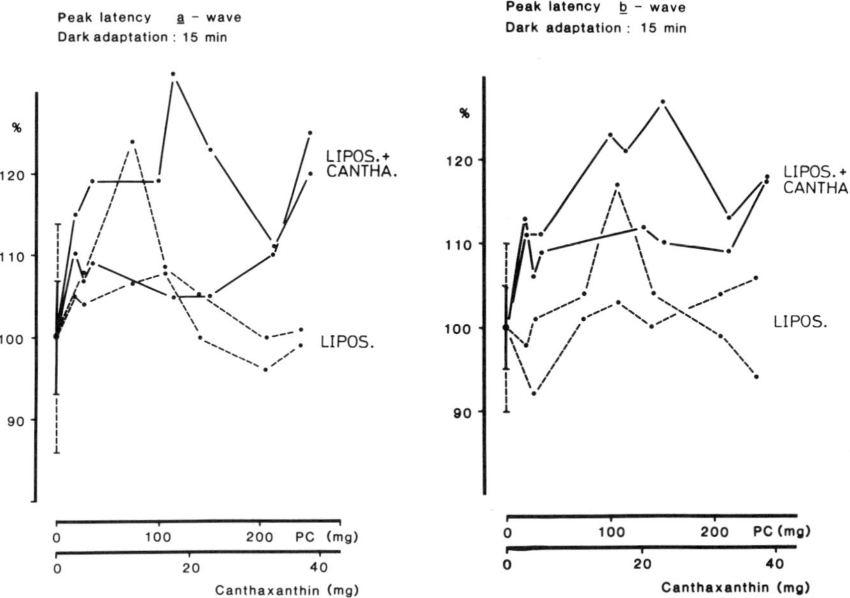

Figure 1. Peak latencies of scotopic a- and b-waves taken from electroretinograms of rabbits treated with phospholipid liposomes with and without canthaxanthin.

REFERENCES

1) Bangham, A.D., Standish, M.M. and Watkins, J.C. (1965) J. Mol. Biol. 13, 228-252.
2) Gregoriadis, G. (1985) Liposome Technology Vol. I-III, CRC Press.
3) Scherphof, G.L. (1985) Abstracts of 13th Intern. Congr. Biochem. TU 377 and TU 380, Elsevier Science Publishers B.V.
4) Dawson, R.M.C. (1959) Biochem. J. 70, 559.
5) Lea, C.H. et al. (1955) Biochem. J. 60, 353.
6) Hanahan et al. (1957) J. Biol. Chem. 228, 685.
7) Mangold, H. (1961) J. Amer. Oil Chem. Soc. 38, 708.
8) Klein, R.A. (1970) Biochim. Biophys. Acta 219, 496.
9) Dittmer, J.C. and Lester, R.L. (1962) J. Lipid Res. 37, 126.
10) Schlieper, P. et al. (1981) Biochim. Biophys. Acta 598, 405.

EASY BREATHING AND THE DESIGN OF ARTIFICIAL LUNG EXPANDING COMPOUNDS (ALEC)

A.D.Bangham

Cambridge Lung Surfactants Ltd., The Cottages, 17 High Green, Great Shelford, Cambridge UK

N.G.A.Miller

Institute of Animal Physiology, Babraham, Cambridge UK

C.J.Morley

Department of Paediatrics, University of Cambridge, Cambridge UK

SUMMARY

A dry protein-free powder consisting of a 7:3 mole/mole mixture of dipalmitoylphosphatidylcholine/phosphatidylglycerol seems to function as a reasonably good substitute for natural lung surfactant in very premature babies. It has the following necessary properties: 1. It spreads rapidly and spontaneously at an air/water interface at $37^{O}C$ reducing the surface tension of water by about 2/3. 2. The unsaturated phosphatidylglycerol (PG) moiety can be squeezed out of the mixed monolayer by rapid overcompression (equivalent to exhalation) and is irreversibly lost to the surface system by reassembly into liposomes. 3. The residual dipalmitoylphosphatidylcholine (DPPC) becomes progressively enriched to the point that, at $37^{O}C$, it condenses out as a solid phase so rigid that it prevents the alveolae from collapsing. 4. The preparation is protein-free.

It is suggested that this simple mixture of two phospholipids exhibits both thermodynamic (equilibrium) and kinetic forces during the course of a compression/decompression cycle on a trough. Likewise, during a respiratory cycle, the alveolae might be kept open (or more precisely, the liquid lining the alveolae is prevented from filling them up) at full expiration by the presence of a permanent residue of almost pure DPPC which, being condensed (solid), is incompressible. The work of extending the uncovered air/water interface, upon inspiration, is reduced by replenishment from the stockpile of dry surfactant as though from a lamellar body. The respiratory cycle is thus seen as a dynamic sequence of ablution and replenishment of phospholipid molecules.

The material has been used in clinical trials involving some 400 or more at-risk babies and the encouraging results will be summarised.

DISCUSSION

Interface physics and chemistry predominate in two aspects of pulmonary physiology. First, at birth, there is an urgent necessity to dry out and expand the lungs by extending an air-water interface, initially no greater than 1 cm^2, to an area of approximately 3 m^2 for a 3 kg infant. This first event clearly requires to be achieved against the smallest force possible and with a minimum total energy, anticipating that the ultimate air cavitites have very small radii of curvature. Subsequently, and for the rest of life, there is an absolute requirement to protect the lungs from refilling with fluid despite adverse gravitational, hydrodynamic, osmotic and capillary forces. Depots of surface active material (lung surfactant) are found in abundance in the lungs of normal, full term infants and consist almost entirely of phospholipids; dipalmitoylphosphatidylcholine (DPPC, approx 40%) being a particularly rich component. Deficiency of lung surfactant in premature and/or immature infants predisposes them to conditions of respiratory distress in the neonatal period. Artificial lung expanding compounds (ALEC's) are being developed in a number of centres throughout the world as replacement therapy, to anticipate and alleviate such conditions (Fujiwara et al. 1980; Morley et al. 1981; Smyth et al. 1983; Hallman et al. 1983; Halliday et al. 1984; Hallman et al. 1985; Enhorning et al. 1985; Kwong et al. 1985).

The approximate pressure of air ($\Delta \underline{P}$ in $N.m^{-2}$) generated by a baby at birth and indeed for most of its life, to extend and sustain the intra-pulmonary air/fluid interface at inspiration is maximally about 20 centimeters of water (1 cm H_2O = 98.3 $N.m^{-2}$), which by the Laplace law (small bubbles emptying into larger ones or $\Delta \underline{P} = 2\gamma_{aw}/\underline{r}$) limits the minimum radius of curvature $\underline{r}$ (m) of any liquid meniscus in the alveolus to about 70 microns if γ_{aw}, the surface tension of the air/water interface of the tracheal fluid, is 70 $mN.m^{-1}$ at 37°C. The radius of curvature $\underline{r}$ reduces to about 25 microns for the same air pressure if γ_{aw} is lowered to 25 $mN.m^{-1}$, a reasonable value for a hydrocarbon-based insoluble surfactant at equilibrium on a water surface. Thus, there appears to be no theoretical reasons to prevent extension of the air/water interface down the baby's trachea, bronchi and smaller branches, to the limit of an advancing air "bubble" of radius 70 microns. The amount of work would be reduced by 2/3 (from 210 mJ or a 240 gm wt falling through 10 cm, to 75 mJ) if soapy water (γ_{aw} = 25 mN.m^{-1}) for example, were present, but the incipient froth would be highly embarrassing. Nature has anticipated this outcome by providing a group of surfactants, the phospholipids we think, which are virtually insoluble in water and can be recruited from crystalline or liquid-crystalline sources, viz. the lamellar bodies of the Type II cells and mixtures of pure phospholipids as formulated in ALEC (Bangham et al. 1984) on the basis of their surface chemical properties (Bangham et al. 1979). Bearing in mind that insoluble surfactants, like their soluble counterparts, lower γ_{aw} by displacing hydrogen-bonded water molecules from the air/water surface by hydrocarbon chains, there is no quibble with the classic concept of surface tension (γ_{aw}). However, it would appear that both man and some animals need to accommodate liquid menisci in their lungs of far smaller radii of curvature than can be supported by the application of the Laplace law and the values discussed above (Hills 1983). The second physico-chemical problem therefore concerns the promotion and stability of these very small menisci, a prerequisite to keep the alveolar space maximally expanded and dry throughout a respiratory cycle and throughout a normal lifespan. Contemporary wisdom suggests, rather naively, that natural lung surfactant (LS) solves this problem by reducing γ_{aw} to whatever value between 72 - 0 mN. m^{-1} is necessary, where necessary (Scarpelli et al. 1983)! This cannot be so (Bangham 1986).

Reminding oneself that surface tension arises because of the extremely short-range but unequal pull experienced by surface molecules, it can and should be argued that a water surface completely covered by a monolayer of dipalmitoylphosphatidylcholine (DPPC) behaves, surface-wise, like a jug full of octane, γ = 20.8 mN.m^{-1} at room temperature! Heat the jug of hydrocarbon up and the surface tension undoubtedly falls until a critical temperature (boiling point) is reached at which point there is no longer an unequal pull, no longer a tension to resolve in the plane of the interface, and therefore, no longer an interface! For octane the temperature is 24°C but for longer chain hydrocarbons such as might be found in phospholipids, in excess of 200°C! Claims, therefore, that LS can achieve near-zero surface tensions (γ_{aw}) are manifestly absurd since a zero surface tension implies no interface! As Comroe relates (Comroe 1977), Pattle very clearly recognised these problems and offered an (almost) entirely satisfactory physical explanation but which one now realises has never been clearly understood (Pattle 1955). Pattle suggested that a material was available in the lungs which, after surface adsorption, denatured to form an incompressible "skin" of molecules similar, in many ways, to the "skin" that forms on hot milk or the tanned protein that sustains the "foam" in the stomach of ruminants suffering (and dying) from "bloat". He based his conclusions on studies made with lung foam, noting that unlike soap bubbles which obey the Laplace law, the extremely small lung bubbles lined with lung surfactant (20 microns or less) remain intact for 20 hours or longer. He concluded: 'It is thus evident that the alveoli are lined with an insoluble protein layer which can abolish the "tension" of the alveolar surface'. He did not go on to explain that an insoluble layer of protein would, in physical terms, have constituted a new (solid and irreversible) phase which now separated the air from the liquid on the alveolar surface. This omission (from his hypothesis) is crucial because his explanation denies, quite correctly, the existence of any true air/liquid interface in the smaller menisci of a lung. In fact, he is saying that there are, indeed, two interfacial regions, liquid/solid and solid/air, and it is the properties of the latter that determine the topography of the underlying liquid. Failure to recognise that a wet alveolar surface might be covered by a solid phase, proteinaceous or phospholipid, has led to the false and quite untenable assumption that the "surface tension" of alveolar liquid can be reduced indefinitely by the presence of a surfactant (Bangham et al. 1979; Bangham 1986). There can be little doubt that this erroneous idea developed from the widely quoted but inappropriately used relationship:-

$$\gamma_{aw} = \gamma_{72} - \Pi$$

where γ_{aw} is the (supposed) surface tension at the air/liquid interface, γ_{72} is the surface tension of clean water (72 $mN.m^{-1}$) and Π is the (osmotic) pressure ($mN.m^{-1}$) of insoluble surfactant molecules relative to a surfactant-free surface; it being wrongly assumed that all values of surface pressure (0 - 72 $mN.m^{-1}$), whether measured in a Langmuir-Adam trough with a barrier balance or a Wilhelmy dipping plate, can be converted to values of γ_{aw}. Indeed, it was a measure of one's familiarity with these matters to be able to invert a tension/area to a pressure/area isotherm without standing on one's head! The confusion arose because of the tradition, on the one hand, of measuring surface concentrations of <u>insoluble</u> surfactants as surface pressures (Π), and of the mistaken belief in the existence of a substance capable of lowering the surface tension (γ_{aw}) to very low values, on the other.

The pressure experienced (although the third dimension is only one molecule thick) by (insoluble) surface-active compounds such as oleic acid, cholesterol, long-chain fatty acids or lung surfactants, DPPC and/or phosphatidylglycerol (PG), when placed on a water surface of defined area is very tangible, equivalent to a force of some 45 mN.m^{-1} (i.e. a gram weight per 22 cm of barrier length), or some 220 atmospheres. George Gaines Jr however, in his book entitled "Insoluble Monolayers" (Gaines 1966), points out that the surface pressure (Π) can, under appropriate circumstances, represent the sum of <u>two</u> energies: an equilibrium (surface spreading) and a kinetic energy. The equilibrium surface energy of an insoluble surfactant corresponds to the spontaneous (reversible) tendency of surface-active molecules to occupy the surface (and by so doing, replace the more highly self-attractive water molecules), and the measured surface pressure genuinely relates to γ_{aw}, as defined thermodynamically, having the same value at all points of a surface; the energy trade-off (or available free-energy) exercised by surfactant molecules being partly a solvation of the (phospholipid) head-groups and partly an entropic gain due to an incongruity between water and hydrocarbon structures. In a Langmuir-Adam trough, the kinetic energy element arises from further compression of the monolayer by an external energy source such as the operators hand or an electrically driven motor, pushing a barrier and squeezing the surface molecules. Since it is, characteristically, insoluble surface-active compounds that are the most likely to react to such kinetic energy, its relevance to LS is obvious. For instance, pure, long-chain, saturated fatty acids, spread and compressed on an acid subphase, condense to become solid films, which sublime to a surface-vapour phase very slowly if and when the pressure is reduced i.e. they exhibit surface-equilibrium hysteresis. Such behaviour becomes understandable for LS when one realises that molecules 20 Å long, constrained by a barrier to a pressure of 70 mN.m^{-1}, (i.e. a typical pressure sustained by a "good" LS), experience a three-dimensional pressure of 352 atm or almost 5000 lb in^{-2}. But it is also likely that the kinetic energy imparted to the surface film is also responsible for the exclusion of minor constituents of LS and, in particular, unsaturated phospholipids (the phosphatidylglycerol, PG, in the case of an artificial lung surfactant). These would tend not to return into the surface film but aggregate in the aqueous phase as liposomes with the result that DPPC, in particular, would be progressively refined. It is the progress and consequence of this surface refinement process which offers an explanation as to how and why small menisci are prevented from filling up into larger ones, particularly upon expiration. If experimentalists with LS were to attempt to rotate their Wilhelmy dipping plates instead of using them to measure meaningless pulls, they would appreciate the stiffness of the compressed surface films and begin to understand what is really taking place on a water surface covered with surfactant!

A monolayer of pure DPPC becomes a condensed (solid) film when $\underline{T}$ < $\underline{T}_c$ = 41°C and such films can sustain very high pressures (Π> 70 mN.m^{-1}) on compression (Watkins 1968). A reason for this property is a high activation energy to the transition from crystalline monolayer to crystalline multilayer or collapse phase so that the transformation is slow and under the condition of over-compression, the monolayer is metastable with respect to the collapse phase and $\Pi \neq \gamma_{72} - \gamma_{aw}$ (Gaines, 1966). Pure DPPC, however, does not have all the appropriate properties to act as the complete substitute for natural lung surfactant because it cannot spread spontaneously over a water surface at $\underline{T}$ < 41°C, i.e. its equilibrium spreading pressure Π_e = 0 mN.m^{-1}. Furthermore, although at $\underline{T} > \underline{T}_c$ dry DPPC can spread spontaneously over a surface (fully hydrated DPPC, as liposomes, does not; see ref. Pattus et al. 1978), Π_e equals approximately 48 mN.m^{-1} (γ is therefore some 24 mN.m^{-1}); compression of the resultant liquid expanded monolayer does not give higher values of Π because over-compression of the liquid monolayer results in rapid removal of phospholipid molecules from the monolayer into either a reversible, dry multilayer collapse phase or an irreversible hydrated collapse phase. A minimum surface tension γ_{min} of 24 mN.m^{-1} is higher than the γ_{min} < 9 mN.m^{-1} observed in contracted lung (Schurch et al. 1976) and would lead to alveolar collapse. Thus, it is apparent that the requirements of rapid, spontaneous spreading at the air-water interface and attainment of zero contractile liquid-force ($\underline{F}_c \approx$ 0 mN.m^{-1}) at high compression of the monolayer are mutually exclusive for pure DPPC at a given temperature (37°C for the lung).

In order to achieve both rapid spreading and $\underline{F}_c \approx 0$ mN.m^{-1} it is necessary to resort to a lipid mixture, as occurs <u>in</u> <u>vivo</u> in the lung surfactant, which contains a heterogeneous mixture of lipids including DPPC and PG (Goerke 1974). A mixed monolayer of DPPC plus a phospholipid with a lower $\underline{T}_c$ behaves differently when compressed repeatedly because it forms an irreversible collapse phase selectively enriched in the least stable component (Phillips et al. 1970) leaving a monolayer progressively enriched in DPPC. By this means, a DPPC monolayer capable of undergoing the transition from liquid-expanded to condensed state on compression is present at the air-water interface at 37°C. Once the DPPC monolayer is condensed, it can be compressed to $\Pi > 70$ mN.m^{-1} thereby eliminating the contractile liquid surface and stabilising the alveoli on expiration. Thus, in order to formulate a lung surfactant replacement based on DPPC (which is normally present), other phospholipids must be added to reduce $\underline{T}_c$ to below 37°C so that the dry mixture can spread spontaneously, but these added lipids must be squeezed out of the mixed monolayer preferentially leaving a condensed

DPPC-enriched monolayer to be compressed to $\underline{F}_c \approx 0$ mN.m^{-1}. Replenishment would be achieved from the resevoir of the dry lipid mixture which spreads into the monolayer when γ falls below γ_e as the alveolus expands.

A 3:1 mol/mol DPPC/PG mixture had the desired properties outlined above (Bangham, et al. 1979) but in practice was found to be extremely difficult to deliver to very small infants as a dry preparation in the form of a dust or smoke. Instead it is dispersed in cold saline (well below the $\underline{T}_c$) which is then placed in the pharynx before or at the time of the first breath, the warmth of the infant promotes spreading whenever the material happens to be at an appropriate site. We have recently completed two clinical trials with ALEC; a randomized trial involving some 341 babies between 23 and 34 weeks gestation (Morley et al. 1986(a)) and most recently, a multicentred, randomized trial within the UK of babies less than 30 weeks gestation (Morley et al. 1986(b)). Both trials report encouraging results which include a reduction of >40% in mortality and to a lesser extent of such crippling complications as intraventricular haemorrhage and reduction in the severity of lung disease.

REFERENCES

Bangham AD (1986) Breathing made easier. (Submitted) "The Lung".

Bangham AD, Miller NGA, Davies RJ, Greenough A, Morley CJ (1984) Colloids and Surfaces 10 : 337-341.

Bangham AD, Morley CJ, Phillips MC (1979) Biochim. Biophys. Acta 573 : 552-556.

Comroe Jr JH (1977) Premature Science and Immature Lungs. Part I: Some premature discoveries pp. 127-135; Part II: Chemical warfare and the newly born pp. 311-323; Part III: The attack on immature lungs pp. 497-518. Am. Rev. Respir. Dis. Volume 116.

Enhorning G, Shennan A, Possmayer F, Dunn M, Chen CP, and Milligan J (1985) Paediatrics 76 : 145-153.

Fujiwara T, Chida S, Watabe Y, Maeta H, Morita T and Abe T (1980) Lancet i:55 -59.

Gaines GL, Jr (1966) Insoluble Monolayers. Interscience Publishers, New York.

Goerke J (1974) Biochim.Biophys.Acta 344 : 241-261.

Halliday HL, McClure G, Reid M, Lappin TRJ, Meban C and Thomas PS (1984) Lancet i : 476-478.

Hallman M, Merritt TA, Schneider H, Epstein BL, Mannino F, Edwards DK and Gluck L (1983) Paediatrics 71 : 473-482.

Hallman M, Merritt TA, Jarvenopaa AL, Boynton B, Mannino F, Gluck L, Moore T and Edwards DK (1985) The Journal of Paediatrics 106 : 963-969.

Hills BA (1983) Pulmonary Surfactant System. Eds. EV Cosmi and EM Scarpelli. Symposia of the Giovanni Lorenzini Foundation Vol. 16 pp. 17-32. Elsevier Science Publications, Amsterdam.

Kwong Sm, Egan EA, Notter RH and Shapiro DL (1985) Paediatrics 76 : 585-592.

Morley CJ, Bangham AD, Miller N and Davis JA (1981) Lancet i : 64-68.

Morley CJ, Gore SM, Greenough A, Miller NGA, Bangham AD, Pool J, Wood SW, South M and Davis JA (a) (Submitted) Randomized trial of artificial surfactant at birth: Pragmatic analysis (1986).

Morley CJ, and 21 collaborators (b) (Submitted). Ten-centred trial of artificial surfactant in very premature babies: Randomized evaluation of artificial surfactant (BREATHE). (1986).

Pattle RE (1955) Nature 175 : 1125-1126.

Pattus F, Desnuelle P and Verger R (1978) Biochim. Biophys. Acta 507 : 62-70.

Phillips MC and Joos P (1970) Kolloid. Z. Z. Polym. 238 : 499-505.

Scarpelli EM, Kumar A, Clutario BC (1983) Pulmonary Surfactant System. Eds. EV Cosmi and EM Scarpelli. Symposia of the Giovanni Lorenzini Foundation Vol. 16 pp. 3-16. Elsevier Science Publications, Amsterdam.

Schurch S, Goerke J and Clements JA (1976) Proc. Natl. Acad. Sci. USA 73 : 4498-4702.

Smyth JA, Metcalfe IL, Duffty P, Possmayer F, Bryan MH and Enhorning G (1983) Paediatrics 71 : 913-917.

Watkins JC (1968) Biochim. Biophys. Acta 152 : 293-306.

THE ROLE OF LIPIDS AND PROTEINS IN THE STRUCTURE AND FUNCTION OF PHOTOSYNTHETIC MEMBRANES : a mechanism for the lateral segregation of thylakoid components and a role for non-bilayer-forming acyl lipids

Denis J. Murphy

Department of Botany, University of Durham, Durham DH1 3LE, UK

INTRODUCTION

The photosynthetic membrane system is by far the most abundant of all the types of biological membrane found on earth. It is also the site of the photosynthetic light reactions, upon which all life on this planet ultimately depend. Despite its evident predominance and biological importance, however, relatively little was known about the organisation and function of this membrane system at the molecular level until very recently. Two of the most fundamental problems in this area are (i) the function and maintenance of lateral segregation and (ii) the role of non-bilayer acyl lipids in photosynthetic membranes. These two topics will be discussed in the context of lipid-lipid, lipid-protein and protein-protein interactions in photosynthetic membranes.

Higher plant photosynthetic membranes, or thylakoids, can be divided into two contiguous lateral regions depending upon whether or not they are appressed to adjacent thylakoids. It is now known that these regions differ dramatically in both their lipid (Murphy & Woodrow, 1983) and protein (Anderson et al., 1978) compositions. Since the two regions are contiguous in the lateral plane, one must account for the evident stability of what at first sight seems to be a thermodynamically unfavourable demixed state.

In a typical crop plant, such as spinach, pea or maize, about 60% of the thylakoid surface is appressed (stacked), while the remainder is non-appressed (unstacked). The appressed regions contain mostly Photosystem II (PSII), Light-harvesting Complex II (LHCII), and probably some Cytochrome, b_6-f Complex (Cyt b_6-f). In contrast, the non-appressed regions contain mostly Photosystem I (PSI), ATP Synthetase and Cyt b_6-f complexes (Murphy, 1986a). There is also a lateral asymmetry of both chlorophyll and acyl lipid distribution between the two regions (Murphy & Woodrow, 1983). Finally there is a large difference in the lipid : protein ratios between the two regions, with appressed regions containing only 15% by weight acyl lipid whereas non-appressed regions contain almost 40% acyl lipid.

LATERAL SEGREGATION

How is it possible to account for such extreme lateral segregation along thylakoid membranes? One mechanism which can produce lateral segregation is phase separation. In this case certain proteins may be excluded from, for example, domains of gel-phase lipid. It is most unlikely, however, that gel-phase domains occur in normal thylakoid membranes since most of the acyl lipids have transition temperatures well below 0°C. The formation of microdomains of gel-phase phosphatidylglycerol (PG) at low temperatures has been proposed to have a role in chilling injury in plants (Raison & Wright, 1983; Murata & Yamaya, 1984) although this is now disputed (Low et al., 1984). In any case, this could not possibly account for the large-scale lateral segregation observed in normal thylakoids.

In the absence of a role for lipid-lipid interactions, one can propose that it is protein-protein interactions which give rise to thylakoid lateral segregation. It is now known with some certainty that the factor responsible for thylakoid appression is the interaction between LHCII complexes on adjacent membranes. The LHCII complexes are transbilayer integral membrane proteins with N-terminal regions projecting from the outer or stromal surface of the thylakoid membrane. Most of the thylakoid membrane surface carries a net negative charge, but the N-terminal region of LHCII, which has the amino acid sequence (Lys-Arg)-Ser-Thr-Thr-Lys-Lys, carries 3-4 positive charges (Mullet, 1983). The localised decrease in the negative surface charge will allow the approach of LHCII complexes on adjacent membranes until the protein surfaces are sufficiently close for the operation of attractive van der Waals forces.

Such an inter-membrane interaction will not of itself necessarily cause lateral segregation but it may favour such a process. For example, given the strong repulsive forces between adjacent thylakoid membranes, it is unlikely that isolated LHCII-LCHII couples would have a long lifetime. There would therefore be a tendency for the LHCII couples to congregate into lateral domains where their inter-membrane interactions would be rendered more stable. The size of such domains would be limited only by the number of LHCII complexes available and the size of the LHCII domain on the adjacent membrane, i.e. the largest possible LHCII domains would be favoured. It is known that the amount of thylakoid appression is generally proportional to the amount of LHCII present in the membrane, e.g., in most crop plants, LHCII accounts for over 50% total thylakoid protein and 55-60% of the thylakoid surface area is appressed, whereas in shade-adapted species like <u>Alocasia</u>, LHCII forms 70% of thylakoid protein and 75% of the thylakoid is appressed (Anderson, 1980). There are a few exceptions to this rule in the case of chlorophyll b-less mutants. Here, the amount of LHCII is sharply reduced but both the PSII content and the extent of thylakoid suppression is only slightly less than in wild-type plants (Miller et al., 1976; Simpson, 1979). This can be explained by assuming that the only change in the mutants is that there are fewer LHCII units surrounding each PSII. When these LHCII units congregate, they will carry along with them their PSII core complexes. Since the size of the PSII-LHCII complexes will only be slightly reduced compared to the wild-type, it may be expected that the extent of thylakoid appression will also be only slightly reduced. One obvious difference in the mutant will be the reduction in the density of LHCII complexes and hence the weakening of the short-range attraction between appressed thylakoids. This is fully consistent with the observation that the maintenance of thylakoid appression in such mutants requires considerably higher cation concentrations than in the wild-type (Burke et al., 1979).

MODEL FOR THE ORGANISATION OF THYLAKOID MEMBRANES

There are many, often very complex, models for the three-dimensional organisation of thylakoid membranes (Staehelin, 1986). More recently, however, a simplified "flattened vesicle" model has been proposed (Murphy, 1986a). According to this model, the thylakoids are envisaged as being a collection of similarly sized overlapping flattened vesicles as shown in Fig. 1. It is in the regions of overlap that membrane appression will occur. This model can be elaborated in order to explain LHCII-induced lateral segregation in the following manner. The flattened thylakoid vesicles contain two flat lengths of membrane connected by a tightly curved margin. It has already been proposed that the extreme curvature in this marginal region will effectively exclude large protein complexes, such as LHCII (Murphy, 1982). The margins should not only be essentially free of bulky protein complexes, they should also provide a very effective barrier to the lateral diffusion of such proteins. Therefore each thylakoid will have two contiguous but non-communicating protein-containing domains. In each of these domains, inter-membrane LHCII-LHCII interactions will lead to the formation of one or two large regions which will be very enriched in LHCII complexes, i.e. appressed (or stacked) thylakoid regions. Most of the LHCII is known to be tightly bound to PSII complexes to form very large, 14-16 nm diameter, PSII-LHCII complexes. One consequence of the segregation of LHCII complexes will therefore be an associated segregation of PSII complexes into appressed regions as shown in Fig. 1.

This may explain the formation of domains enriched in LHCII-PSII complexes, but why are the other protein complexes also asymmetrically distributed along thylakoid membranes? Again, it is possible to explain all of these phenomena simply on the basis of inter-membrane LHCII-LHCII interactions. As was argued above, these interactions will result in the congregation of PSII-LHCII complexes into large domains, which will be separated by only 3-4 nm from similar domains on an adjacent appressed thylakoid membrane. The other protein complexes would tend to be excluded from such domains since they play no role in membrane appression (Ryrie, 1983). The ATP synthetase complex would be excluded from appressed regions anyway due to steric factors, since its bulky CF_1 headpiece protrudes some 9-14 nm from the membrane surface. There is also evidence that LHCII complexes may interact laterally with other LHCII complexes in the same membrane (Murphy, 1986a). This will lead to a reinforcement of their segregation into regions very enriched in PSII-LHCII complexes. The entropic forces which favour randomisation of protein distribution would be more than compensated for by the electrostatic and van der Waals forces responsible for both inter-membrane and the intra-membrane LHCII-LHCII interactions.

To summarise, the attractive forces between LHCII complexes will cause not only the appression of thylakoid membranes, but also an extreme lateral segregation of their lipid and protein components between the appressed and non-appressed regions. Each thylakoid may be considered as a flattened vesicle. Along each planar surface of the thylakoid, essentially all of the PSII-LHCII complexes will congregate into large regions which will be closely appressed to similarly sized regions, also rich in PSII-LHCII complexes, on neighbouring thylakoids. The functional role of the lateral segregation of thylakoid membrane components is beyond the scope of this article but has recently been reviewed (Murphy, 1986a). The reason for the lateral segregation of acyl lipids will now be considered in the context of the role of these lipids in the thylakoid membrane.

ROLE OF DIACYL DIGALACTOSYL GLYCEROL

Thylakoids are very unusual membranes, not least by virtue of their very high proportions of uncharged lipids. Some 75% of the total acyl lipid is uncharged galactolipid, in the ratio of 2:1 diacylgalactosylglycerol (DGG): diacyldigalactosylglycerol (DDG). The major phospholipid is PG, which forms about 10% and most of the remainder is sulpholipid (< 15%). It is well known that DGG will not form lamellar bilayer structures, except in the presence of an excess of bilayer-forming lipids, such as DDG or PG (Quinn and Williams, 1983; Sprague and Staehelin, 1983).

One of the major unresolved questions confronting researchers in this area is, "what is the role of DGG?". This galactolipid contains two α-linolenate moieties and is therefore highly polyunsaturated. While this will ensure the existence of a liquid-crystalline phase down to temperatures as low as $-30^{\circ}C$, it represents a considerable "overkill" in terms of the maintenance of bilayer fluidity under physiological conditions. It also means that the lipid will tend to form non-bilayer or at least non-planar surfaces (Murphy, 1982; Williams et al., 1984; Murphy 1986b). Finally, the trienoic acyl groups will be highly sensitive to oxidative damage, more especially since the thylakoid is also the site of the oxygenic reactions of water photolysis. It seems unlikely therefore that DGG, which makes up over half of the thylakoid acyl lipid, is simply an "ordinary" bilayer component in thylakoid membranes.

In general, two possible roles for DGG may be considered: (i) the stabilisation of the tightly curved margins of thylakoid membranes and (ii) the packaging of the large transbilayer protein complexes of thylakoids. The shape of the DGG molecule resembles that of a truncated cone, with a

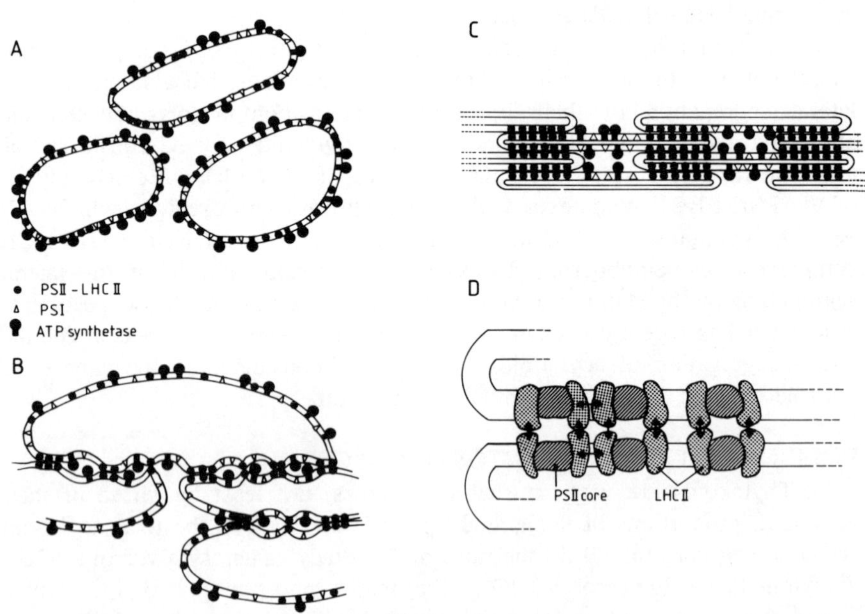

Figure 1

Figure 1. Mechanism of lateral segregation in thylakoid membranes.

In this schematic diagram, the lateral segregation of three of the major intrinsic membrane protein complexes of thylakoids is shown, i.e. the Photosystem I, Photosystem II - Light-harvesting Complex II (PSII-LHCII) and ATP synthetase complexes. The location of the Cytochrome b_6-f complex has yet to be positively resolved (Murphy, 1986a) and is not shown here.

(A). When thylakoids are destacked, they form a population of vesicles in which the location of all of the major protein complexes is completely randomised in the lateral plane.

(B). Under the appropriate conditions of high cation concentrations, LHCII complexes on adjacent membranes are attracted together. This will cause a flattening of the membrane surface in the region of LHCII-LHCII interaction. The domains containing such regions would tend to grow and to fuse in order to exclude protein complexes, such as PSI and ATP synthetase. The latter would tend to repel each other and/or project out of the membrane, both processes tending to act against the attractive LHCII-LHCII interactions.

(C). The most stable end-state would be one in which each thylakoid had four large appressed domains - two on its upper-facing stromal surface and two on its lower-facing stromal surface. Therefore each thylakoid would be connected to four neighbouring thylakoids to form a network of thylakoid "strings" which would ramify throughout the chloroplast. Such an arrangement would give rise to the alternate stacks of appressed and non-appressed commonly seen in electron micrographs. It is also easy to see how thylakoids could readily and reversibly destack and restack in response to cation fluctuations in such a system.

(D). The inter-membrane LHCII-LHCII interactions cause not only thylakoid appression, but also an extreme lateral segregation into domains very enriched in PSII-LHCII complexes and domains containing the other thylakoid protein complexes. This lateral segregation may be further augmented by intra-membrane attraction between adjacent LHCII complexes. If different protein complexes tended to have different lipid populations in their immediate vicinity, then the lateral segregation of proteins would also lead to the lateral segregation of lipids that has been observed in thylakoid membrane fractions (Murphy & Woodrow, 1983).

---

narrow galactosyl head-group and a bulky polyunsaturated acyl chain region (Murphy, 1982, 1986b). Such a molecule will tend to induce a concave curvature in any membrane into which it is inserted. Thylakoid membranes are kept flat by the LHCII-LHCII interactions responsible for membrane appression as discussed above, and also by trans-lumenal interactions as discussed in Murphy (1986a). Since thylakoids are flattened vesicles, this means that there will be a region of very tight curvature connecting the two flattened regions, i.e. the thylakoid margin. Such a region will have an extreme curvature imposed upon it by the forces maintaining a planar configuration in the rest of the thylakoid. A molecule like DGG would favourably partition into the concave region of such a domain and hence stabilise it more effectively than more cylindrical lipids like PG or DGG. It is possible that phosphatidylethanolamine, which has similar non-bilayer forming tendencies to DGG, may perform a similar role in tightly curved regions of extra-chloroplastic membrane systems, such as mitochondria or endoplasmic reticulum. This effect may account for up to 25% of the DGG found in thylakoids but cannot explain the presence of the remainder. For this, one has to propose a role for DGG in the packaging and/or the function of the thylakoid protein complexes.

A protein packaging function for DGG has been proposed by several authors (Murphy, 1982; Williams et al., 1984; Gounaris and Barber, 1983), but it has often been uncertain as to why these specific proteins would require such special "packaging" by DGG, while other large membrane proteins, e.g. those of the mitochondria, evidently do not. It was proposed by Israelachvili (1977, 1980) that large integral protein complexes may induce deformations in the planar conformation of the lipid bilayer, but no evidence for this has yet been presented.

STRUCTURE OF THYLAKOID PROTEINS

Recent progress in the elucidation of the amino acid sequences of most of the integral thylakoid protein components has allowed for more concrete predictions of their secondary and tertiary structures than was hitherto possible. The length of the probable membrane-spanning regions of thylakoid proteins is quite variable. About 25 α-helically orientated amino acids are required to span a lipid bilayer but the size of the putative, α-helical regions of thylakoid polypeptides ranges from less than 20 to about 30 amino acids. Since these hydrophobic regions would be excluded from the aqueous phase, a localised curvature would be imposed upon the lipid bilayer in order to match up as closely as possible the adjacent lipid and protein hydrophobic and hydrophilic surfaces. It has been calculated that this would result in a significant increase in elastic energy of the lipid molecules in such a domain (Murphy, 1986a). This would provide a force which would favour the partitioning of non-cylindrical lipid molecules into such a domain. For example, if a membrane protein had several transbilayer α-helical spans of 28 amino acids, this would induce a concave curvature into the lipid bilayer (Fig. 2A) and hence favour the presence adjacent to the protein of a population of cone-shaped lipid molecules, such as DGG.

The reason for the extra length of the transbilayer α-helical regions of some thylakoid proteins may be related to their pigment-coordinating functions. The size of a chlorophyll molecule is such that it is just possible to fit one on each side of the lipid bilayer. There is increasing evidence, however, that in many cases the tetrapyrrole ring may be tilted by as much as 30° with respect to the bilayer normal (Nabedryk et al., 1984a,b). It is possible that this may require a hydrophobic protein domain slightly thicker than the average bilayer width. There is now evidence that the α-helical domains of some pigment proteins are also tilted by 30° (Breton and Nabedryk, 1984; Nabedryk et al., 1984b) and this would favour the presence of non-cylindrical lipid molecules in the adjacent region, as shown in Fig. 2B. This effect would probably be localised to the first few shells of acyl lipid molecules surrounding a large protein complex. There would be a tendency for the entire assembly to diffuse as a unit and this would lead to the lateral segregation of certain lipid classes in the wake of the proteins with which they would tend to associate.

LIPID-PROTEIN INTERACTIONS

Since acyl lipids are normally free to diffuse laterally in a bilayer in the liquid-crystalline state, the lateral segregation of lipids in thylakoids strongly implies some sort of lipid-protein interaction of the kind mentioned above. This would not be a strong interaction of the sort between the photosynthetic pigments and their apoproteins. Other bilayer lipids would be free to diffuse in and out of the area around such a protein complex but, due to their molecular geometry and/or other factors, their residence times in the region adjacent to the protein would be considerably shorter than that of the favoured lipid population. This would lead to a relatively specific lipid-protein association but, since it would not involve strong lipid-protein binding, it would not necessarily be detectable following purification of the protein complexes.

A more promising approach to the detection of lipid-protein interactions in thylakoid membranes is by ESR spectroscopy. Preliminary studies in which spin-labelled PG was used have implied that, at least in the case of this lipid, there is a great deal of lipid-protein interaction in thylakoid membranes (Murphy & Knowles, 1984). It is obviously of great

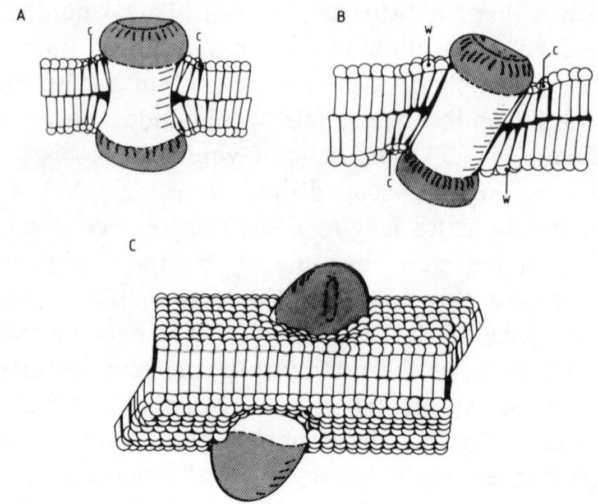

Figure 2. Role of non-bilayer-forming acyl lipids in thylakoid membranes. In this schematic diagram, a possible protein packaging role for DGG is shown. A role for diacylgalactosylglycerol (DGG) in the stabilisation of thylakoid margins has already been proposed (Murphy, 1982). (A). The presence of a large protein molecule with a hydrophobic domain wider than that of the lipid bilayer will tend to induce a localised concave curvature in the bilayer. This is because the hydrophobic surfaces of both lipids and protein will tend to be excluded from the aqueous phase. A cone-shaped lipid (c), such as DGG, would favourably partition into such a region of concave membrane curvature. Regions of low acyl chain density in the bilayer midplane may favour the presence of plastoquinone as previously discussed (Murphy, 1986a). This could lead to extremely rapid rates of lateral plastoquinone diffusion if the protein density of the membrane were sufficiently high, e.g., in appressed thylakoid regions, and would favour the action of plastoquinone as the principal long-range electron transport intermediate in photosynthesis. (B). Some thylakoid proteins are tilted by at least 30-35° with respect to the bilayer normal. This will create packing defects in the bilayer which could be appropriately sealed by the presence of cone-shaped (c) lipids (DGG?) and wedge-shaped (w) lipids (DDG?). (C). Schematic representation of the localised deformation in a lipid bilayer region adjacent to a large integral protein. The increase in elastic energy of the bilayer components in this region could supply a force for the lateral segregation of certain lipids into such a region. In the case of thylakoid membranes, integral proteins may constitute up to 80% of the membrane (by weight) and the regions of bilayer deformation may therefore involve a substantial proportion of the total acyl lipid population.

interest to look at galacto-lipid-protein interactions in the same way and such studies are now in progress in our laboratories. Reconstitution studies have also demonstrated the importance of phospholipids and, in some cases, galactolipids in many aspects of thylakoid protein function, as reviewed by Murphy (1986a). One must, however, be cautious in the interpretation of such studies since the most effective reconstitution agent may simply form better liposomes than the other lipids or package the proteins better in the reconstitution system. Such interactions do not necessarily imply the existence of an identical lipid-protein interaction in vivo.

Thylakoids are very rich in protein, particularly in their appressed regions where proteins constitute about 85% by weight of the membrane (Murphy, 1986a). This extreme crowding of large protein complexes means that most acyl lipid molecules are in the immediate vicinity of proteins. It is therefore likely that, at least in appressed regions, almost the entire acyl lipid population undergoes perforce a degree of interaction, either directly or indirectly with intrinsic protein complexes. This would principally involve acyl chain dynamics rather than headgroup interactions. Given the complexity of such a system and the evident widespread extent of the lipid-protein interactions, it seems unlikely that it will be possible to say that, while certain acyl lipids definitely interact with certain proteins, other lipids do not. It is probably more realistic, in this context, to state that all thylakoid acyl lipids are able to interact with the membrane proteins in their midst but that some lipid populations may have a greater tendency to associate with certain proteins than others. It should be pointed out that if these associations are based upon lipid geometry rather than upon headgroup interactions, then acyl molecular species may well be more important than the lipid class. Lipid-protein interactions based upon packing geometry would be relatively weak in comparison to pigment-protein binding, and may only be manifest in specific domains of intact thylakoid membranes. While the interactions between the neutral galactolipids and proteins are more probably based upon packing geometry, it is quite possible that the anionic sulpholipid and PG undergo headgroup interactions with thylakoid proteins. It remains for research workers in this field to elucidate, preferably by means of relatively non-intrusive techniques, such as NMR or ESR spectroscopy, the exact nature and scale of lipid-protein interactions in this fascinating and important membrane system.

REFERENCES

Anderson, J.M. (1980) FEBS Lett. 117, 327-331
Anderson, J.M., Waldron, J.C. and Thorne, S.W. (1978) FEBS Lett. 92, 227-233

Burke, J.J., Steinback, K.E. & Arntzen, C.J. (1979) Plant Physiol. 63, 237-243

Breton, J. & Nabedryk, E. (1984) FEBS Lett. 176, 355-359

Gounaris, K. & Barber, J. (1983) Trends in Biochem. Sci. 8, 378-381

Israelachvili, J.N. (1977) Biochim. Biophys. Acta 469, 222-225

Israelachvili, J.N., Marcelja, S. and Horn, R. (1980) Q. Rev. Biophys. 13, 121-200

Low, P.S., Ort, D.R., Cramer, W.A., Whitmarsh, J. and Martin, B. (1984) Arch. Biochem. Biophys. 231, 336-344

Miller, K.R., Miller, G.J. & McIntyre, K.R. (1976) J. Cell Biol. 71, 624-638

Mullet, J.E. (1983) J. Biol. Chem. 258, 9941-9948

Murata, N. & Yamaya, J. (1984) Plant Physiol. 74, 1016-1024

Murphy, D.J. (1982) FEBS Lett. 150, 19-26

Murphy, D.J. (1986a) Biochem. Biophys. Acta 864, 33-95

Murphy, D.J. (1986b) in: Encyclopaedia of Plant Physiology, Vol. 16 (Staehelin, L.A. & Arntzen,C.J.,eds) Springer, Berlin, pp. 713-725

Murphy, D.J. & Knowles, P.F. (1984) in Structure, Function and Metabolism of Plant Lipids (Siegenthaler, P.A. & Eichenberger, W., eds) pp. 425-428, Elsevier, Amsterdam

Murphy, D.J. & Woodrow, I.E. (1983) Biochim. Biophys. Acta 725, 104-112

Nabedryk, E., Andrianambinintsoa, S. & Breton, J. (1984a) Biochim. Biophys. Acta 765, 380-387

Nabedryk, E., Biondet, P., Darr, S., Arntzen, C.J. & Breton, D. (1984b) Biochim. Biophys. Acta 767, 640-647

Quinn, P.J. and Williams, I.E. (1983) Biochim. Biophys. Acta 737, 223-266

Raison, J.K. & Wright, L.C. (1983) Biochim. Biophys. Acta 731, 69-78

Ryrie, I.J. (1983) Eur. J. Biochim. 137, 203-213

Simpson, D.J. (1979) Carlsberg Res. Commun. 44, 235-254

Sprague, S.G. & Staehelin, L.A. (1983) in Biosynthesis and Function of Plant Lipids (Thomson, W.W., Mudd, J.B. & Gibbs, M., eds) pp. 144-159, Am. Soc. Plant Physiol. Rockville, MD

Staehelin, L.A. (1986) in: Encyclopaedia of Plant Physiology, Vol. 16, pp. 1-84 (Staehelin, L.A. & Arntzen, C.J., eds.), Springer, Berlin

Williams, W.P., Goumaris, K. and Quinn, P.J. (1984) in "Proc. 6th Int. Congr. on Photosynthesis", Vol. III, (ed. E. Sybesma) pp. 123-130, Dr. Junk, The Hague

ENVIRONMENTAL ADAPTATION

ALTERNATIVE MEMBRANE LIPIDS - ADAPTATION TO THE ENVIRONMENT OR EVOLUTIONARY DIVERGENCE?

W.D. Grant

Department of Microbiology, University of Leicester, Leicester LE1 7RH

INTRODUCTION

The Precambrian fossil record indicates that discrete cellular forms existed 3.5×10^9 years ago. No-one would dispute that the enclosure of a population of organic molecules separate and distinct from those in the surrounding environment, i.e. the appearance of an efficient membrane structure, constituted a major event in the evolution of life forms on the early earth. Furthermore, a consideration of a wide range of present day life forms makes it apparent that the model of membrane structure based on fatty acid glycerol ester lipids won the war of selection amongst the presumed range of 'protocells' precariously coexisting at that time.

The virtual universality of the fatty acid-ester model implies a remarkable degree of functional success and adaptability evidenced in phylogenetic terms by its presence in the widely different kingdoms of plants, animals and bacteria, and by the capacity of the system to cope with shorter-term changes in the environment. Microorganisms have proved to be useful model systems in which to test the latter capacity, in view of the wide range of so-called 'extreme' habitats occupied by bacteria in particular. Extreme environments are usually defined in relation to our subjective view of what constitutes an equable environment - in truth, a taxonomic definition is more appropriate, since the 'stressed' environment is simply one in which most organisms cannot grow and survive. In experimental terms, the most fashionable stressed environments include those produced by extremes of temperature, water activity (a_w) and pH, although equally interesting stresses are produced by high pressure or high radiation flux.

Microorganisms have a remarkable capacity to colonize stressed environments, bacteria being often the only inhabitants as conditions become more and more demanding. It is possible, therefore, to test the adaptability of the fatty acid-ester model under a wide range of conditions, both in organisms specifically adapted to particular extreme niches, and also in those organisms capable of growing over a range of stress from the equable to the extreme (a more rare but potentially more interesting situation).

The stress imposed by extremes of temperature is most studied, and for those organisms capable of growth over a wide range of temperature, lower growth temperatures generally lead to an increase in fatty acid unsaturation, the converse occurring at high temperatures, sometimes along with variations in chain length and the type of branching. Occasionally, thermophilic microorganisms have large amounts of cyclized fatty acids (Langworthy, 1978). Those organisms specifically adapted to particular temperatures show a permanent increase in the appropriate fatty acids. These temperature-dependent changes in fatty acid composition are usually explained in terms of membrane fluidity and its maintenance.

Water activity ($\underline{a}_w$) effects have been largely imposed by high salt (NaCl) concentrations and it has often not been possible to distinguish between $\underline{a}_w$ <u>per</u> <u>se</u> and the stress due to high concentrations of ions. However, it is clear that in those bacteria such as <u>Vibrio</u> <u>costicola</u> that grow over a wide range of salt concentrations, the main effect is a shift towards higher concentrations of more negatively charged phospholipids and a reduction in amino lipids (Kates, 1986). There is also some evidence for an increase in the proportion of branched chain and cyclopropane fatty acids. These changes may be required to neutralize counter-ions, so again the overall effect can be explained in terms of the necessity to maintain hydrocarbon fluidity over a range of conditions.

The effect of extremes of pH is much less studied, but there is some evidence that cells modulate the proportions of particular charged phospholipids in response to pH. These changes may be involved in H^+ transport across the membrane and thus internal pH homeostasis. Some acidophiles contain large amounts of aminophospholipids (Langworthy, 1978), whereas those few alkaliphiles studied this far only rarely possess amino-lipids and lack glycolipids (Grant & Tindall, 1986) although the significance of the latter is not clear. There is no conclusive evidence for changes in fatty acid composition since despite the fact that certain acidophiles contain large amounts of cyclized fatty acids, these organisms are also thermophiles and the adaptation may be to high temperature rather than low pH.

SPECIAL EXTREMOPHILES FROM EXTREME ENVIRONMENTS

Although many microorganisms with membranes based on the fatty acid ester model are to be found living and growing in extreme environments, there are a few odd bacteria that have a different kind of membrane architecture living in the most extreme examples of these stressed environments.

(a) Obligate halophiles: In salt lakes in tropical and semi-tropical parts of the world, when the salt (NaCl) concentration approaches saturation (approximately 5M), characteristic populations of red-coloured bacteria develop. These red bacteria (halobacteria) have an obligate requirement for 2M NaCl in order to grow, imparting the overall striking red colouration to the environment due to the possession of C_{50} carotenoids known as bacterioruberins that have a protective function against the high light intensities experienced by the cells. There are several different kinds of salt lake, each harbouring characteristic and distinct halobacteria (Grant & Ross, 1986), but at the moment only four genera are recognized, the rod-shaped or pleomorphic isolates (<u>Halobacterium</u> spp.) and the coccoid forms (<u>Halococcus</u> spp.) from neutral or slightly acid salt lakes, and the more recently described equivalent <u>Natronobacterium</u> and <u>Natronococcus</u> spp. found in highly alkaline (soda) salt lakes (Tindall et al., 1984).

More than 25 years ago the classic work of Kates (see Kates, 1978) established that halobacteria (in this case <u>Halobacterium</u> spp.) lacked significant amounts of fatty acids, and moreover had membranes based on lipids that had core structures resistant to saponification procedures. Acid methanolysis of halobacterial biomass, followed by hexane extraction and tlc in appropriate solvents (Ross et al., 1981) reveals only low amounts of fatty acid methyl esters, but large amounts of other lipid material. This acid-resistant material was first characterised by Kates as a glycerol diether with two C_{20} saturated branched isoprenoid chains in place of the usual fatty acids (Figure 1). Moreover, the glycerol has the <u>sn</u>-2,3 configuration rather than the <u>sn</u>-1,2 configuration characteristic of glycerol fatty acid esters. It will be noted that the molecular dimensions of this C_{20},C_{20} lipid are similar to a C_{16},C_{16} diester lipid, and this odd structure is clearly functionally directly analogous. Thin sections of halobacteria indicate a bilayer structure similar in thickness to that in other bacteria (Stoeckenius & Rowen, 1967). A family of phospholipids and glycolipids similar to those found in other bacteria, but based on this diether core is also to be found, again with the opposite configuration for glycerols in head groups (Kates, 1986).

In more recent times, two other diether core lipids have been found in halobacteria, a C_{20},C_{25} diether and a C_{25},C_{25} diether (De Rosa et al., 1982, 1983)(Figure 2) particularly common in haloalkaliphilic isolates. Again, there are complex lipids based on these cores, and the presence or absence of particular complex lipids has proved of use in the taxonomy of the group (Ross & Grant, 1985; Grant & Ross, 1986).

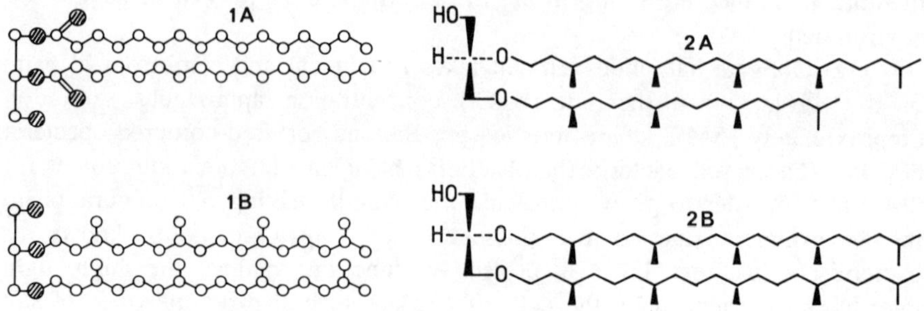

Figure 1. Diagrammatic representation of: (A) 1,2-di-O-palmitoyl-sn-glycerol diester;(B) 2,3-di-O-phytanyl-sn-glycerol diether.

Figure 2. (A) 2-O-Sesterterpanyl-3-O-phytanyl-sn-glycerol diether;(B) 2,3-di-O-sesterterpanyl-sn-glycerol diether.

(b) Thermophiles: In extremely hot (and sometimes acidic) environments at temperatures close to $100^{\circ}C$ (and occasionally exceeding $100^{\circ}C$ in pressurized sea-floor environments), sulphur metabolizing bacteria are to be found that have an important similarity to halobacteria. These sulphur-dependent thermophilic bacteria, presently comprising eight genera, Sulfolobus, Thermoplasma, Desulfurococcus, Thermodiscus, Thermococcus, Pyrodictium, Thermoproteus and Thermofilum (Stetter & Zillig, 1985), like halobacteria, lack saponifiable core lipids, most of the core lipid being a C_{40},C_{40} dibiphytanyl tetraether diglycerol, again with the sn-2,3 configuration (Figure 3). This compound is clearly two C_{20},C_{20} diethers linked head to head. In at least one genus of these bacteria, a certain proportion of the tetraethers have one of the glycerols replaced by a nine carbon polyol (De Rosa et al.,1980a). There are complex lipids based on the tetraether core (De Rosa et al., 1980b) although these have not been completely characterized for any of the isolates, and a certain amount of "fine-tuning" of membrane fluidity occurs in some isolates, in that cyclopentanediyl rings are inserted as the growth temperature rises (De Rosa et al., 1980c) (Figure 3).

METHANOGENIC BACTERIA

Initially, the possession of curious tetraether lipids by certain thermophiles, and the possession of diethers by halobacteria, were considered to be the ultimate expression of membrane adaptation to the environment. A membrane based on C_{40},C_{40} lipids in particular, seemed to be the perfect answer to the high-temperature environment, spanning the

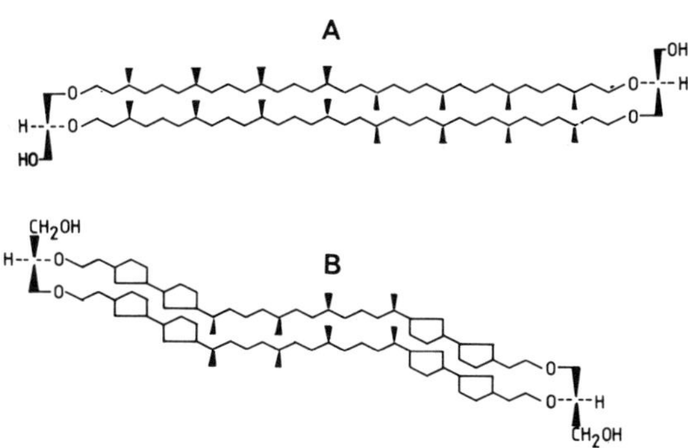

Figure 3. (A) 2,3,2',3'-Tetra-*O*-dibiphytanyl-di-<u>sn</u>-glycerol tetraether; (B) as (A) with four cyclopentane rings - one to four rings may be present symmetrically distributed in each C_{40} chain.

membrane to produce a rigid monolayer, not subject to the same problems of fluidity maintenance experienced by bilayers. It was also assumed that phytanyl chains had characteristic and necessary fluidity properties in high salt.

However, it is not possible to argue that methanogenic bacteria occupy an extreme niche. These bacteria live in the main in complex, consortial associations with other bacteria in highly anaerobic environments, usually at 'normal' temperature, pH, $\underline{a}_w$ and pressure. There are thermophilic representatives, but these are the exception (Whitman, 1985). The group is extremely diverse comprising three orders and eleven genera (Archer & Harris, 1986), and remarkably the members of the group also have lipids based on glycerol isopranyl ethers of comparable structure to those seen in thermophiles and halophiles.

Most methanogens have a mixture of C_{20},C_{20} and C_{40},C_{40} core ether lipids (Langworthy et al., 1982), a few have only C_{20},C_{20} diethers (e.g., <u>Methanococcus</u> <u>voltae</u>). However, the picture is more complex than in the halophiles and thermophiles in that a C_{40} macrocyclic diether is known (Figure 4) for one <u>Methanococcus</u> sp. (Comita & Gagosian, 1983) and more recently <u>Methanosarcina</u> spp. have been shown to have C_{40},C_{40} tetraethers, glycerol C_{20} monoethers, C_{20},C_{25} diethers and C_{20},C_{20} tetritol diethers (Figure 4)(De Rosa et al., 1986). It is no longer possible to ascribe cyclopentane rings only to the thermophilic environment, since mesophilic <u>Methanosarcina</u> spp. also possess 1-3 rings in C_{40},C_{40} chains (De Rosa et al., 1986). There is clearly a tremendous range of complex phospholipids and glycolipids based on different cores and these have value in the taxonomy of

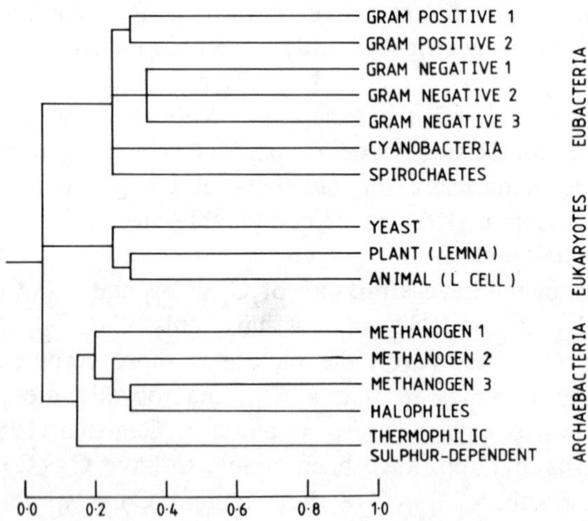

Figure 4. (A) Macrocyclic diether; (B) tetritol-diphytanyl-diether; (C) 3-O-phytanyl-sn-glycerol monoether.

Figure 5. Schematic representation of phylogenetic relationships redrawn according to the evolutionary tree constructed by Fox et al. (1980) and Woese and Fox (1977) after comparative analysis of oligonucleotide sequences.

the group (Grant et al., 1985), although the complex lipids have only been characterised for one species (Kushwaha et al., 1981). There is no doubt that other ether core lipids remain to be characterized in this group (Grant et al., 1985).

HALOPHILES, THERMOPHILES, METHANOGENS AND ARCHAE-BACTERIA

Classification of prokaryotes are in the main phenetic (based on present-day phenotypic characters), and unlike the situation with eukaryotes, cannot draw on cell structure or the fossil record in any major way since there is insufficient detail available to construct any real phylogenies. However, it can be argued that there is an accessible fossil record in gene structure i.e. an organism's evolutionary history is enshrined in its genetic material, the sequence of which is a function of a mutational clock that has been running since the origins of cell lines some 3.5-4 x 10^9 years ago. It should therefore be possible to deduce phylogenetic relationships amongst prokaryotes by comparing genetic sequence in appropriate genes (either directly by sequencing these, or indirectly by sequencing gene products such as ribosomal RNA or protein). Comparisons of these 'chronometric' macromolecules have been extensively carried out over recent years, notably by Woese and colleagues (Fox et al., 1980) with a view to constructing a prokaryote phylogeny and also with a view to establishing the origins of the eukaryotic cell line, since it is possible to compare genes of chloroplasts and mitochondria with present day prokaryotes and with the equivalent genes located in the eukaryote nuclear genome.

In particular, phylogenetic analysis of prokaryote 16S rRNA and equivalent eukaryote 18S rRNA by partial sequence comparisons (the S_{AB} coefficient)(Fox et al., 1977) have proved particularly useful. As predicted, mitochondria and chloroplasts share common origins with present day prokaryotes, whereas the eukaryote nuclear genome appears to be a separate and distinct line of evolution, supporting the ancient endosynbiont theory for the origin of the eukaryote cell (which supposes a now extinct "18S" cell line engulfed ancestor cells of the present-day "16S" line which subsequently became chloroplasts and mitochondria). A phylogeny of present-day prokaryotes can also be constructed by this means (Fox et al., 1980; Stackebrandt & Woese 1981), generating taxa, some of which would be expected based on present day phenotypic characteristics, others less so - in particular, cell shape, a character traditionally given great emphasis in prokaryote taxonomy, seems to be a poor phylogenetic marker.

When the methanogenic, halophilic and sulphur-dependent thermophilic bacteria were analysed by the S_{AB} coefficient (which is an indirect quantitation of common oligonucleotides in the RNAs of organisms A and B), surprisingly, these prokaryotes were as unrelated in phylogenetic terms from other bacteria (now named the eubacteria) as they were from eukaryotes, although related amongst themselves at a comparable level as that seen amongst the major groups within the eubacteria (Figure 5). These curious prokaryotes have been given the name archaebacteria by Woese (Woese & Fox, 1977; Woese et al., 1978), the name denoting supposed antiquity to underline the hypothesis that these organisms were once the dominant inhabitants of the early earth, each phenotype today coping with an environmental stress that would once have been common on the early earth.

There is now a wealth of independent biochemical information in addition to the possession of glycerol isopranyl ether lipids supporting the phylogenetic distinctness of archaebacteria (see Kandler, 1982). Recently it has been possible to sequence completely genes from different archaebacteria and such sequence comparisons with eubacteria have confirmed the earlier RNA partial sequence comparison conclusions (Yang et al., 1985). Archaebacteria are characterized by wide metabolic and structural diversity within the three major phenotypes comparable to that seen amongst the major phenotypes of eubacteria. Some features more closely resemble eukaryotic characteristics, leading to the suggestion that the original eukaryote was a chimaera of ancient eubacteria, 18S cell and archaebacteria (Woese, 1982).

Given the phylogenetic coherence of these extraordinarily phenotypically diverse prokaryotes, one is drawn inexorably to the conclusion that the possession of ether-linked core lipids is not a function of different environmental adaptations, but rather a consequence of evolution from a common archaebacterial ancestor with this odd membrane model. Precisely why present-day archaebacteria seem confined to extreme environments is not clear.

BIOSYNTHESIS OF ARCHAEBACTERIAL LIPIDS

Despite work carried out over a period of more than twenty years, it is still unclear in detail how certain features of the basic isopranyl ether structures are achieved, although a general consensus of the overall process is now emerging. Reviews of progress in the field include Kates (1978; 1986), Langworthy (1985), and very recently, De Rosa et al. (1986) have considered biosynthetic aspects along with structural and physicochemical properties.

Labelling studies make it beyond doubt that the hydrocarbon chains are synthesized exclusively by the mevalonate pathway for isoprenoid chains, which must be modified to include hydrogenation steps for the formation of saturated phytanyl groups. Formation of the ether linkage generates the sn-2,3 configuration and it is still not clear how this is achieved, although labelling studies indicate that it is likely that the configuration of the glycerol C2 is a consequence of the interconversion of glycerol (or glycerol phosphate) and dihydroxyacetone (or dihydroxyacetone phosphate). Since allyl pyrophosphates are known to function as alkylating agents in other biosynthetic mechanisms, there is no problem in imagining a stereospecific alkylation step. This alkylation ability is considerably reduced in saturated non-allyl pyrophosphates, so De Rosa et al.(1986) consider this as circumstantial evidence for the hydrogenation stage in the formation of the phytanyl chains occurring after ether linkage to glycerol.

In methanogens and thermophilic archaebacteria C-C bonds are formed within or between pranyl chains, in the head-to-head coupling of phytanyls to form C_{40} biphytanyls and in cyclization within residues to form 5-membered cyclic rings in biphytanyls. At present, it is not known as to whether the head-to-head coupling is between ether-linked C_{20} chains or between precursors. Neither is there any evidence for free C_{40} chains, nor any known biological precedent for such a coupling. The structural regularity of the cyclized forms, however, suggests cyclization within the tetraether.

Both the head-to-head linkage and ether linkages are remarkably stable, and it is of interest that isopranoid hydrocarbons and hydrocarbon ethers of similar structure to archaebacterial lipids have been found in ancient sediments as old as the Precambrian period. These chemical fossils have been adduced as supportive evidence of an archaebacterial era on the early earth (Hahn & Haug, 1985).

YET MORE MEMBRANE MODELS

One might be forgiven for assuming that the survival of two distinct membrane types is remarkable enough given natural selection. To find other models coexisting with these is extraordinary indeed. However, two other types, based on core lipids different from either the fatty acid ester model or the isopranyl ether model have come to light in recent years.

Thermodesulfotobacterium commune is a thermophilic, obligately anaerobic, sulphate-reducing bacterium isolated from thermal muds, springs and algal mats associated with volcanic activity (i.e. environments similar to those occupied by thermophilic and sulphur-dependent archaebacteria)(Zeikus et al. 1983). Analyses of lipids have revealed core lipids based on glycerol ethers, but in this case sn-1,2 dialkyl glycerols (Langworthy et al. 1983), the principal glycerol diether being 1,2-di-Oanteisoheptadecyl-<u>sn</u>-glycerol (Figure 6). No phytanyl chains are present, but this is the only organism apart from archaebacteria with diether core lipids, albeit with the <u>sn</u>-1,2 configuration for glycerol.

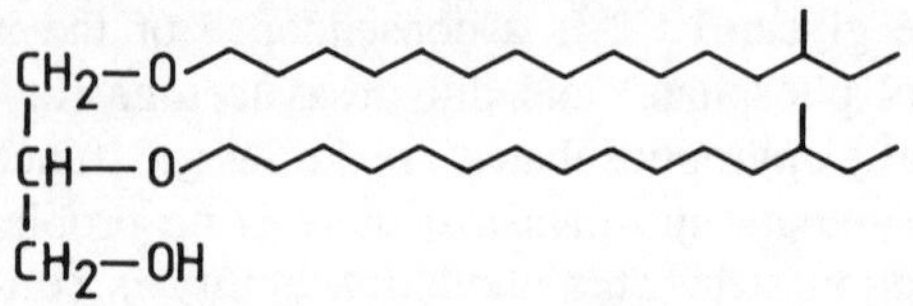

$$CH_2-O\sim\sim\sim\sim$$
$$CH-O\sim\sim\sim\sim$$
$$CH_2-OH$$

Figure 6. 1,2-Di-<u>O</u>-anteisoheptadecyl-<u>sn</u>-glycerol.

<u>Thermomicrobium</u> <u>roseus</u> is a thermophilic bacterium lacking a normal cell wall, and lipid analyses indicate the total absence of glycerol-derived lipids. A series of C_{18}-C_{23} straight chain and branched 1,2-diols replace glycerolipids, n-C_{21} being present in largest amount (Pond et al., 1986)(Figure 7). They can be viewed as structural analogues of glycerolipids if there is a bend in their configuration at C3. These odd bacteria also have fatty acids which are ester-linked to the diols and it can be imagined that when esterified at C2, the diols would provide a double chain form capable of forming a bilayer.

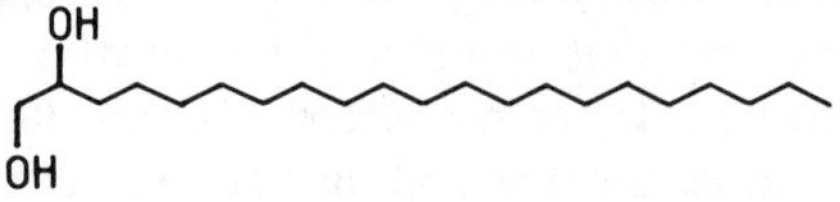

Figure 7. 1,2-Eicosanediol.

Both <u>T. roseus</u> and <u>T. commune</u> do not appear to have archaebacterial cell wall polymers, or characteristic archaebacterial transcriptional or translational machinery (as determined by antibiotic sensitivities), but rather represent separate and distinct early eubacterial lines (Gibson et al., 1985). However, the precise branching order of eubacterial and archaebacterial branches is difficult to determine from the existing data and its exact positioning will require extensive gene sequence comparisons from a variety of different genes.

If one accepts membrane architecture as a primary phylogenetic marker, one is left with the intriguing possibility that five or more distinct cell lines co-evolved and competed on the early earth, and that, remarkably, more than one type is around today.

ACKNOWLEDGEMENT
The diagrams were drawn by Gary Pinch and his work is greatly appreciated.

REFERENCES

Archer, D.B. & Harris, J.F. (1986). In: Anaerobic Bacteria in Habitats Other than Man (eds. E.M.Barnes & G.C.Mead) pp. 185-224. Blackwell Scientific Publications.

Comita, P.B. & Gagosian, R.B. (1983). Science 222 : 1329-1331.

De Rosa, M., De Rosa, S., Gambacorta, A. & Bu'lock, J.D. (1980a). Phytochemistry 19 : 249-254.

De Rosa, M., Gambacorta, A., Nicolaus, P. & Bu'lock, J.D. (1980b). Phytochemistry 19 : 821-825.

De Rosa, M., Esposito, E., Gambacorta, A., Nicolaus, B. & Bu'lock, J.D. (1980c). Phytochemistry 19 : 827-831.

De Rosa, M., Gambacorta, A., Nicolaus, B., Ross, H.N.M., Grant, W.D. & Bu'lock, J.D. (1982). J. Gen. Microbiol. 128 : 343-348.

De Rosa, M., Gambacorta, A., Nicolaus, B. & Grant, W.D. (1983). J. Gen. Microbiol. 129 : 2333-2337.

De Rosa, M., Gambacorta, A. & Gliozzi, A. (1986). Microbiol. Rev. 50 : 70-80.

De Rosa, M., Gambacorta, A., Lanzotti, V., Trincone, A., Harris, J.E. & Grant, W.D. (1986). Biochim. Biophys. Acta 875 : 487-492.

Fox, G.E., Pechman, K.R. & Woese, C.R. (1977). Int. J. System. Bact. 27 : 44-57.

Fox, G.E., Stackebrandt, E., Hespell, R.B., Gibson, J., Maniloff, J., Dyer, T.A., Wolfe, R.S., Blach, W.E., Tanner, R.S., Magrum, L.J., Zablen, L.B., Blakemore, R., Gupta, R., Bonen, L., Lewis, B.J., Stahl, D.A., Luehrsen, K.R., Chen, C.N. & Woese, C.R. (1980). Science 209 : 257-463.

Gibson, J., Ludwig, W., Stackebrandt, E. & Woese, C.R, (1985). System. App. Microbiol. 6 : 152-156.

Grant, W.D. & Ross, H.N.M. (1986). FEMS Microbiol. Rev. 39 : 9-15.

Grant, W.D. & Tindall, B.J. (1986). In: Microbes in Extreme Environments. (Eds. G.A. Codd & R.A. Herbert). pp. 24-53. Academic Press, London, New York, San Francisco.

Grant, W.D., Pinch, G., Harris, J.E., De Rosa, M. & Gambacorta, A. (1985). J. Gen. Microbiol. 131 : 3277-3286.

Hahn, J. & Haug, P. (1985) In: The Bacteria, Volume 8. (Eds. C.R. Woese & R.S. Wolfe) pp. 215-256. Academic Press. New York & London.

Kandler, O. (1982) (ed.). Archaebacteria. Fischer, Stuttgart.

Kates, M. (1978). Prog. Chem. Fats and other Lipids. 15 : 301-342.

Kates, M. (1986). FEMS Microbiol. Rev. 39 : 95-101.

Kushwaha, S.C., Kates, M., Sprott, G.D. & Smith, I.C.P. (1981). Biochim. Biophys. Acta 664 : 228-244.

Langworthy, T.A. (1978). In: Microbial Life in Extreme Environments (ed. D. Kushner) pp. 279-317. Academic Press, London, New York, San Francisco.

Langworthy, T. (1985). In: The Bacteria, Volume 8. (Eds. C.R. Woese & R.S. Wolfe) pp. 459-498. Academic Press, New York & London.

Langworthy, T.A., Tornabene, T.G. & Holzer, G. (1982). Zentralbl. Bakteriol. Mikrobiol. Hyg. Abt. 1 Orig. C3 : 228-244.

Langworthy, T.A., Holzer, G. & Zeikus, J.G. (1983). System App. Microbiol. 4 : 1-17.

Pond, J.L., Langworthy, T.A. & Holzer, G. (1986). Science 231 : 1134-1136.

Ross, H.N.M. & Grant, W.D. (1985). J.Gen.Microbiol. 131 : 165-173.

Ross, H.N.M., Collins, M.D., Tindall, B.J. & Grant, W.D. (1981). J. Gen. Microbiol. 123 : 75-80.

Stackebrandt, E. & Woese, C.R. (1981). In: Molecular and Cellular Aspects of Microbial Evolution. (Eds. M.J. Carlile, J.F. Collins & B.C.B. Moseley). pp. 1-31. Cambridge University Press.

Stetter, K.O. & Zillig, W. (1985). In: The Bacteria, Volume 8. (Eds. C.R. Woese & R.S. Wolfe) pp. 85-170. Academic Press, New York, London.

Stoeckenius, W. & Rowen, R. (1967). J.Cell Biol. 34 : 365-393.

Tindall, B.J., Ross, H.N.M. & Grant, W.D. (1984). Sys. App. Microbiol. 5 : 41-57.

Whitman, W.B. (1985). In: The Bacteria, Volume 8. (Eds. C.R. Woese & R.S. Wolfe). pp. 3-84. Academic Press, London, New York.

Woese, C.R. (1982). Zentralbl. Bakteriol. Mikrobiol. Hyg. Abt. 1. Orig. C3 : 1-17.

Woese, C.R. & Fox, G.E. (1977). Proc. Nat. Acad. Sci. (Wash.) 74: 5088-5090.

Woese, C.R., Magrum, L.J. & Fox, G.E. (1978). J.Molec.Evol. 11 : 245-252.

Yang, D., Kaine, B.P. & Woese, C.R. (1985). System. Appl. Microbiol. 6 : 251-257.

Zeikus, J.G., Dawson, M.A., Thompson, T.G., Ingvorsen, K. & Hatchikian, E.C. (1983). J.Gen.Microbiol. 129 : 1159-1169.

THE THERMODYNAMICS OF ENVIRONMENTAL ADAPTATION

R. A. Klein

Medical Research Council, Molteno Institute, University of Cambridge, Downing Street, Cambridge, CB2 3EE, UK

INTRODUCTION

Adaptation towards the environment experienced by an organism may be viewed in terms of the physical and chemical features of the environment. The most important aspects of the physical environment, in terms of adaptive strategy, are the intensive variables of temperature and pressure. Visible light and ionizing radiation, such as ultraviolet light, may also be important but more because of the chemical consequences of, for example, free radical generation. The chemical nature of the environment, the presence or lack of oxygen or organic matter or the presence of salt, may have very profound effects on the strategy for survival adopted by the organism; the use of alternate electron acceptors other than oxygen is just one such area in which micro-organisms are highly specialised.

It is important to distinguish clearly between those adaptive strategies which have occurred on an evolutionary time scale (this is relative to the life span of the organism), that is genotypic change, and those which have taken place in a particular organism adapting to a change in environment over hours, days or months, that is phenotypic.

EVOLUTIONARY ADAPTATION

Micro-organisms offer the clearest examples of adaptation on an evolutionary time scale. In the beginning the atmosphere of the earth was hotter than today and contained not only little or no oxygen, but hydrogen, nitrogen and H_2S or SO_2. These primaeval selective conditions brought about adaptation in various micro-organisms which may still be seen today in, amongst others, the obligate anaerobes, the thermophiles, the acidophiles and the halophiles. Changes in the atmosphere, with increasing oxygen content and more moderate temperatures, brought about adaptation to a much more oxidative environment. Bacteria are known which survive at temperatures as high as +90°C in 2N acid, at pressures of thousands of atmospheres in oil wells, or at subzero temperatures (Kates, 1986). Studies of the similarities in the genetic composition of organisms, particularly by Woese and his colleagues (Fox et al., 1980; Woese and Fox, 1977; Woese et al., 1978; Woese, 1982; Stackebrandt and Woese, 1981), using the similarity index S_{AB} - see the discussion by Grant in this book - have led to the important identification of a third evolutionary line, the ancestral Archaebacteria, as well as the ancestral Eubacteria and Urkaryotes, and an understanding of the evolutionary relationship between species.

The Archaebacteria are especially interesting because of their occurence in extreme adaptive niches. Grant (q.v.) has discussed the variety of lipid structures found in these organisms, and I will restrict my discussion to the thermodynamic strategies that the structures represent.

ADAPTATION BY INDIVIDUALS

Individuals of a species may adapt over relatively short periods to extremes of temperature and pressure. The higher homeotherms, for example man, have a limited environmental niche. But one should remember that even man is able to adapt to extremes of temperature (for example Eskimos and the arterio-venous shunts found in Aborigines), or to reduced pressure such as is found in high altitude acclimatization. Some animals are able to reduce body temperature for long periods of time, and some of the features of hibernators are discussed by Ellory et al. in this volume. Far and away the most fascinating forms of adaptation are found in those animals which live in the oceans. Temperatures may range from approximately -2°C to +40°C, and pressures may be as high as 1000 atmospheres in the deep sea trenches found in deep sea tectonic plate subduction zones.

Volcanic activity at constructive plate margins in mid-ocean ridges may give rise to "black smokers" where water temperatures may be high (+300°C) and the chemical composition quite bizarre with a high preponderance of sulphides in solution (Jannash, 1984).

The oceans show thermoclines which are seasonal and depend on latitude, with temperature gradients which range from close to zero at the poles to as high as 20°C/500 metres in the tropics. Pressure, of course, increases uniformly by about 1 ATA per 10 metres as one gets deeper.

Two completely separate forms of animal living in the oceans must be distinguished. There are those poikilotherms which live at fairly constant temperature and pressure, whether at the surface or on the abyssal floor, and then there are those species which have adapted to moving in a vertical water column representing a few hundred atmospheres change in pressure together with a change in temperature of perhaps 20°C-30°C. The whale is a good example of an animal able to change its environmental temperature and pressure rapidly by diving.

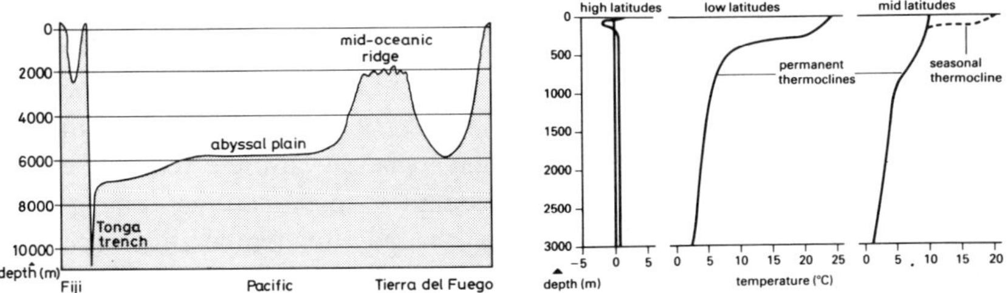

Figure 1. Schematic representation showing the morphology of the ocean basins and temperature profiles at different latitudes (from Friday, A. and Ingram, D.S. (1985), The Cambridge Encyclopaedia of Life Sciences, Cambridge University Press).

I will discuss the possible forms of thermodynamic compensatory phenomena that are available to organisms as adaptive strategies to changing pressure and temperature.

TEMPERATURE ADAPTATIONS

For technical reasons adaptation to temperature, rather than to pressure, has been worked on for a very long time, and the effects of pressure on membranes were not seriously considered until around 1960 (except for the comments of Ebbecke in 1936 - Macdonald, 1986). One should distinguish the effects of environmental temperature on enzymes and other proteins from its effects on the composition of the membrane lipids.

Much of the work on acclimatization to temperature has been done with fish, or the protozoan <u>Tetrahymena</u> <u>pyriformis</u>, or bacteria (for example see Hazel 1984). The generalization that comes out of this large body of work is that fatty acids of the membrane lipids tend to become more unsaturated, or of shorter chain length, as the acclimitazation temperature is lowered. This applies to many organisms (see for example Litchfield and Morales, 1976; Joseph, 1979). Increasing unsaturation, especially with double bonds, leads to increasing membrane disorder, i.e., an increase in fluidity. This effectively increases the entropy, ΔS, of the membrane system and since the transition temperature, T_m, is given by

$$T_m = \Delta H_{cal}/\Delta S \qquad (1)$$

for the lipid liquid-crystalline to gel first-order phase transition, increasing the entropic term effectively reduces the transition temperature, $\underline{T}_m$. Shorter acyl chains with lower values of $\Delta\underline{H}$ and $\underline{T}_m$ produce the same effect on fluidity (McElhaney, 1982).

It would appear that the essential requirement is for the membrane to be above its $\underline{T}_m$ at normal environmental temperatures (see discussion by McElhaney, 1986). So long as this is so, a particular value for the hydrocarbon chain fluidity does not appear to be an absolute requirement for continuous function. The clearest examples of adaptive changes in membrane composition resulting form changes in temperature occur in relatively simple organisms such as <u>Acholeplasma</u> <u>laidlawii</u> (Silvius and McElhaney, 1980; McElhaney, 1986).

Cossins (1986) has recently demonstrated in a series of specialized marine fish that the temperature compensation is never complete if one examines the shifts in the fluorescence polarization for DPH versus temperature profiles for the various species.

The introduction of a <u>cis</u>-double bond increases membrane disorder, i.e., entropy. Extreme thermophiles, or organisms living in highly oxidizing or reducing atmospheres, are unable to utilize this strategy and must rely upon another. As described by Grant (q.v.), these organisms may use branched methyl-substituted chains such as the phytanyl ethers, cyclopentane-substituted or cyclopropane-substituted hydrocarbons (Kates, 1986), as a mechanism for maintaining membrane fluidity by inserting packing defects, i.e., increasing entropy. Some years ago we observed the presence of <u>vic</u>-dimethyl-branched dicarboxylic acids (diabolic acids) in a rumen bacterium subject to a highly reducing environment (Klein et al., 1979), thus utilizing the steric rigidity of the vicinal dimethyl group for the same purpose.

All chemical reactions show a dependency on the temperature. One may express the effect of temperature on equilibria, or on rates using the free energy, $\Delta\underline{G}$, in the following expressions - for example in the equilibrium partitioning of a solute between two solvent phases, the partition coefficient, $^1\underline{P}_2$ is given by

$$\ln(^1\underline{P}_2) = -\Delta^1\underline{G}_2/\underline{RT} \tag{2}$$

In this example of partitioning the free energy term may be usefully factored into contributions for solvation, van der Waals, electrostatic, configurational, cavitation and hydrogen bond donor/acceptor contributions.

$$\underline{F} = \underline{F}_{solv} + \underline{F}_{vdw} + \underline{F}_{es} + \underline{F}_{conf} + \underline{F}_{cav} + \underline{F}_{hba} + \underline{F}_{hbd} \tag{3}$$

Reaction rates are usually expressed through the formalism of the Eyring absolute rate theory, in which the activated–state enthalpy and entropy are used

$$\ln(\text{rate}) = \underline{A} - \Delta\underline{H}^{\ddagger}/\underline{RT} + \Delta\underline{S}^{\ddagger}/\underline{R} \qquad (4)$$

The equilibrium and activated-state enthalpy both include the pressure-dependent term $\underline{P}.\Delta\underline{V}^{O}$ or $\underline{P}.\Delta\underline{V}^{\ddagger}$; this will be discussed further in the section on pressure adaptation.

The full analytical expression for the velocity of an enzyme reaction in which the catalytic rate constant and the Michaelis constant are replaced by the appropriate thermodynamic expressions may become so complicated that extraction of the parameters may be difficult experimentally even in the simplest case, and quite meaningless in more complicated cases, unless a very great deal is known about the mechanism of the reaction (for a discussion of the problems see Segel, 1975 or Roberts and Elmore 1974; Roberts, 1977). The naive measurement of the "activation enthalpy" without knowing if the enzyme is substrate-saturated at all temperatures at which measurements are made, tells one little or nothing, since both the catalytic rate constant $\underline{V}_{max}$ and the $\underline{K}_{m}$ terms have separate temperature dependences (Klein, 1982; Hall et al., 1982).

Temperature modulation of $\underline{V}_{max}$ and $\underline{K}_{m}$ can provide an extremely important short-term adaptive mechanism (Hochachka and Somero, 1973). 'Positive' modulation occurs when the substrate affinity decreases with temperature, i.e., $\underline{K}_{m}$ rises as does the temperature. This process can stabilize the reaction rate if the substrate concentrations are not saturating, and in a particular range as shown in Figure 2.

Many examples of temperature modulation, both 'positive' and 'negative', are known to exist (see references quoted in Klein, 1982). With negative modulation, $\underline{K}_{m}$ increases as the temperature drops, or the $\Delta\underline{H}^{O}$ is positive, and this is thought to be a possible mechanism for limiting environmental habitat.

PRESSURE ADAPTATIONS

The sensitivity of processes to pressure depends on the value of $\Delta\underline{V}^{\ddagger}$ or $\Delta\underline{V}^{O}$ in the free energy term

$$\Delta\underline{G}^{\ddagger} = \Delta\underline{E}^{\ddagger} + \underline{P}.\Delta\underline{V}^{\ddagger} - \underline{T}.\Delta\underline{S}^{\ddagger} \qquad (5)$$

Interspecies differences in the molar volume term may have significance in adaptation to high pressures as found in the oceans (up to

1000 atmospheres), as well as determining species zonation (Siebenaller and Somero, 1978). As a general proposition deep-sea species show values of $\Delta\underline{V}$ that are more negative than surface-living species or land animals (Hochachka and Somero, 1973; etc) with consequent advantages.

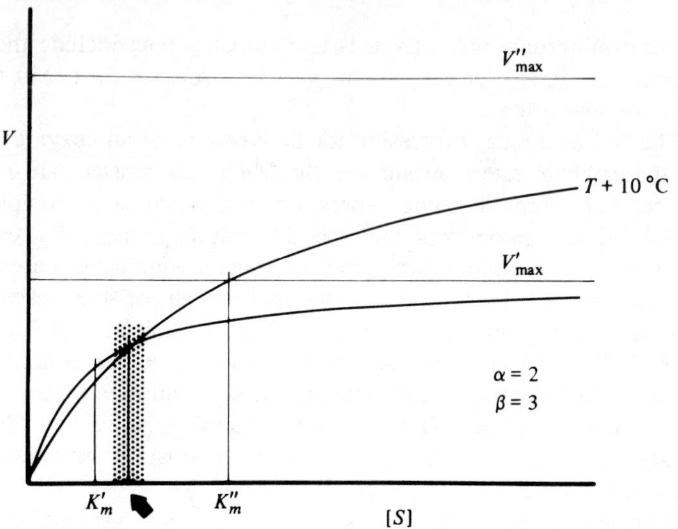

Figure 2. The effect of temperature on the rate of an enzymic reaction showing positive thermal modulation of the Michaelis constant $\underline{K}_m$. Compensatory rate behaviour will occur over a limited temperature range in the region of the substrate concentration $[\underline{S}]$ shown by the stippled area. If this substrate concentration, which is given by the expression

$$[\underline{S}] = \underline{K}_m.(\beta-\alpha)/(\alpha-1)$$

where 'α' and 'β' are the ratios of the maximal velocities and Michaelis constants at the two extremes of temperature respectively, lies within the physiological range then the compensatory behaviour has functional significance. For crossing over to occur as shown in the figure, $\alpha < \beta$ (from Klein, 1982).

Not only does pressure affect the rate at which enzymes operate per se, but it can also increase the 'micro-viscosity', or order parameter, of the lipid comprising the membrane thus affecting enzyme rates indirectly. The dependence of the phase transition temperature $\underline{T}_m$ on the pressure is given by the expression

$$\delta T_m/\delta P = T_m \cdot \Delta V_m/\Delta H_{vH} \tag{6}$$

For DPPC vesicles in the range of 1 to 300 ATA, this expression has a value of about +0.024 K atm^{-1}, and a volume change associated with the transition of 0.035 ml/g (Epand and Epand, 1980; Wilkinson and Nagle, 1981). The membrane of <u>Halobacter</u> <u>halobium</u>, a halophile containing predominantly phytanyl ether lipids, has a compressibility somewhat less than DPPC liposomes, indicating an effect of the methyl-branched phytanyl chains (Lin and Kay, 1977; Marque et al., 1984).

In recent discussions of the combined effects of temperature and pressure on the composition of the membrane lipids, in which increasing pressure and decreasing temperature result in an elevation of the proportion of unsaturated fatty acids present, the equivalence in terms of membrane ordering of temperature and pressure has been recognized (Macdonald, q.v.). 100 ATA has an ordering effect equivalent to around 2°-8°C decrease in temperature at atmospheric pressure (Macdonald 1984, 1986). Why should this be so, and on what does it depend? An equivalent ordering of between 2°-3°C per 100 ATA (see Chong and Cossins, 1984) is satisfactorily accounted for by the temperature dependence of the lipid transition temperature - viz. DPPC with +0.0238 K/ATA. However, higher values for the pressure-temperature equivalence suggest processes other than just an effect on hydrocarbon core ordering, as in the simple case of homogeneous DPPC liposomes. One should remember, however, that as the transition broadens, i.e., ΔH_{vH} becomes less, thus the coefficient, which basically depends on ΔV, becomes larger. Anomalous <u>increases</u>, however, of rotational rate with increasing pressure have been reported (Chong and Cossins, 1984), somewhat resembling the paradoxical increase in K$^+$ permeability with a reduction in temperature seen in red cells (Stewart et al., 1980).

Recent work from Cossins and Macdonald (1984) has shown that adaptive acclimatization to temperature and pressure is never 100% efficient. That is if one measures membrane 'fluidity', using a fluorescent probe, as a function of temperature for a species adapted to, say, 5°C and 25°C, or for different species from different depths, the explainable adaptation expressed as an interpolated difference on the temperature (or pressure) axis is always less than the actual environmental difference.

COMPENSATORY PHENOMENA

Enthalpy-entropy compensation occurs when changes in the enthalpy of a process are offset by changes in the entropic term, thus

$$\Delta H = \alpha + \beta \Delta S \tag{7}$$

The constant β (= $\delta\underline{H}/\delta\underline{S}$) has the form of a temperature (Giese, 1977), and is often referred to as the 'compensation temperature', $\underline{T}_c$. Compensation occurs in many systems with changes in the environment such as solvent composition, pH or ionic composition. It is of interest because it results in minimizing changes in the overall free energy, $\Delta\underline{F}$, and hence reaction rate. There has been a great deal of argument concerning the meaning of the compensation temperature, $\underline{T}_c$ (Lumry and Rajender, 1970; Katz and Diamond, 1974a,b). The compensation temperature is often around 290 K. It seems that the actual value of $\underline{T}_c$, however, is not a property of the reaction or process but of the solvent, most often water (Saenger, 1980). This underlines the point that is often forgotten when discussing enzyme reaction rates or membrane permeability: that the properties of the solvent (water) are important and may be peculiar, especially so at high pressures (Franks, 1975; Stillinger, 1980). Solute hydration effects may be highly significant (Hall et al., 1982).

Compensation in biological systems may occur trivially because of (i) a limited temperature interval being available for experiments (($\underline{T}_1$-$\underline{T}_2$) / ($\underline{T}_1$+$\underline{T}_2$)<0.05), and (ii) a strong correlation between $\Delta\underline{H}^{\ddagger}$ and $\Delta\underline{S}^{\ddagger}$ because

$$(\delta\underline{S}/\delta\underline{T})_p = \underline{C}_p/\underline{T} \tag{8}$$

since $\ln(1+\delta\underline{x}) = \delta\underline{x}$ for $\delta\underline{x}<<0.1$.

A strong correlation between $\Delta\underline{H}^{\ddagger}$ and $\Delta\underline{S}^{\ddagger}$ is often unavoidable because of the way in which these terms are derived experimentally. If one determines the value for $\Delta\underline{H}^{\ddagger}$ from the slope of an Arrhenius plot, and the $\Delta\underline{S}^{\ddagger}$ term from the y-intercept, the consequence is this: since the slope, $\underline{m}$, has an inherent error, determined by the spread in the data points, of $\pm\delta\underline{m}$ which may quite reasonably be between 1% and 5% (= $\delta\underline{m}/\underline{m}$), the entropic term, measured as an intercept, is in error by the approximate amount

$$\delta\Delta\underline{S} = (\underline{R}/<\underline{T}>).\delta\underline{m} \tag{9}$$

where $<\underline{T}>$ is the mean absolute temperature of the measurements. Reasonable values for $\Delta\underline{H}$ and the range of temperature would give an error in $\Delta\underline{S}$ of between 0.1 and 1.0 (e.u.) cal mol^{-1} K^{-1}; a quite considerable error.

Appreciable enthalpy-entropy compensation is seen for lactate dehydrogenase, D-glyceraldehyde-3-phosphate dehydrogenase and muscle glycogen phosphorylase-b in animals as diverse as rabbits and halibut (Low et al., 1973; Somero and Siebenaller, 1979). Values for the free energy of activation, $\Delta\underline{G}^{\ddagger}$, are sensibly the same for different species, without noticeable systematic differences contrary to the authors' claim. This form of compensation is not obviously of adaptive significance since the enzyme rates should have been compared not at the same temperature but at the normal environmental temperature. On the other hand the ectothermic species do show a marked reduction (up to 25%) in the enthalpy term. Again, however, these measurements are only meaningful for the comparison of rate data if the term $\underline{P}.\Delta\underline{V}^{\ddagger}$, is also known: in general these have not been determined. Oxamate binding to the muscle LDH also shows lower $\Delta\underline{H}^{\ddagger}$ values for ectothermic species (Hochachka et al., 1976). Adaptively the reduction of the enthalpy term is important because this reduces temperature sensitivity. One of the most dramatic examples I am aware of is for tuna compared to mammalian haemoglobin (Rossi-Fanelli and Antonini, 1960 - 1.8 kcal mole^{-1} compared with 13 kcal mole^{-1}).

Another form of compensation which could have considerable significance for those species that move through the water column with changes in temperature (20-30°C) and pressure (100-400 ATA), involves offsetting the enthalpy with the molar volume term. The temperature sensitivity (Arrhenius slope) is given by the term

$$\text{slope} = -(\Delta\underline{H}^{\ddagger}/\underline{R}) = -(\Delta\underline{E}^{\ddagger} + \underline{P}.\Delta\underline{V}^{\ddagger})/\underline{R} \tag{10}$$

Values of $\Delta\underline{H}^{\ddagger}$, and $\Delta\underline{E}^{\ddagger}$, for many processes at atmospheric pressure are typically between 20 and 80 kJ mole^{-1}. Values for the molar volume, $\Delta\underline{V}^{\ddagger}$, are usually positive (but see discussion in Macdonald, 1984; Hall et al., 1982) and around 30 cm^3 mole^{-1}. However, protein conformational changes are often associated with much higher values - 300 cm^3 mole^{-1} for myosin polymer (Dreizen and Kim, 1971; Lukasiewicz and Dreizen, 1977).

Consider the enthalpy term at temperatures $\underline{T}_1$ and $\underline{T}_2$, and pressures $\underline{P}_1$ and $\underline{P}_2$, where $\underline{T}_1 > \underline{T}_2$ and $\underline{P}_1 < \underline{P}_2$. The difference in the exponent in the rate equation due to the enthalpy will be

$$-(\Delta\underline{E}^{\ddagger}+\underline{P}_1.\Delta\underline{V}^{\ddagger})/\underline{R}\underline{T}_1 + (\Delta\underline{E}^{\ddagger}+\underline{P}_2.\Delta\underline{V}^{\ddagger})/\underline{R}\underline{T}_2 \tag{11a}$$

$$= -\Delta\underline{E}^{\ddagger}/\underline{R}.(1/\underline{T}_1 - 1/\underline{T}_2) - \Delta\underline{V}^{\ddagger}/\underline{R}.(\underline{P}_1/\underline{T}_1 - \underline{P}_2/\underline{T}_2) \tag{11b}$$

For there to be complete compensation

$$\Delta\underline{E}^{\ddagger}/\Delta\underline{V}^{\ddagger} = (\underline{P}_2/\underline{T}_2 - \underline{P}_1/\underline{T}_1)/(1/\underline{T}_1 - 1/\underline{T}_2) \tag{12a}$$

$$= (\underline{P}_2\underline{T}_1 - \underline{P}_1\underline{T}_2)/(\underline{T}_2 - \underline{T}_1) \tag{12b}$$

For a thermocline of 27°C to 7°C and a change of 400 ATA

$$\Delta\underline{E}^{\ddagger} = -798.6 \times \Delta\underline{V}^{\ddagger} \qquad (13)$$

in joules mole^{-1} and cm^3 mole^{-1}. Typical values for K$^+$ leakage in the human red cell are $\Delta\underline{E}^{\ddagger} = 78$ kJ mole^{-1} and $\Delta\underline{V}^{\ddagger} = -84$ cm^3 mole^{-1} (Hall et al., 1982), suggesting that such compensation is at least possible.

It is perhaps a truism to point out that compensation in one process, as measured experimentally, may be irrelevant and an example of Macdonald's reductionism (see the Round Table discussion in this volume). What matters in a living organism is that overall compensation takes place. In the example of the red cell the equilibrium between energy-dependent (that is 'active') pumps and passive diffusion - the leak - must be compensated in an overall manner for adaptation to occur.

REFERENCES

Chong, P.L-G. and Cossins, A.R. (1984) Biochim. Biophys. Acta 772: 197-201

Cossins, A.R. (1986) Biochem. Soc. Trans 14: in press

Cossins, A.R. and Macdonald, A.G. (1982) J. Physiol. Lond. 332: 64-65

Cossins, A.R. and Macdonald, A.G. (1984) Biochim. Biophys. Acta 776: 144-150

Dreizen, P. and Kim, H.D. (1971) Amer. Zool. 11: 513-521

Epand, R.M. and Epand, R.F. (1980) Chem. Phys. Lipids 27: 139-150

Fox, G.E., Stackebrandt, E., Hespell, R.B., Gibson, J., Maniloff, J., Dyer, T.A., Wolfe, R.S., Blach, W.K., Tanner, R.S., Magrum, L.J., Zablen, L.B., Blakemore, R., Gupta, R., Bonen, L., Lewis, B.J., Stahl, D.A., Luehrsen, K.R., Chen, C.N. and Woese, C.R. (1980) Science 209: 457-463

Franks, F. (1975) in "Water", Vol.2 pp.1-54, Plenum Press, New York and London

Giese, B. (1977) Angew. Chem., Int. Ed. 16: 125-136

Grant, W.D. (1986) in "Topics in Lipid Research - from Structural Elucidation to Biological Function" (eds. B. Schmitz and R.A. Klein), Royal Society of Chemistry, London

Hall, A.C., Ellory, J.C. and Klein, R.A. (1982) J. Membrane Biol. 68: 47-56

Hazel, J.R. (1984) Amer. J. Physiol. 246: R460-R470

Hochachka, P.W. and Somero, G.N. (1973) "Strategies of Biochemical Adaptation", pp. 179-303, W.B. Saunders, Philadelphia, London and Toronto

Hochachka, P.W., Norberg, C., Baldwin, J. and Fields, J.H.A. (1976) Nature Lond. 260: 648-650

Jannash, H.W. (1984) in "Hydrothermal processes at sea floor spreading centers", eds. P.A. Rona, K. Bostrom, and L. Laubier, Plenum Press, New York

Joseph, J.D. (1979) Prog. Lipid Res. 18: 1-30

Kates, M. (1986) F.E.M.S. Microbiology Reviews 39: 95-101

Katz, Y. and Diamond, J.M. (1974a) J. Membrane Biol. 17: 69-86

Katz, Y. and Diamond, J.M. (1974b) J. Membrane Biol. 17: 101-120

Klein, R.A. (1982) Quart. Rev. Biophys. 15: 667-757

Klein, R.A., Hazlewod, G.P., Kemp, P. and Dawson, R.M.C. (1979) Biochemical Journal 183: 691-700

Lin, N. and Kay, R.L. (1977) Biochemistry 16: 3484-3486

Litchfield, C. and Morales, R.W. (1976) in "Aspects of Sponge Biology", pp.183-200 (eds. F. Harrison and R.C. Cowden) Academic Press, New York

Low, P.S., Bada, J.L. and Somero, G.N. (1973) Proc. Natl. Acad. Sci. U.S.A. 70: 430-432

Lukasiewicz, R.J. and Dreizen, P. (1977) Biophys. J. 17: 37A

Lumry, R. and Rajender, S. (1970) Biopolymers 9: 1125-1227

Macdonald, A.G. (1984) Phil. Trans. Roy. Soc. Lond. B304: 47-68

Macdonald, A.G. (1986) in "Topics in Lipid Research - from Structural Elucidation to Biological Function", eds. B. Schmitz and R.A. Klein, Royal Society of Chemistry, London

Macdonald, A.G. and Cossins, A.R. (1983) Biochim. Biophys. Acta 730: 239-244

McElhaney, R.N. (1982) Chem. Phys. Lipids 30: 229-259

McElhaney, R.N. (1986) Biochem. Soc. Trans. 14: in press

Marque, J., Eisenstein, L., Gratton, E., Sturtevant, J.M. and Hardy, G.J. (1984) Biophys. J. 46: 567-572

Roberts, D.V. (1977) "Enzyme Kinetics", Cambridge University Press

Roberts, D.V. and Elmore, D.T. (1974) Biochem. J. 141: 545-554

Rossi-Fanelli, A. and Antonini, O. (1960) Nature Lond. 186: 895-896

Saenger, W. (1980) Angew. Chem., Int. Ed. 19: 344-362

Segel, I.H. (1975) "Enzyme kinetics", John Wiley, New York

Siebenaller, J. and Somero, G.N. (1978) Science 201: 255-257

Silvius, J.R. and McElhaney, R.N. (1980) Proc. Natl. Acad. Sci. U.S.A. 77: 1255-1259

Somero, G.N. and Siebenaller, J.F. (1979) Nature Lond. 282: 100-102

Stackebrandt, E. and Woese, C.R. (1981) in "Molecular and Cellular Aspects of Microbial Evolution", eds. M.J. Carlisle, J.F. Collins and B.C.B. Moseley, pp.1-31, Cambridge University Press

Stewart, G.W., Ellory, J.C. and Klein, R.A. (1980) Nature Lond. 286: 403-404

Stillinger, F. (1980) Science 209: 451-457
Wilkinson, D.A. and Nagle, J.F. (1981) Biochemistry Philadelphia 20: 187-192
Woese, C.R. (1982) Zentralblatt Bakteriol. Mikrobiol. Hyg. (Abt. 1) C3: 1-17
Woese, C.R. and Fox, G.E. (1977) Proc. Natl. Acad. Sci. U.S.A. 74: 5088-5090
Woese, C.R., Magrum, L.J. and Fox, G.E. (1978) J. Molecular Evol. 11: 245-252

THE TEMPERATURE DEPENDENCE OF MEMBRANE TRANSPORT IN HIBERNATOR AND NON-HIBERNATOR SPECIES' RED CELLS

J. Clive Ellory, Andrew C. Hall[+], Michael W. Wolowyk[*] and Lawrence C.H. Wang[x]

University Laboratory of Physiology, Parks Road, Oxford OX1 3PT, [+]Physiological Laboratory, Downing Street, Cambridge CB2 3EG, [*]Faculty of Pharmacy and Pharmaceutical Sciences, and [x]Department of Zoology, University of Alberta, Edmonton, Alberta, Canada T6G 2N8

INTRODUCTION

The ability of hibernating mammals to survive for long periods at low temperatures ($0^{\circ}C$-$4^{\circ}C$) is thought to be due mainly to the persistence of transport processes and the maintenance of the functional integrity of cell membranes (Willis, 1978). Thus, with cooling, the balance between active and passive fluxes is maintained and the steady-state ensured. In cells from non-hibernators, the greater temperature-sensitivity of active compared to passive transport leads to the collapse of established gradients and ultimately cell death. Mammalian hibernators also show the phenomenon of periodic arousal in which body temperature rises rapidly to normal ($37^{\circ}C$) and after a short period (hours) the animal returns to the hibernating state (Willis, 1978). Clearly, the regulation of cellular composition both at low and normal body temperatures in hibernators' cells relies on modifications to transport systems which are presumably not present or insufficient in the non-hibernators' cells.

There are many reviews on membrane transport in mammalian cells and in particular the adaptations of hibernators' transport pathways to function at low temperatures (Willis, 1978,1979; Willis et al., 1981). We have, therefore, omitted the Na^+/K^+ pump and amino acid transport pathways (see Ellory & Willis, 1981,1983), and confined ourselves to studies on cation transport in erythrocytes (for review see Ellory & Tucker, 1983). Red cells are a valuable model cell and, furthermore, can be obtained from animals during hibernation, and compared with the transport properties following arousal. Initially, we describe some transport pathways in erythrocytes, and attempt to determine whether they play a significant role in hibernators' red cells at low temperatures. Subsequently, we focus on Ca^{2+} transport, in red cells from a non-hibernator, and from hibernators in the awake and hibernating states.

RESIDUAL CATION PERMEABILITY

All biological membranes exhibit a ground leak permeability (i.e., that which remains after all known carrier-mediated pathways have been inhibited). In this context, K^+ permeability is probably the best understood. However, experiments on the passive (i.e., (ouabain + bumetanide + Ca^{2+})-insensitive) K^+ transport in red cells of a wide variety of mammalian species (hibernators: thirteen-lined ground squirrel (TLGS), Columbian ground squirrel, woodchuck, hamster; non-hibernators: guinea pig, gerbil, rat, lemming, dog, rabbit, human) have failed to show any correlation of passive K^+ permeability with the ability to hibernate (Hall & Willis, 1984 a,b; 1986).

Passive Na^+ transport is complex because of the possible contribution of certain pathways (e.g. Na^+-Li^+ exchange, (Pandey, et al., 1978); $NaCO_3^-$ via capnophorin (Funder, 1980)) at low temperatures which have yet to be assessed. Experiments on passive Ca^{2+} permeability are similarly complicated by the presence of the ATP-fuelled Ca^{2+} pump (Sarkadi, 1980) which must be abolished by ATP-depletion, a procedure which might inadvertently alter passive Ca^{2+} permeability. Despite this, the passive influx or efflux of Na^+ or Ca^{2+} is not significantly different between hibernators and non-hibernators (Ellory & Hall, 1983; Hall et al., 1986; Hall & Willis, unpublished and see below). Of interest was the finding that the apparent activation energies (E_a) for passive cation (K^+, Na^+, Ca^{2+}) permeability below 18.5°C (computed from Arrhenius plots, see Hochachka & Somero, 1973) are very low, and for some species are close to that for free diffusion in water (Hall & Willis, 1984 b; Hall et al., 1986).

Ca^{2+}-ACTIVATED K^+ CHANNEL

A rise in $[Ca^{2+}]_i$ stimulates the quinine-sensitive K^+ channel, which leads to a loss of cell K^+ (Lew & Ferreira, 1978). Studies on the temperature dependence of this pathway have utilised the Ca^{2+} ionophore A23187 to raise $[Ca^{2+}]_i$ thereby fully activating the K^+ channel. The difference in temperature sensitivity of K^+ channels in a hibernator (TLGS) and a non-hibernator (guinea pig) was slight (about x2) considering the enormously stimulated K^+ flux (Hall & Willis, 1984a). It is conceivable that the disruption of the membrane abolished a subtle difference in the hibernators' cells. Indeed, simple ghosting of red cells markedly alters the temperature dependence of the Na^+/K^+ pump (Willis et al., 1978).

VOLUME-SENSITIVE KCl PATHWAY

In many species' erythrocytes, swelling activates a Cl^--dependent K^+ flux leading to a loss of cell $[K^+]_i$ (Dunham & Ellory, 1981; Ellory et al., 1985a). Such a pathway may have an important regulatory function by limiting the swelling caused by ionic imbalance at low temperature although this flux has been shown to be temperature-sensitive (Ellory et al., 1985b). Both hibernators' and non-hibernators' red cells possess this transporter (Ellory et al., 1985a); however at reduced temperatures there does not appear to be a marked difference between the potency of the pathway in a hibernator (TLGS) and a non-hibernator (guinea pig) (Hall & Willis, unpublished observations). Clearly more work on this system and the amiloride-sensitive Na^+-H^+ pathway (Benos, 1982) at low temperature is needed before their role can be fully determined.

(Na^++K^+)-COTRANSPORT

Previous studies on the TLGS and the guinea pig (Hall & Willis, 1984a) showed the persistence of cotransport at 5^oC in the hibernator and its abolition in the non-hibernator. Recent work on a wider range of species suggests that there is only a poor correlation between cotransport and the ability to hibernate (Hall & Willis, unpublished observations). Nevertheless, the role of this pathway in volume regulation is thought to be important in many cell types (e.g. Kregenow, 1981). Thus, although there may not be a universal adaptation to this pathway in hibernators' cells, cotransport may for some species have a significant supplementary role to compensate for deficiencies in other transport pathways.

Ca^{2+} PUMP

In the presence of external Ca^{2+}, K^+ permeability mediated by the Ca^{2+}-activated K^+ channel is markedly stimulated in red cells from a non-hibernator (guinea pig) whereas in a hibernator (TLGS) there is little effect at low temperature (Hall & Willis, 1984a). This suggests that the regulation of a low $[Ca^{2+}]_i$ in non-hibernators' cells fails with cooling thereby causing a rise in $[Ca^{2+}]_i$ (Hall & Willis, 1984c). Since the maintenance of a low $[Ca^{2+}]_i$ is known to be essential for cell viability, a cold-resistant form of the pump may be an important adaptive strategy for survival at low temperature in hibernators' cells.

Fig. 1 shows ATP-dependent, and ATP-independent Ca^{2+} efflux from guinea pig or Richardson's ground squirrel erythrocytes at 5^oC. In cells from the non-hibernator, depletion only slightly reduced Ca^{2+} efflux indicating that the Ca^{2+} pump was almost completely inhibited at low

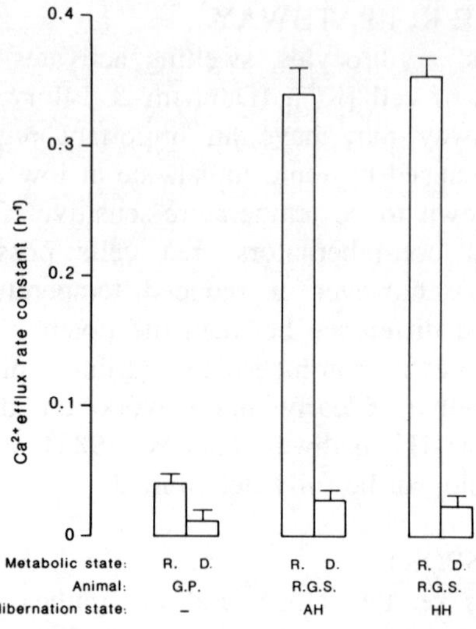

Figure 1. The effects of low temperature on Ca^{2+} efflux from rodent red cells. Erythrocytes were obtained from guinea pig (GP; a non-hibernator), or Richardson's ground squirrel (RGS; a hibernator) in the awake (AH) or hibernating (HH) states, and maintained replete (R) or depleted (D) of ATP by incubation with iodoacetamide. The cells were then loaded with ^{45}Ca using the ionophore A23187 under 'chemical clamp' conditions (see Hall & Willis, 1984a for experimental details). The ionophore was then washed away using albumin, and the appearance of ^{45}Ca measured at 5°C as the efflux rate constant under 'chemical clamp' conditions. Results are means ($\pm$ s.e.m.) of three experiments on blood from different donors.

temperature. In contrast, red cells from ground squirrels, whether obtained from awake or hibernating individuals, exhibited a large active Ca^{2+} efflux. There was no significant difference between the capacity of the Ca^{2+} pump in AH or HH cells suggesting that the temperature resistance of the transport pathway is inherent and not altered during a bout of hibernation. Notice also that in depleted cells passive Ca^{2+} permeability was similar under all conditions.

CONCLUSIONS

In the present paper, we have described several specific cation transport pathways in mammalian erythrocytes, and attempted to determine if transport capacity correlates with the ability to hibernate. Although there are many measurements of specific fluxes, crucial experiments on the long-term role of specific transport pathways in the control of cellular composition similar to those already performed (Kimzey & Willis, 1971) are lacking. Nevertheless, the temperature resistance of the Na^+/K^+ and Ca^{2+} pumps are significant and important adaptations in hibernators' red cells. The data do not , however, support the hypothesis that the adaptation of transport systems in hibernators' cells relies on a general membrane effect as has been observed in poikilotherms (Cossins, 1983). Rather, the reduced temperature sensitivity of certain transport pathways and the ability to function for long periods both at low and normal body temperatures must reflect specific adaptations to those transport systems. This is emphasized by the results for passive cation (K^+, Na^+, Ca^{2+}) transport systems which show no apparent correlation with hibernating ability.

ACKNOWLEDGEMENTS

This work was supported in part by the Alberta Heritage Foundation for Medical Research, and a University of Alberta C.R.F. grant. A.C.H. and J.C.E. received Wellcome Research Travel Grants.

REFERENCES

Benos, D.J. (1982). Amer. J. Physiol. 242, C131-C145.

Cossins, A.R. (1983). Biochem. Soc. Trans. 11, 332-333.

Dunham, P.B. & Ellory, J.C. (1981). J. Physiol. 318, 511-530.

Ellory, J.C. & Hall, A.C. (1983). J. Physiol. 345, 148P.

Ellory, J.C., Hall, A.C. & Stewart, G.W. (1985a). Molecular Physiol. 8, 235-246.

Ellory, J.C., Hall, A.C. & Stewart, G.W. (1985b). In "Transport Processes, Iono- and Osmoregulation." Eds. R. Gilles & M. Gilles-Baillien. Springer-Verlag, Heidelberg. pp. 401-410.

Ellory, J.C. & Tucker, E.M. (1983). In "Red Blood Cells of Domestic Mammals." Eds. N.S. Agar & P.G. Board. Elsevier. pp. 291-314.

Ellory, J.C. & Willis, J.S. (1981). In "Effects of Low Temperature on Biological Membranes." Eds. G.J. Morris & A. Clarke. Academic Press, London. pp. 107-119.

Ellory, J.C. & Willis, J.S. (1983). Biochem. Soc. Trans. 11, 330-332.

Funder, J. (1980). Acta Physiol. Scand. 108, 31-37.

Hall, A.C., Ellory, J.C., Wolowyk, M.W. & Wang, L.C.H. (1986). J. Therm. Biol: In Press.

Hall, A.C. & Willis, J.S. (1984a). J. Physiol. 348, 629-643.

Hall, A.C. & Willis, J.S. (1984b). J. Physiol. 354, 49P.

Hall, A.C. & Willis, J.S. (1984c). Biochem. Soc. Trans. 12, 312-314.

Hall, A.C. & Willis, J.S. (1986). Cryobiology: In Press.

Hochachka, P.W. & Somero, G.N. (1973). "Strategies of Biochemical Adaptation." W.B. Saunders, Philadelphia.

Kimzey, S.L. & Willis, J.S. (1971). J. Gen. Physiol. 58, 620-633.

Kregenow, F.M. (1981). Ann. Rev. Physiol. 43, 493-505.

Lew, V.L. & Ferreira, H.G. (1978). Curr. Topics Membr. Trans. 10, 217-277.

Pandey, G.N., Sarkadi, B., Haas, M., Gunn, R.B., Davis, J.M. & Tosteson, D.C. (1978). J. Gen. Physiol. 72, 233-247.

Sarkadi, B. (1980). Biochim. Biophys. Acta 604, 159-190.

Willis, J.S. (1978). In "Strategies in Cold: Natural Torpidity and Thermogenesis." Eds. L.C.H. Wang & J.W. Hudson. Academic Press, New York. pp. 417-460.

Willis, J.S. (1979). Ann. Rev. Physiol. 41, 275-286.

Willis, J.S., Ellory, J.C. & Becker, J.H. (1978). Amer. J. Physiol. 235, 159-167.

Willis, J.S., Ellory, J.C. & Cossins, A.R. (1981). In "Effects of Low Temperature on Biological Membranes." Eds. G.J. Morris & A. Clarke. Academic Press, London. pp. 124-142.

THE ROLE OF MEMBRANE LIPIDS IN THE ADAPTATION OF DEEP SEA ORGANISMS TO HIGH PRESSURE

A.G. Macdonald

Physiology Department, Marischal College, University of Aberdeen, Aberdeen AB9 1AS, Scotland

INTRODUCTION

High pressure affects rate constants and equilibria, through its contribution to the free energy change $\Delta F = (\Delta E + P\Delta V) - T\Delta S$ (ΔF, free energy; ΔE, internal energy; P, pressure; ΔV, volume; T, temperature and ΔS, entropy). The magnitude of the pressure in the deep sea, several hundred atmospheres, is sufficient to affect many physiological processes. In this paper the effects of deep sea pressures on the structure and function of lipid bilayers are outlined. Predictions are then made of the nature of pressure-adapted membranes in deep sea organisms. This is an extension of the theory of homeoviscous adaptation of membranes and it will be shown that certain membranes, prepared from the tissues of deep sea fish, are consistent with the predictions. The first section of the paper however, considers the deep sea environment and the evidence that deep sea organisms are indeed adapted to their ambient high pressure.

DEEP SEA ADAPTATION

For present purposes the deep sea is a very simple environment. It begins at about 1000 m depth, off the continental slope, and comprises the vast bulk of the world's oceans, whose mean depth is 3800 m. Highly localized deep sea trenches extend to approximately 10,000 m depths. It is by far the largest of biological environments, inhabited by animals and bacteria in biomass densities which reflect the attenuated flux of organic material through the system, which is far removed from the photosynthetic surface layer (Macdonald 1975; Somero 1982). Local "hot spots" where hot brines leak through vents in the ocean floor create oases with a distinctive flora and fauna, and emphasise the desert-like nature of the rest of the abyssal plain (Jannash 1984). Generally, temperatures in the deep sea are below 4°C and dissolved oxygen is present in abundance. Pressure increases by 1 atm for each 10 m of the water column and so the pressure range of the deep sea is 100-1000 atm.

Commercial (i.e. human) diving is concerned with the much smaller depth and pressure range of 1-100 atm which is also the range occupied by commercially important fish and the spectacular marine mammals and, for research purposes, the two complementary fields of high pressure biology conveniently overlap. The pressure range of interest in physical science is, of course, virtually infinite, but in the study of aqueous solutions, of protein structure and of chemical kinetics and thermodynamics, several thousand atm pressure is commonly used. In that sense deep sea pressures are not particularly high and the physical perturbations they cause are subtle rather than extreme.

What is the evidence that deep sea organisms are specifically adapted to high pressure? At the level of the integrated whole organism (animal or bacterial cell) the evidence comes from two types of study. One is of the effect of deep sea pressure on shallow water organisms otherwise resembling deep sea species, and the second is of the effect of hydraulically decompressing deep sea organisms to normal, surface, atmospheric pressure. Both approaches demonstrate that pressure has profound effects, and that shallow and deep living creatures depend on their normal ambient pressure.

In the case of shallow water animals, including primates and humans, pressures over the 1-100 atm range cause major disturbances in the central and peripheral nervous systems, manifest as motor abnormalities such as limb tremor, convulsions, and paralysis. Only the paralysis which inhibits respiration or circulation causes death; the tremor and convulsions are entirely reversible phenomena whether in invertebrates, fish or mammals (Brauer 1984; Bennet & Elliott 1982).

Experiments with deep sea animals are not common largely because of the specialised nature and high cost of the ships required to collect them.

Invertebrates collected from the relatively shallow depth of 680 m, such as the isopod <u>Munnopsis typica</u>, the scaphopod <u>Dentalium sp</u>, the hydroid <u>Aglaophenia bispinosa</u>, and the crinoid <u>Rhizocrinus lofotensis</u>, have been observed to recover movement when restored to their normal ambient temperature and pressure (Menzies George and Paul, 1974). Crabs (<u>Bythograea thermydron</u>) collected from hydrothermal vents at 2500 m in the Pacific were found to survive preferentially at elevated pressure and their heart rate and rhythm were impaired when decompressed (Mickel & Childress, 1982).

A systematic set of observations on the motor activity of benthic amphipods, collected over the depth range 1000-4000 m in the north

Atlantic, yielded the following conclusions. Amphipods from 2000 m were deleteriously affected by exposure to normal atmospheric pressure but they resumed normal activity at their normal pressure of 200 atm. Amphipods from 1300 m were not seriously affected by decompression, but those from 4000 m were totally immobilised, and appeared dead. On recompression to their normal pressure (400 atm) locomotor activity was gradually restored to the majority (Macdonald & Gilchrist 1978, Macdonald & Gilchrist, 1980). Using a specially developed pressure-retaining trap the same authors demonstrated a sustained level of normal activity in amphipods collected from 4000 m, and observed at 400 atm. During controlled decompression to atmospheric pressure the amphipods became immobilized (Macdonald & Gilchrist, 1982). Furthermore the tolerance of these abyssal amphipods to higher than normal pressure was very marked. Whereas amphipods from less than 4000 m depths show a graded increase in the pressure required to elicit convulsions, proportional to the pressure of their habitat, those from 4000 m failed to exhibit convulsions at pressures up to 900 atm, which immobilized them (Macdonald & Gilchrist 1982; Macdonald et al.1980).

Yayanos (1978) has also observed that amphipods in a pressure-retaining trap lose their motor activity on decompression.

Amphipods collected from 900-1400 m depths in the fresh water Lake Baikal revealed a loss of Na^+ whilst kept at atmospheric pressure, but not when held at their normal pressure (Brauer et al.1980).

The case for deep water invertebrates, and crustacea in particular, being dependent on their normal high pressure and therefore highly adapted to pressure, seems well established. More recently fish have been investigated but these are generally more susceptible than small amphipods to mechanical damage when collected. Using the pressure required to elicit convulsions, Brauer et al.(1984) found that cottoid fish from various depths in Lake Baikal comprised two groups. Shallow water fish convulsed at low pressures, whilst those from significant depths convulsed at rather higher pressures. Deep sea fish appear to show comparable tolerance. In experiments carried out on Challenger cruise 6B/85 and yet to be published, Macdonald et al. (in preparation) pressurised <u>Trachyscorpia cristulata chinata</u> trawled from 900 m in the Atlantic, and observed the partial restoration of motor activity and convulsions at pressures higher than those required to elicit convulsion in shallow water fish.

The dependence of the whole, integrated, animal on its high ambient pressure is thus experimentally established and that even the primitive methods used reveal significant adaptation to 100 atm (1000 m). There is little doubt that the phenomenon is graded over the 1-1000 atm pressure range in the oceans. Equally, there is little doubt that the nervous system is particularly sensitive to pressure and the adaptation of a pressure-stable system is a major factor in deep sea adaptation.

The practical question is now, how best to investigate these properties of deep sea animals?

Adaptations to high pressure, manifest at cellular and molecular levels, are in some ways more convenient to study than those in the integrated animal. A practical distinction for those who work at sea is between biochemical experiments, typically on enzyme properties, which can use "dead" material and physiological experiments which require "living" preparations to be used on board ship. Examples of the former are Swezey & Somero (1982), Hochachka (1974, 1975), Low & Somero (1975) and represent an important body of knowledge and ideas. Examples of the latter are confined to a limited number of more recent experiments, including ECG recordings from crabs (Mickel & Childress 1982), the Na^+ balance studies on amphipods from Lake Baikal (Brauer et al. 1980), and the resuscitation by pressure of isolated nerve, skeletal muscle and hearts from fish collected at 900 and 4000 m depths (Forbes et al. 1985, 1986a, 1986b).

This paper is concerned with the adaptation of membranes from deep sea fish to high pressure, and bridges the divide between "biochemical" and "physiological" work. Although some measurements use deep frozen samples, much depends on the preparation of fresh, good quality isolated membranes, at sea.

DEEP SEA PRESSURE

The significance of deep sea pressures on the structure and function of membranes emerged from work primarily concerned with the application of pressure as an analytical variable, like temperature, to understand the nature of lipid bilayers. It is a remarkable historical fact that pressure was not considered to be a variable of any particular interest in bilayer structure and function until about the late 1960's. It is remarkable for two reasons. The first is the extensive and fascinating body of knowledge which had grown from pressure studies of cellular processes, such as cytoplasmic and muscular contraction (Zimmerman 1970), might have prompted someone to consider pressurised membranes as potentially rewarding. The second reason is that sufficient knowledge of isotropic solvents under pressure existed in chemistry to initiate a natural progression to membrane bilayers. That progression did not occur until about 20 years ago, despite the brilliant speculative comments made by Ebbecke (1936) whose main experimental interest was with animal and nerve-preparations under high pressure.

The present wave of interest in the effects of pressure on membranes largely came from the observation that general anaesthesia could be reversed by pressure in certain animal experiments, reinforced by the established field of high pressure cell biology. Two publications mark the event: Macdonald & Miller 1976; Zimmerman & Zimmerman 1977. Both are reviews in multi-author books and both emphasised the gap which then existed in our knowledge of membranes and of bilayers in particular in pressurised cells.

The properties of isotropic solvents under pressure provide a basis for comparable studies on lipid bilayers. Three major interrelated properties are (i) bulk compressibility (ii) viscosity (iii) self-diffusion.

A typical isotropic solvent undergoes a bulk compression of about 1%/100 atm, and its viscosity may be more than doubled by a few hundred atmospheres (Isaacs 1981; Utracki 1983). The self-diffusion coefficients of its molecules are also significantly reduced by a hundred atmospheres. All this is intuitively understandable, and clearly consistent with Le Chatelier's principle. Interestingly, isotropic solvents are in marked contrast to protein gels (of actin and other important multimeric proteins such as fibrin or tubulin), all of which are depolymerized by pressure, and rendered more fluid (Swezey & Somero 1982; Collen et al.1970; Jaenicke 1983; Weber & Drickamer 1983).

So the behaviour of isotropic solvents under pressure strongly suggests, as Ebbecke pointed out 50 years ago, that the lipid bilayer, in so far as it is isotropic, should respond likewise. The anisotropic properties of bilayers, however, make them more interesting.

The bulk isothermal compressibility of DPPC liposomes is approximately 1.2% 100 atm^{-1} (Line & Kay 1977) and that of the membrane of <u>Halobacterium</u> <u>halobium</u> 0.26% 100 atm^{-1} (Marque et al. 1984). X-ray analysis of partially hydrated vesicles subjected to pressure shows an increase in bilayer thickness, due to the more ordered packing of the acyl chains, and a marked lateral compression, by which the volume of the bilayer as a whole compresses (Stamatoff et al. 1978). There is no reason to suppose that these observations should not apply to membrane bilayers in general.

The effect of pressure on the motion and dynamic structure of membrane bilayers has been studied using various spectroscopic methods: ESR, fluorescence methods (depolarization of diphenyl hexatriene and excimer probes) and Raman spectroscopy. These are reviewed in Macdonald (1984), and for present purposes the important conclusion is that pressure mimics a reduction in temperature, ordering bilayers to the extent that 100 atm is equivalent to a decrease of between 2° and 8°C, depending on the specific parameter.

Raman spectroscopy confirms the idea from X-ray analyses that acyl chains in a bilayer straighten under pressure, with the <u>trans</u> conformer being favoured at the expense of the <u>gauche</u> form (Yager & Peticolas 1980; Wong et al. 1982). In short chain isotropic solvents, pressure acts in the reverse way, favouring the <u>gauche</u> form whose volume "swept clear" is less than that of the <u>trans</u> conformers (Schoen et al. 1977, but see also Schwickert et al. 1982).

An isothermal phase transition may be brought about if sufficient pressure is applied to an appropriate bilayer. The important thermodynamic factors are $\Delta \underline{V}$, the volume change associated with the transition and $\Delta \underline{H}$, the enthalpy change. The Clausius-Clapeyron relationship $\delta \underline{T}/\delta \underline{P} = \underline{T}.\Delta \underline{V}/\Delta \underline{H}$ applies, with typical transition temperatures increasing by approximately $2^{\circ}C$ per 100 atm. (Reviewed in Macdonald 1984 and Wong 1984.)

From this outline it is clear that deep sea pressures are sufficient to exert a significant ordering effect on membrane bilayers, acting in the same direction as low temperature. Accordingly a bilayer at the average ocean bottom depth of 4000 m (400 atm) and at the typical bottom temperature of, for example $2^{\circ}C$, may be regarded as at the equivalent of at least $(2 - 8)^{\circ}C$ = $-6^{\circ}C$. On this basis it may be predicted that deep sea membrane bilayers require compensating changes in fluidity, i.e. homeoviscous adaptation to high pressure, in addition to low temperature.

The fluidity (order, motional freedom) of a bilayer in a natural membrane is largely determined by the number of <u>cis</u> double bonds in the fatty acid acyl chains. The <u>cis</u> bond imposes a kink in the linearity of the acyl chain, creating significant free volume in the bilayer. The first <u>cis</u> double bond to be inserted in an acyl chain pair attached to a phospholipid exerts a much bigger influence on the order of the molecule than second or third double bonds; hence the relationship between order and number of <u>cis</u> double bonds is non-linear (Stubbs et al. 1981). Furthermore cholesterol and the inclusion of other large molecules reduce fluidity and mimic pressure (Chong & Cossins 1984). Homeoviscous adaptation to high pressure therefore predicts increased <u>cis</u> double bonds and diminished cholesterol content to be among the chief means by which fluidity compensation is achieved.

The prediction assumes that the pressure-ordering effect imposes constraints on the performance of membrane processes. This is a reasonable assumption and specific examples of membrane-bound ATPases in which this is claimed are to be found in Champeil et al. (1982), Heremans (1982) and Chong et al. (1982). The general question of rate-limitation by pressure-mediated changes in bilayer properties is discussed in Macdonald (1986). For the present however, this paper proceeds on the above assumption, and now turns to an examination of membrane bilayers in membranes which normally function in organisms living at high pressures.

DEEP SEA MEMBRANES

Because the membranes extracted from various fish tissues show marked homeoviscous adaptation to temperature changes, deep sea fish were chosen to test for the existence of homeoviscous adaptation to pressure. The work has involved collaboration with Dr. Cossins of Liverpool University. Access to freshly caught, deep sea fish is severely restricted by the high operating cost of research ships capable of collecting them, and of accommodating the associated scientific work. From the U.K. the most convenient ocean depths, where flat topography facilitates bottom trawling, lie at 4000 m depths in and around the Porcupine Sea Bight (circa 49°30'N, 11-14°W). A typical research cruise might provide 8 working days. The best source of deep sea fish is the ocean floor, as mid water species are small and sparse. Fish trawled from the ocean floor are moribund on arrival at the surface, but tissues such as liver and brain can be rapidly dissected, chilled and subjected to standard membrane isolating procedures without undue deterioration. Centrifugation at sea is necessarily limited to rather low $\underline{g}$ forces; in practice 15000 g is achieved using an IEC Centra 3R-S bench-top centrifuge, or slightly more with a bigger machine fitted with an "excess motion" cut out. The ship requires to be hove-to in order to minimise motion and of course calm weather is also necessary. These practical limitations might one day be circumvented by freezing whole tissues at sea and subsequently isolating membranes from them.

In collaboration with Dr. Cossins two independent tests of homeo-viscous adaptation have been undertaken: one is based on measurements of membrane (bilayer) fluidity, determined at sea, using the steady - state fluorescence polarisation of diphenyl hexatriene (DPH) and the second involved the fatty acid analysis of membrane lipids, specifically the fatty acids of the major phospholipids. The favourable membrane preparation for these tests were, for (i) brain myelin-rich and synaptosomal-rich fractions and for (ii) (requiring more material), liver mitochondria.

Steady-state fluorescence polarisation measurements of DPH yield the steady-state anistropy (r_s), which is an index of bilayer order, rather than viscosity. (The term homeoviscous theory was coined at a time when bilayer fluidity was addressed in viscosity terms.) According to Jähnig (1979), Heyn (1979) and Van Blitterswicjk et al. (1981), r_s is approximately equal to r_{∞}, which in turn is proportional to the square of the order parameter obtained by NMR. The steady-state anisotropy r_s may thus be used to compare "fluidities", with due regard for its limitations.

Myelin-rich and synaptosomal-rich membrane fractions were prepared at sea from fish freshly trawled from bottom depths of from 200 to 4000 m; r_s values were determined at 4°C and normal atmospheric pressure. The results showed that in the case of the myelin-rich fraction, r_s varied with the depth of origin of the donor fish by -0.0027 units km^{-1} (Cossins & Macdonald 1984). This offsets more than half the pressure ordering effect, (0.0034 units 100 atm^{-1}) over the depth range of 4000 m. Since the measurements were all made at the temperature ($4^{o}C$) which is close to that prevailing at depth, any variation of temperature within the 4000 m vertical range does not affect the conclusion, and these are clearly consistent with the predicted homeoviscous adaptation to pressure.

Liver mitochondria from a variety of deep sea fish yielded the following phospholipids in practical amounts: phosphatidylcholine (PC), phosphatidylethanolamine (PE), cardiolipin, phosphatidylserine and phosphatidylinositol. Only the first two were available in sufficient quantity from fish collected over a range of depths to provide an adequate set of data. Fatty acid methyl esters were analysed in the convential way, yielding composition "scores" from different depths. The most illuminating "score" was found to be the saturation ratio, which is the ratio of weight % saturated:unsaturated fatty acids. The regression of the saturation ratio against the depth of the origin of the donor fish produced a significant trend in the fatty acids from PC and PE. The saturation ratio decreases with the normal living depth of the fish (Cossins & Macdonald 1986). This is consistent with prediction. However, several factors other than pressure vary with depth, of which temperature is the most significant. Over the depth range in question, 200-4000 m, the sea temperature varies by approximately $8^{o}C$ (we are not concerned with surface temperatures). Could the change in saturation ratio be caused by a decrease of $8^{o}C$? Comparison with the fresh water sunfish (Cossins et al. 1980) shows that a lesser reduction in the saturation ratio occurs over the temperature range of $25-5^{o}C$, i.e. $20^{o}C$. This suggests that the $8^{o}C$ decrease in the sea temperature is insufficient to account for the observed decrease in the ratio. Pressure could well account for the "excess saturation ratio", that is, the excess temperature decrease which the saturation appears to be compensating for, namely $8^{o}C$ minus $20^{o}C$, i.e. $-12^{o}C$. Is 400 ATA pressure sufficient to match the $-12^{o}C$ "worth" of bilayer ordering? According to the spectroscopic analyses 400 ATA is equivalent to a decrease of $8^{o}C$, in reasonable agreement. The conclusion from this is that the fatty acid analyses are quantitatively consistent with the predicted homeoviscous adaptation to pressure. Furthermore the data of Avrova (1984) for ganglioside fatty acids extracted from fish brains are similar, showing a reduced saturation ratio in the case of one deep sea species <u>Antimora rostrata</u>, from 2300 m.

Two laboratory acclimation experiments using high pressure have been undertaken to elicit changes in membrane lipid composition. One used the ciliated protozoan <u>Tetrahymena</u> <u>pyriformis</u> NT-1, renowned for its homeoviscous adaptation to temperature. The results were the converse of what was predicted for adaptation to high pressure, and more consistent with pressure actually inhibiting desaturase activity which is required to change the saturation ratio. The other experiment (DeLong & Yayanos 1985) used the deep sea bacterium CNPT3, which was grown at $2^{\circ}C$ over the pressure range 1-690 atm. The saturation ratio, to use the same criterion as the fish work described above, decreased over the whole pressure range, with one minor anomalous value at 517 atm, which is unimportant. Clearly this result is also consistent with homeoviscous theory, although there was no correlation between saturation ratio and the generation time of the cells.

CONCLUSION

The tests using deep sea fish membranes to search for homeoviscous adaptation to high pressure produced encouraging results. In those cases where the results were obscure, and not reported here, technical difficulties probably played a major part, (e.g. quenching in some of the fluorescence determinations using mitochondria, and a shortage of material in some of the fatty acid analyses). None of the results seriously conflicted with the predictions. Future measurements are undoubtedly desirable, including assays of cholesterol and other membrane lipids. Whilst it is gratifying to find the results are consistent with prediction, it is clearly important, and a much bigger task, to establish the nature of the inter-relationships between pressure, membrane lipids, bilayer order parameters and the rates of membrane-based processes. There are many limitations in this research, particularly the availability of ship time and the refinement of laboratory methods suited to the shipboard conditions. For the present, a practical way forward with deep sea fish material would be to measure specific membrane bilayer order parameters, lipid composition, and functional activity, such as that of ATP-ases, and relate them all to the living depth of the donor animal. Deep sea bacteria offer special possibilities particularly in understanding how pressure is "sensed" and how the ensuing change in the saturation ratio of membrane phospholipids is achieved.

ACKNOWLEDGEMENT
The work was supported by grants from the NERC.

REFERENCES

Avrova N.F. (1984) Comp. Biochem. Physiol. 78B: 903-909.

Bennett, P.B. & Elliott, D.H. (1982) The Physiology and Medicine of Diving. Bailliere Tindall. London.

Brauer, R.W., Beckman, M.Y., Keyser, J.B., Nesbitt, D.L., Shvetzob, S.G., Sidelev, G.N. & Wright, S.L. (1980) Comp. Biochem. Physiol. 65A: 119-127.

Brauer, R.W. (1984) Phil. Trans. Roy. Soc. London B 304:17-30.

Brauer, R.W., Sidelyova, V.C., Dail, M.B., Galazii, G.I. & Roer, R.D. (1984) Comp. Biochem. Physiol. 77A: 699-705.

Champeil, P., Buschlen, S. & Guillain, F. (1981) Biochemistry 20: 1520-1524.

Chong, P.L-G., Jameson, D.M., Fortes, P.A.G. & Weber, G. (1982) Biophys. J. 148a.

Chong, P.L-G. & Cossins, A.R. (1984) Biochim. Biophys. Acta 772: 197-201.

Collen, D., Vandereycken, G. & De Maeyer, L. (1970) Nature, London 288: 669-671.

Cossins, A.R., Kent, J. & Prosser, C.L. (1980) Biochim. Biophys. Acta. 599: 341-358.

Cossins, A.R. & Macdonald, A.G. (1984) Biochim. Biophys. Acta. 776: 144-150.

Cossins, A.R. & Macdonald, A.G. (1986) Biochim. Biophys. Acta 860: 325-335.

DeLong, E.F. & Yayanos, A.A. (1985) Science 288: 1101-1103.

Ebbecke, U. (1936) Pflügers Arch. ges. Physiol. 238: 441-451.

Forbes, A., Gilchrist, I., Green, T., Harper, A.A., Macdonald, A.G., McDiarmid, V., Pennec, J-P., Tetteh-Lartey, N. & Wardle, C. (1985) J. Physiol. 372: 57P.

Forbes, A., Gilchrist, I., Green, T., Harper, A.A., Macdonald, A.G., McDiarmid, V., Pennec, J-P., Tetteh-Lartey, N. & Wardle, C. (1986a) J. Physiol. in press.

Forbes, A., Gilchrist, I., Green, T., Harper, A.A., Macdonald, A.G., McDiarmid, V., Pennec, J-P., Tetteh-Lartey, N. & Wardle, C. (1986b) J. Physiol. in preparation.

Heremans, K. (1982) Ann. Rev. Biophys. Bioeng. 11: 1-21.

Heyn, M. (1979) FEBS Letters, 108: 359-364.

Hochachka, P.W. (1974) Biochem. J. 143: 535-539.

Hochachka, P.W. (1975) Comp. Biochem. Physiol. 52B: 39-41.

Isaacs, N.S. (1981) Liquid-phase High Pressure Chemistry. Wiley & Sons. Chichester.

Jaenicke, R. (1983) Naturwissenschaften 70: 332-341.

Jähnig, F. (1979) Proc. Natl. Acad. Sci. U.S.A. 76: 6361-6365.

Jannash, H.W. (1984) In: Hydrothermal processes at sea floor spreading centers. Rona, P.A., Bostrom, K. and Laubier, L. (eds). Plenum Publishing Corp. N.Y.

Line, N. & Kay, R.L. (1977) Biochemistry N.Y. 16: 3484-3486.

Low, P.S. & Somero, G.N. (1975) Comp. Biochem. Physiol. 52B: 67-74.

Macdonald, A.G. (1975) Physiological Aspects of Deep Sea Biology. Monograph of the Physiological Society. Cambridge University Press. Cambridge.

Macdonald, A.G. and Miller, K.W. (1976) in "Biochemical and Biophysical Perspectives in Marine Biology", eds. D.C. Malins and J.R. Sargent, pp. 117-147, Academic Press, New York and London.

Macdonald, A.G. & Gilchrist, I. (1978) Marine Biol. 45: 9-21.

Macdonald, A.G. & Gilchrist, I. (1980) Comp. Biochem. Physiol. 67A:149-153.

Macdonald, A.G., Gilchrist, I., Wann, K.T. and Wilcock, S.E. (1980) In: Animals and Environmental Fitness. ed.Gilles, R. Pergamon Press. Oxford.

Macdonald, A.G. & Gilchrist, I. (1982) Comp. Biochem. Physiol. 71A: 349-352.

Macdonald, A.G. (1984) Phil. Trans. Roy. Soc. London B 304:47-68.

Macdonald, A.G. (1986) In preparation.

Marque, J., Eisenstein, L., Gratton, E., Sturtevant, J.M. & Hardy, C.J. (1984) Biophys. J. 46: 567-572.

Menzies, R.J., George, R.Y. & Paul, A.Z. (1974) Int. Revue. ges. Hydrobiol. 59: 187-197.

Mickel, T.J. & Childress, J.J. (1982) Biol. Bull. 162: 70-82.

Schoen, P.E., Priest, R.G., Sheridan, J.P. & Schnur, J.M. (1977) Nature, London 270: 412-414.

Schwickert, H., Strobl, G.R. & Eckel, R. (1982) Colloid & Polymer Sci. 260: 588-593.

Somero, G.N. (1982) In: The Environment of the Deep Sea. Ernst, W.G. & Morin, J.G. (eds). Prentice Hall Inc. New Jersey.

Stamatoff, J., Guillon, D., Powers, L. & Cladis, P. (1978) Biochem. Biophys. Res. Commun. 85: 724-788.

Stubbs, C.G., Konyama, T., Kinosita, K. & Ikegami, A. (1981) Biochemistry 20: 4257-4262.

Swezey, R.R. & Somero, G.N. (1982) Biochemistry 21: 4496-4503.

Utracki, L.A. (1983) Polym. Engin. Sci. 23: 446-451.

Van Blitterswicjk, W.J., Van Haeven, R.P. & Van Deer Meer, B.W. (1981) Biochim. Biophys. Acta 644: 323-332.

Weber, G. & Drickamer, H.G. (1983) Quart. Rev. Biophys. 16: 89-112.

Wong, P.T.T., Murphy, W.F. & Mantsch, H.H. (1982) J. Chem. Phys. 76: 5230-5237.

Wong, P.T.T. (1984) Ann. Rev. Biophys. Bioeng. 13: 1-24.

Yager, P. & Peticolas, W.J. (1980) Biophys. J. 31: 359-370.

Yayanos, A.A. (1978) Science 200: 1056-1059.

Zimmerman, A.M. (1970) ed. High Pressure Effects on Cellular Processes. Academic Press. New York.

Zimmerman, S.B. & Zimmerman, A.M. (1977) The effects of hydrostatic pressure on cell membranes. In: Mammalian cell membranes. 5. Responses of Plasma Membranes. ed. Jamieson, G.A. & Robinson, D.M. Butterworth. London. pp.47-71.

HYDROSTATIC PRESSURE EFFECTS ON MEMBRANE TRANSPORT IN RED CELLS

Andrew C. Hall and J. Clive Ellory*

Physiological Laboratory, Downing Street, Cambridge CB2 3EG, U.K., and
*University Laboratory of Physiology, Parks Road, Oxford OX1 3PT, U.K.

INTRODUCTION

High hydrostatic pressure is well known for its ability to reduce the motional freedom of the acyl chains of phospholipids in both artificial and natural membranes (e.g. Boggs et al., 1976; Finch & Kiesow, 1979). In addition, the inhibitory effects of pressure on enzymes and other biochemical systems have been extensively studied (Macdonald, 1975; Zimmerman & Zimmerman, 1977). One might therefore reasonably propose that raising pressure would inhibit permeability; however, there have been suprisingly few studies of pressure effects on transport with which to test this hypothesis (reviewed by Macdonald, 1984). In this paper we present data on well-characterised erythrocyte cation transport systems, and demonstrate that pressure effects on permeability are complex and difficult to predict on the basis of model systems.

METHODS

The methodology used for flux studies at high pressure has already been documented (Hall et al., 1982) and therefore only an outline will be given here. In brief, radioisotope K^+ (using ^{86}Rb as a congener) and ^{22}Na fluxes were measured over the range 1-400 ATA (ATA = atmospheres absolute; 1 ATA = 0.101 MPa) at $37^{\circ}C$ in fresh, washed human red cells. Ouabain and bumetanide were used as required to inhibit Na^+/K^+ pump and $(Na^+ + K^+)$ cotransport systems respectively (see Hall & Willis, 1984), and the Ca^{2+}-chelating agent EGTA prevented the activation of the Ca^{2+}-sensitive K^+ channel (Lew & Ferreira, 1978). Cell volume was varied by altering medium osmolarity, and anion substitution performed as described (Dunham et al., 1980).

RESULTS AND DISCUSSION

Raising pressure inhibits the Na^+/K^+ pump and (Na^++K^+) cotransport systems; however, in contrast, the K^+ flux remaining in the presence of ouabain and bumetanide is increased and dependent on the anion composition of the medium (Hall et al., 1982). Since we might expect that increasing pressure would inhibit transport, this result was unexpected; however, in this report we describe experiments which enable us to dissect out two significant effects of pressure on erythrocyte permeability.

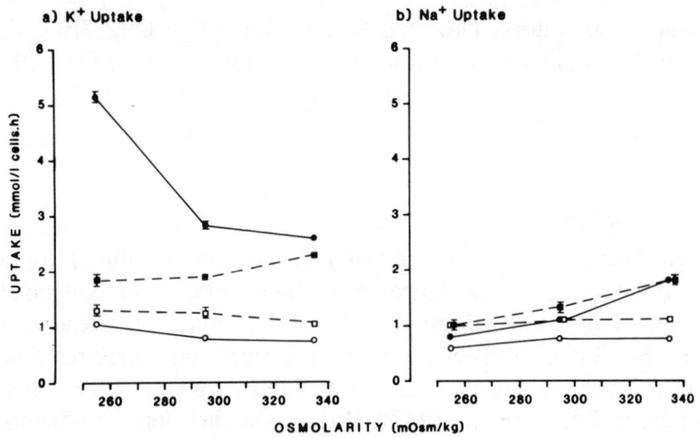

Figure 1. The effect of varying medium tonicity on passive (a) K^+ and (b) Na^+ uptake at 1 and 400 ATA in human red cells. Radioisotope uptake was measured in Cl^- or $CH_3SO_4^-$-containing media in the presence of the inhibitors ouabain, bumetanide, and EGTA as previously described (Hall et al., 1982; Ellory et al., 1985). Symbols represent uptake in a Cl^- medium ($\circ$,$\bullet$) at 1 and 400 ATA respectively, and in a $CH_3SO_4^-$ medium ($\square$,$\blacksquare$) at 1 and 400 ATA respectively. $[K]_o$ and $[Na]_o$ were maintained constant at 62.5 mM. Results are means $\pm$ s.e.m. of triplicate determinations.

Fig. 1 shows an experiment where the influence of medium osmolarity and anion composition were tested on (a) K^+ and (b) Na^+ influx at 1 ATA and 400 ATA. At atmospheric pressure, varying medium osmolarity (range 255-335 mOsm/kg) had little effect on passive (i.e., ouabain + bumetanide + EGTA - insensitive) K^+ or Na^+ uptake. At 400 ATA there was a marked increase in permeability to both cations, although K^+ uptake in a Cl^- medium was increased with decreasing medium osmolarity (i.e. cell swelling) whereas Na^+ permeability in either medium, or K^+ influx in $CH_3SO_4^-$ medium was reduced. In further studies, the effect of increasing $[K]_o$ on K^+ uptake in a Cl^- medium was studied at 1 and 400 ATA over the range 2.5-130 mM. The results demonstrated the presence of a saturable flux in swollen cells ($\underline{K}_m$ about 40 mM, $\underline{V}_{max}$ about 6 mmol per (litre cells per hour)) which was abolished in shrunken cells (i.e., uptake was a linear function of $[K]_o$). In swollen cells, when Cl^- was replaced by $CH_3SO_4^-$, no saturable flux was observed.

These data therefore suggest that pressure has at least two effects on passive monovalent cation permeability: (1) a generalised increase in Na^+ and K^+ transport and (2) the activation of a specific KCl pathway. These two effects may be separated because cation permeation via system (1) is unaffected by the anion present (Cl^- $\underline{vs}$ $CH_3SO_4^-$), is reduced with cell swelling and demonstrates non-saturable uptake kinetics. In contrast, only K^+ (or Rb^+) is transported by system (2), and this flux shows the kinetic characteristics of a low–affinity, high-capacity system dependent on Cl^- for activity and increased by cell swelling in a hypotonic medium.

The nature of the pressure-induced permeability increase ((1) above)) is obscure because we would expect that with all known transport pathways inhibited (including the KCl pathway) the effects of pressure on cation transport would resemble those previously reported in liposomes (i.e. artificial lipid membranes) in which raising pressure decreased permeability (Johnson & Miller, 1975). It is likely that membrane-bound proteins are a critical determinant of the effects of pressure on erythrocyte permeability.

An explanation for the activation of the otherwise latent volume-sensitive KCl system is slightly easier since there are conditions other than pressure which may reveal this pathway. These are (1) the addition of the thiol-reactive agent NEM (Lauf & Theg, 1980), and (2) in certain haemolytic anaemias, for example hereditary spherocytosis (HS) (Ellory et al., 1985). The question which then arises is, do pressure, NEM and HS share a common mode(s) of action? In an attempt to clarify this, we have fixed red cells treated with NEM (2 mM, 10% haematocrit, 10 min) or at high hydrostatic pressure (400 ATA, 30 min) and examined their morphology by scanning electron microscopy (SCEM). The shape of a significant fraction (15-50% NEM-treated cells; 5-35% pressurised cells) of the cell population was found to be cup-shaped i.e., similar to the morphology of HS cells at 1 ATA (Dacie & Lewis, 1975). A small decrease (about 4%) in cell volume accompanied these shape changes. Not all cells participated in the morphological transformation suggesting a considerable degree of heterogeneity in the cell population. The importance of this observation is underlined by experiments on HS erythrocytes which show large pressure-induced, volume-sensitive KCl fluxes (Hall et al., 1982; Gardiner et al., 1983; Ellory et al., 1985). Interpretation of these data is, however, complicated by the presence of large numbers of reticulocytes and young red cells characteristic of haemolytic anaemias. If young but not mature cells possessed functional volume-sensitive KCl transporters this might explain the marked effect of pressure on K^+ permeability in HS red cells.

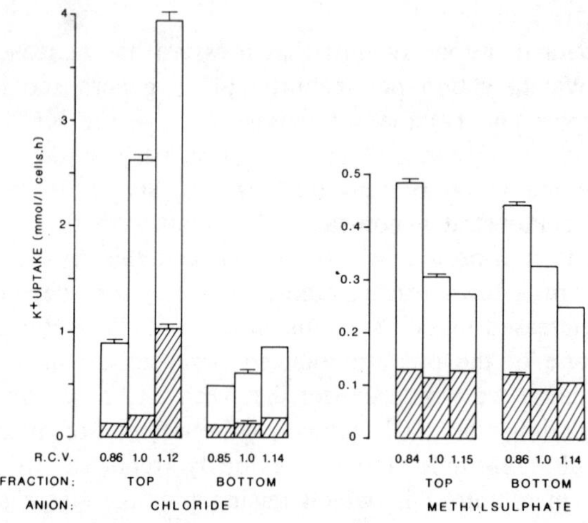

Figure 2. The effect of cell volume, anions and hydrostatic pressure on passive K^+ uptake in red cells from different fractions of human blood. Passive K^+ influx was measured in cells of differing volume (r.c.v. = relative cell volume; cells of normal volume, r.c.v. = 1.00) at 1 ATA (cross-hatched) and 400 ATA (total area at each condition). Results are means $\pm$ s.e.m. of triplicate determinations.

In order to determine whether young but not mature cells possess this KCl pathway, we have used red cells from normal donors because interpretation would not then be compromised by morphologically abnormal cells. Fractions of red cells were enriched with young or mature cells by the centrifugation technique of Tucker and Young, (1982). Fig. 2 shows that at 1 ATA, cell swelling markedly increased the Cl⁻-dependent K^+ uptake of red cells from the top (i.e., less dense) fraction, but had little effect on cells from the bottom fraction. At pressure, a large volume-sensitive Cl⁻-dependent increase in K^+ uptake was observed in the young red cell-enriched population, with a similar, although smaller effect being detected in mature cells. In the presence of $CH_3SO_4^-$, cell swelling <u>reduced</u> K^+ uptake indicating the generalised effect of pressure on cation permeability, (pathway (1) above). Comparable experiments on Na^+ fluxes showed no effect of altering cell volume at 1 ATA, an elevation in flux at pressure, but cell swelling <u>reduced</u> permeability and there was no significant effect when Cl^- was replaced by $CH_3SO_4^-$. Thus, the pressure effect on Na^+ permeation is quite different compared to K^+ in a Cl^- medium, but resembles K^+ uptake in a $CH_3SO_4^-$ medium. These data therefore support the hypothesis that young red cells possess functional volume-sensitive KCl transporters and that the pathway is present in mature cells but latent under normal experimental conditions.

Thus, pressure may alter red-cell shape from the normal discoid to cup-shaped forms and that this activates the otherwise latent volume-sensitive KCl system. Our studies suggest that cell volume <u>per se</u> is unlikely to be the sole determinant of the activity of the KCl transporter. Nevertheless, alterations to cell volume undoubtably have a major influence in cells where morphology is compromised and clearly the details of the relationship between these two factors is essential before we will have a full understanding of the mechanisms which activate the pathway.

Despite these uncertainties, it seems highly likely that the major determinant of red cell shape, the cytoskeleton (Lux, 1979), is involved. It has been known for many years that pressure disrupts structural proteins in a wide variety of cell types (Zimmerman & Zimmerman, 1977) and that HS erythrocytes are thought to be defective in certain critical structural protein components (Lux, 1979; Shotton, 1983). It is not unreasonable to propose that the receptor complex for the KCl transporter is situated in the cytoskeleton in intimate contact with the membrane-bound transport protein(s). Factors which strain (e.g. cell swelling) or disrupt (e.g. pressure) the cytoskeleton may therefore profoundly influence the activity of the KCl pathway.

In conclusion, our experiments demonstrate that increasing hydrostatic pressure has complex effects on erythrocyte cation transport. Pressure inhibits the permeability mediated by some systems but markedly stimulates transport by others. The activation of the carrier-mediated KCl pathway may be related to the integrity of the protein cytoskeleton which underlies the membrane. Clearly, a simple hypothesis which proposes that pressure influences permeability by a direct effect solely on the order of membrane lipids is untenable. Thus, we still have much to learn about the basic factors which underlie and regulate the permeability of biological membranes and for these studies pressure may be a very useful tool.

ACKNOWLEDGEMENTS

We would like to thank Mr A. Hockaday for the scanning electron microscopy of red cells, and the Department of Anatomy, University of Cambridge for the use of the microscope. This work was supported by the M.R.C.

REFERENCES

Boggs, J.M., Yoong, T. & Hsia, J.C. (1976). Molec. Pharmac. 12: 127-135.

Dacie, J.V. & Lewis, S.M. (1975). Practical Haematology. 5th Edn. Churchill Livingstone, Edinburgh.

Dunham, P.B., Stewart, G.W. & Ellory, J.C. (1980). Proc.Natl.Acad.Sci. U.S.A. 77: 1711-1715.

Ellory, J.C., Hall, A.C. & Stewart, G.W. (1985). Molec. Physiol. 8:235-246.

Finch, E.D. & Kiesow, L.A. (1979). Undersea Biomed. Res. 6: 41-45.

Gardiner, R.M., Barnes, N.D. & Ellory, J.C. (1983). Arch. Disease Childhood 58: 547-549.

Hall, A.C., Ellory, J.C. & Klein, R.A. (1982). J. Membrane Biol. 68:47-56.

Hall, A.C. & Willis, J.S. (1984). J. Physiol. 348: 629-643.

Johnson, S.M. & Miller, K.W. (1975). Biochim. Biophys. Acta 375: 286-291.

Lauf, P.K. & Theg, B.E. (1980). Biochem. Biophys. Res. Comm. 92: 1422-1428.

Lew, V.L. & Ferreira, H.G. (1978). Curr. Topics Membranes & Transport 10: 217-277.

Lux, S.E. (1979). Semin. Haematol. 16: 21-51.

Macdonald, A.G. (1975). Monographs of the Physiological Society. No. 31. Cambridge University Press, Cambridge.

Macdonald, A.G. (1984). Phil. Trans. Roy. Soc. Lond. B 304: 47-68.

Tucker, E.M. & Young, J.D. (1982). In:Red Cell Membranes; A Methodological Approach. Eds. J.C. Ellory & J.D. Young. Academic Press, London. pp.31-41.

Shotton, D. (1983). In: Electron Microscopy of Proteins. Ed. J.R. Harris. Academic Press, London. pp.205-330.

Zimmerman, S.B. & Zimmerman, A.M. (1977). In:Mammalian Cell Membranes. Eds. G.A. Jamieson & D.M. Robinson. Butterworth, London.